T0214274

Lecture Notes in Business Information Processing 377

Series Editors

Wil van der Aalst ⓘ
RWTH Aachen University, Aachen, Germany
John Mylopoulos ⓘ
University of Trento, Trento, Italy
Michael Rosemann ⓘ
Queensland University of Technology, Brisbane, QLD, Australia
Michael J. Shaw
University of Illinois, Urbana-Champaign, IL, USA
Clemens Szyperski
Microsoft Research, Redmond, WA, USA

More information about this series at http://www.springer.com/series/7911

Henriqueta Nóvoa · Monica Drăgoicea ·
Niklas Kühl (Eds.)

Exploring
Service Science

10th International Conference, IESS 2020
Porto, Portugal, February 5–7, 2020
Proceedings

 Springer

Editors
Henriqueta Nóvoa (iD)
University of Porto
Porto, Portugal

Monica Drăgoicea (iD)
University Politehnica of Bucharest
Bucharest, Romania

Niklas Kühl (iD)
Karlsruhe Institute of Technology (KIT)
Karlsruhe, Germany

ISSN 1865-1348   ISSN 1865-1356   (electronic)
Lecture Notes in Business Information Processing
ISBN 978-3-030-38723-5   ISBN 978-3-030-38724-2   (eBook)
https://doi.org/10.1007/978-3-030-38724-2

This Springer imprint is published by the registered company Springer Nature Switzerland AG
The registered company address is: Gewerbestrasse 11, 6330 Cham, Switzerland

Preface

Service Science has been the object of a remarkable development since initial calls for action at its own creation, more than ten years ago. Since then, the Service Science field has started to take shape by building a significant body of research, gathering researchers with different backgrounds in the pursuit of a better definition of its main constructs, as well as understanding the research boundaries of this new discipline.

The International Conference on Exploring Service Science (IESS), celebrating its 10th edition, has been clearly one of the primordial efforts to foster this new research community, gathering academics from all over the world in a stimulating environment.

In this volume of "Exploring Service Science," we have collected the peer-reviewed papers of IESS 2.0, organized and held at the Faculty of Engineering of the University of Porto (FEUP), during February 5–7, 2020. The book includes papers that extend the view on different concepts related to the development of the Service Science domain of study, applying them to frameworks, advanced technologies, and tools for the design of new, digitally-enabled service systems.

Twenty-eight full papers were selected in a blind review process, from authors of fourteen different nationalities. All submissions were reviewed by at least two members of the International Program Committee, composed of Service Science experts from over 20 countries.

This book is structured in six parts, based on the six main conference themes, covering a broad range of relevant challenges for Service Science: Customer Experience, Data Analytics in Service, Emerging Service Technologies, Service Design and Innovation, Service Ecosystems and Service Management.

The Customer Experience theme encompasses articles that review the literature on the topic and highlight the customer role and perceptions regarding their interactions with service providers. The Data Analytics in Service theme presents a set of articles focused on predicting service-related outcomes, load and service level planning, and a recommendation engine for service personalization. The Emerging Service Technologies theme brings together articles focused on a diverse and rich set of innovative technologies applied in a service setting. These technologies include blockchain, artificial intelligence, service robots, and innovative technology-enabled services, such as smart services and fintechs. The Service Design and Innovation theme explores how to overcome the resistance to innovation adoption, presenting a new service design method and studing the connection between service design and design thinking, and the application of service design to business process reengineering and the printing industry. The Service Management theme adopts an organizational perspective, with an emphasis on the characteristics of management and service organizations, and modeling service processes. Finally, the Service Ecosystem theme discusses societal and macro-level challenges, such as public safety, societal progression, country-wide benchmarking, and service ecosystem simulation. It also includes an article that reviews the ten years of the IESS conference and sets a research agenda for the Service Science field.

This publication is the result of teamwork where many people have contributed. The review process actively involved all International Program Committee members to ensure theoretical and methodological rigor. We owe to the entire Program Committee a deep thank you for being exigent and scrupulous with demanding deadlines. The entire review process, as well as the compilation of the proceedings, was supported through the Easychair platform, and it is never too much to say how indebted we are to the IS/IT developers of Easychair, who have facilitated the life of countless researchers in the world.

We would also like to extend our thanks to the IESS Steering Committee for their support and the opportunity to host this conference once more in Porto. A good conference must have inspiring keynote speakers: we are grateful to Jim Spohrer and Tuure Tuunanen, for immediately accepting our invitation and for contributing undoubtedly to the success of this event. Finally, the conference could not have been held without the continuous commitment and enthusiasm of the local organizing team at FEUP – Jorge Teixeira, Marta Ferreira, and Vera Miguéis. Thank you all!

We hope you have enjoyed the conference and we wish you a pleasant and fruitful reading of the IESS 2.0 proceedings.

November 2019

Henriqueta Nóvoa
Monica Drăgoicea
Niklas Kühl

Organization

Program Committee

Jesus Alcoba	La Salle University Center, Spain
Theodor Borangiu	University Politehnica of Bucharest, Romania
Marta Campos Ferreira	University of Porto, Portugal
Luca Carrubbo	University of Salerno, Italy
Fabrizio D'Ascenzo	Sapienza University of Rome, Italy
María Valeria De Castro	Universidad Rey Juan Carlos, Spain
Marco De Marco	Uninettuno University, Italy
Monica Dragoicea	University Politehnica of Bucharest, Romania
João Falcão E Cunha	University of Porto, Portugal
Jose Faria	University of Porto, Portugal
Teresa Fernandes	University of Porto, Portugal
Jorge Grenha Teixeira	University of Porto, Portugal
Peter Hottum	Karlsruhe Institute of Technology, Germany
Manuele Kirsch-Pinheiro	Université Paris 1 Panthéon-Sorbonne, France
Natalia Kryvinska	University of Vienna, Austria
Niklas Kühl	Karlsruhe Institute of Technology, Germany
Thang Le Dinh	Université du Québec à Trois-Rivières, Canada
Michel Leonard	University of Geneva, Switzerland
Weiping Li	Peking University, China
Paul Lillrank	Aalto University, Finland
Paul Maglio	University of California, Merced, USA
Maria Maleshkova	University of Bonn, Germany
Vera Migueis	University of Porto, Portugal
Miguel Mira Da Silva	Universidade de Lisboa, Portugal
Jean-Henry Morin	University of Geneva, CUI, Switzerland
Henriqueta Nóvoa	University of Porto, Portugal
Lia Patrício	University of Porto, Portugal
Geert Poels	Ghent University, Belgium
Francesco Polese	University of Salerno, Italy
Jolita Ralyté	University of Geneva, Switzerland
Melanie Reuter	Karlsruhe Institute of Technology, Germany
Shai Rozenes	Afeka College of Engineering, Israel
Gerhard Satzger	Karlsruhe Institute of Technology, Germany
Mehdi Snene	University of Geneva, Switzerland
Maddalena Sorrentino	University of Milan, Italy
Tuure Tuunanen	University of Jyväskylä, Finland
Leonard Walletzký	Masaryk University, Czech Republic

Zhongjie Wang   Harbin Institute of Technology, China
Stefano Za   D'Annunzio University of Chieti–Pescara, Italy

Additional Reviewers

Barbosa, Flávia   University of Porto, Portugal
Benz, Carina   Karlsruhe Institute of Technology, Germany
Enders, Tobias   Karlsruhe Institute of Technology, Germany
Guimarães, Luis   University of Porto, Portugal
Hunke, Fabian   Karlsruhe Institute of Technology, Germany
Martin, Dominik   Karlsruhe Institute of Technology, Germany
Vössing, Michael   Karlsruhe Institute of Technology, Germany
Walk, Jannis   Karlsruhe Institute of Technology, Germany
Wolff, Clemens   Karlsruhe Institute of Technology, Germany

Contents

Service Ecosystems

Service Management

Customer Experience

Customer Experience Literature Analysis Based on Bibliometry

Jorge Henrique O. Silva[1], Glauco Henrique S. Mendes[1],
Paulo A. Cauchick-Miguel[2(✉)], and Marlene Amorim[3]

[1] Universidade Federal de São Carlos, São Carlos, Brazil
jorgeh.oliveira@hotmail.com, glauco@dep.ufscar.br
[2] Universidade Federal de Santa Catarina, Florianópolis, Brazil
paulo.cauchick@ufsc.br
[3] Universidade de Aveiro, Aveiro, Portugal
mamorim@ua.pt

Abstract. Customer experience (CX) has attracted an increasing interest of practitioners and scholars, since it has become an important basis for the competitiveness of organizations. To better understand how the literature in the customer experience field has evolved, this study presents a bibliometric study of 613 articles published in peer-reviewed journals from 1991 to 2018 using citation and reference co-citation analyses. The results show that service quality, service encounter, and service-dominant logic literature encompass the foundations (intellectual roots) of the CX research field. This work brings four main contributions. Firstly, it demonstrates the emergence of customer-centric theories that have contributed to the development of the CX field. Secondly, initial studies emphasized the hedonic aspects of consumption, although subsequent studies have broadened the CX scope. Thirdly, customer experience has evolved from a static to a dynamic view. Lastly, as the field consolidates, the role of technology has gained prominence.

Keywords: Customer experience · Service experience · Customer journey · Bibliometry

1 Introduction

According to a study by Gartner [1], 81% of marketers say their companies compete mostly on the basis of customer experience (CX). Multiple international brands have job titles such as chief experience officer, vice-president of customer experience, and customer experience manager, who are in charge of managing CX [2]. CX is crucial for B2B companies [3] and has the great strategic potential for small businesses [4]. The practitioner interest has strongly stimulated customer experience research, which has become a key focus of service marketing and service science literature [5].

Although there is no consensus about how to define the CX concept [6], the major accepted view defines CX as a multidimensional construct focusing on the customer's cognitive, emotional, behavioural, sensorial, and social responses to all direct or indirect interactions with the firm during the customer's entire journey [6, 7].

H. Nóvoa et al. (Eds.): IESS 2020, LNBIP 377, pp. 3–20, 2020.
https://doi.org/10.1007/978-3-030-38724-2_1

Regarding CX consequences, both practitioners and scholars agree that CX significantly determines the customer behaviour, affecting relevant marketing outcomes such as customer satisfaction, loyalty, and word-of-mouth communications [8–11]. In light of the increasing interest in CX, this promising research field should be reviewed to a better understanding of its underlying theories and related themes. In the context of services, the concept of CX is also determined by the characteristics of the production processes, because customers play an active role in the delivery activities, and their actions can have a substantial impact in the process experience. Despite a number of efforts to review CX literature, this study aims to analyze the customer experience research field through bibliometry. Thus, the primary goal of this article is to provide a map of CX research in terms of intellectual roots (intellectual structure) that should be of help to scholars seeking to understand the various streams of research in CX and the historical development of the field. Based on 613 CX-related articles (1991–2018), a bibliometric analysis is developed, grounded in two methods, namely: citation analysis and co-citation.

The remainder of the paper is structured as follows. Section 2 outlines a brief overview of the customer experience concept and how other literature reviews have dealt with this subject. This is followed by Sect. 3 that highlights the methodological research procedures. The results in Sect. 4 provide a descriptive analysis based on an overview of customer experience field and co-citation analysis. Finally, Sect. 5 draws some concluding remarks and research implications of this study.

2   Customer Experience Concept

Researchers have agreed that the customer perceives his/her experience in a holistic way which involves multiple subjective responses to all interactions with an organization [2], across multiple touchpoints along the stages of the customer journey (pre-purchase, purchase, and post-purchase) [7]. However, despite the major acceptance of this holistic view, CX has been conceptualized in multiples ways through different lenses [12]. In addition, customer experience can be characterized as an umbrella construct which means a broad concept used to encompass a diverse set of phenomena [13]. Furthermore, the literature has shown that 'customer experience', 'service experience' and 'customer service experience' are common terms used interchangeably to designate the same concept, despite the fact that some authors pointed conceptual differences among them as e.g. [6, 14, 15]. This study draws on research that uses either term to depict customers' experiences in the service context. Nevertheless, for clarity purposes, this study adopts the term customer experience in line with relevant sources in marketing (e.g. [7, 11]) and service science (e.g. [16, 17]).

For the understanding of customer experience, some literature review papers have dealt with this subject. Helkkulla [18] presented a systematic review of the concept by characterizing CX research in three categories: 'phenomenon-based'; 'process-based'; and 'outcome-based'. Vasconcelos et al. [15] integrated CX in a conceptual framework containing three dimensions: 'predispositions', 'interactions', and 'reactions'. De Keyser et al. [2] proposed a framework for understanding and managing CX that considers insights from marketing, philosophy, psychology, and sociology literature.

Hwang and Seo [19] integrated the CX research in a conceptual framework describing the antecedents (internal and external factors) and consequences (emotional, behavioral and brand-related outcomes) of CX management. Lipkin [20] posited CX research by applying three different perspectives: 'stimulus-based'; 'interaction-based'; and 'sense making-based'. Jain et al. [6] summarize the CX literature in terms of the emergence and development of the concept, proposing a CX research agenda at three different levels: conceptual, management and measurement. Kranzbühler et al. [13] integrate CX research from organizational and consumer perspectives, connecting insights from both angles. Bueno et al. [8] identified how CX has been measured in relevant publications in the marketing field. Mahr et al. [12] synthesize contributions about CX across its five dimensions (physical, social, cognitive, affective and sensorial) in the services and marketing domains.

While complimenting the previous studies, this work differs from them in its objective and coverage. Specifically, it addresses the main theoretical basis of customer experience research field by combining co-citation analysis with a qualitative content analysis based on 613 articles. The next section provides more details on the research methods.

3 Research Methods

Bibliometry involves an analysis of scientific publications to structure and bridge-related research within a field [21]. When conducting bibliometric analysis, an often used method is co-citation analysis, which aims to identify researchers who are cited together, and, therefore, represent similar ideas, unveiling the intellectual structure of a research field [21]. Co-citation has been widely used in services, marketing, and business research as e.g. [22–24]. In order to build and visualize the results, VOS-viewer software is usually adopted [25].

The data for this study were obtained through the Social Science Citation Index (SSCI) accessible online through the Web of Science (WoS), which is one of the largest multidisciplinary databases of peer-reviewed literature. It is the most frequently used database for bibliometric studies in management and organizational areas [21]. Moreover, the WoS provides a unique feature of citation counts and metadata, which allows the use of influence objective measures to qualify the relative importance of the articles [26]. Therefore, several scholars have used only the WoS database to perform bibliometric studies as e.g. [24, 27].

The sample used in this study is limited to articles written in English and published in peer-reviewed journals. The search was performed in January 2019 with the following terms: 'customer experience' AND 'service'; OR 'service experience'. The terms 'customer journey' AND 'service' have also been included due to the close link between the customer journey perspective and the notion of customer experience [28]. The search was performed on the titles, abstracts, and keywords (time period from 1945 to 2018), resulting in 1,112 publications. To ensure that the final sample was in accordance with our aim of addressing only articles related to customer experience several steps were taken. One of the authors checked all the articles in the initial sample by reading articles' titles, keywords, and abstracts. Since decisions regarding inclusion and exclusion remain relatively subjective, when an abstract content was ambiguous, the paper was read in full by two authors to make a decision about whether it should be included. This screening process involved two researchers working in parallel and comparing their judgments. After excluding 499 articles, the final sample comprised 613 articles (focal sample) and 25,535 cited references (reference sample), which were used to perform the analysis. The final step consisted in interpreting and summarizing the findings. Thus, the bibliometric analysis was complemented with a qualitative content review. Specifically, a thematic approach [29] was adopted to reading and categorizing the articles, identifying themes and, extracting relevant data about the CX research field under investigation, from which results are presented in the next section.

4   Results: Mapping the Research on Customer Experience

The first part of this section discusses the focal sample of articles in terms of frequency and time of publication, most prominent journals, and citation statistics. Figure 1 shows that the number of articles published per year has increased during the review period: 27 between 1991 and 2004 (14 years); 43 between 2005 and 2008 (4 years); 161 between 2009 and 2014 (6 years); 382 between 2015 and 2018 (4 years). The milestones in the time period were associated with the observation of sharp increases in the number of publications (e.g., 2004–2005 from 4 to 9 articles). Although there were few articles since the early 90s, there was a significant increase in the number of publications in the past 10 years (2009–2018). Actually, since 2015 a sharp increase in the number of papers published was evident, with almost than twice as many publications as in any of the six preceding years. It is not possible to infer about a decrease in 2018, since the search was performed in January 2019 and databases are frequently updated during the following year. In summary, customer experience can be considered a recent research field, which has significantly grown since 2009, and whose frequency has intensified in the past four years (2015–2018).

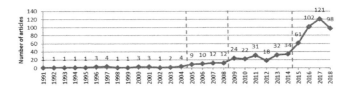

Fig. 1. Frequency of publications

The 613 articles were published in 229 peer-reviewed journals. However, only 15 journals had more than eight focal papers during the review period, which corresponds to 250 papers (approximately 41%). The leading journals were: Journal of Service Management (42 papers), Journal of Service Research (36), Journal of Services Marketing (36), Journal of Service Theory and Practice (17), and Journal of Business Research (17). In fact, CX research has mainly been published in service and marketing research journals. Service journals comprise 108 papers, while marketing journals totalize 70, as exhibited in Table 1. Moreover, 35 papers were published in hospitality journals and 37 in other journals (e.g., management and information technology). Furthermore, there are no high-ranked outlets entirely devoted to research on CX, which was suggested earlier as a young field [30]. Structures in customer experience research have developed slowly, particularly through specific conferences and journal special issues. For instance, some special issues have been published on this research subject (e.g., volume 26, issue 2 of Journal of Service Management: 'Co-creating Service Experience').

Table 1. Top 10 journals publishing the core articles (Note: Journal's field followed the Academic Journal Guide's categorization)

Rank	Source	Field	#
1	Journal of Service Management	Sector-Service	42
2	Journal of Service Research	Sector-Service	36
	Journal of Services Marketing	Marketing	36
3	Journal of Service Theory and Practice	Sector-Service	17
	Journal of Business Research	Management	17
4	International Journal of Contemporary Hospitality Management	Sector-Hospitality	14
5	Service Industries Journal	Sector-Service	13
6	International Journal of Hospitality Management	Sector-Hospitality	12
7	BT Technology Journal	Info Management	11
8	Journal of Retailing and Consumer Services	Marketing	10
9	Computers in Human Behavior	Info Management	9
	Cornell Hospitality Quarterly	Sector-Hospitality	9
10	European Journal of Marketing	Marketing	8
	Journal of Retailing	Marketing	8
	Journal of the Academy of Marketing Science	Marketing	8

Table 2 lists the most-cited focal articles and most-cited references. Those articles may be considered the most influential publications in the field of customer experience. The citations of each focal article refer to its citations in the SSCI-WoS, while the citations of a reference mean the number of focal articles citing that specific reference. To avoid bias towards older publications – as recommended elsewhere [21] – Table 2 also shows the impact of each focal article and reference based on their citation average per year (criterion used to rank both focal articles and references). This allows having a comparison among articles published at different times.

As can be seen in Table 2, Ostrom et al. [16, 17] are the most cited articles. Although they are not CX-centered publications, both discuss priorities for service research and point out CX as one of them. Certainly, they have influenced scholars to engage with customer experience research topics. Brodie et al. [31] ranked in second

Table 2. Top 20 most-cited focal articles and their references

#	Focal articles	Citation	Average	#	References	Citation	Average
1	Ostrom et al. (2010)	580	58.0	1	Lemon and Verhoef (2016)	30	15.0
2	Ostrom et al. (2015)	290	58.0	2	Verhoef et al. (2009)	125	13.9
3	Brodie et al. (2011)	509	56.5	3	Klaus and Maklan (2013)	40	8.0
4	Verhoef et al. (2009)	598	54.3	4	Vargo and Lusch (2004)	110	7.9
5	Lemon and Verhoef (2016)	153	38.2	5	Lemke et al. (2011)	54	7.7
6	Grewal et al. (2009)	278	25.2	6	Jaakkola et al. (2015)	23	7.7
7	Puccinelli et al. (2009)	260	23.6	7	Vargo and Lusch (2016)	15	7.5
8	Zomerdijk and Voss (2010)	228	22.8	8	Meyer and Schwager (2007)	81	7.4
9	Heinonen et al. (2010)	222	22.2	9	Zomerdijk and Voss (2010)	58	7.3
10	Lemke et al. (2011)	193	21.4	10	Vargo and Lusch (2008)	71	7.1
11	Wu et al. (2014)	123	20.5	11	Gentile et al. (2007)	77	7.0
12	Payne et al. (2009)	224	20.3	12	Ostrom et al. (2015)	21	7.0
13	Meyer and Schwager (2017)	263	20.2	13	Brakus et al. (2009)	57	6.3
14	Chandler and Lusch (2015)	93	18.6	14	Helkkula et al. (2011)	38	6.3
15	Patricio et al. (2011)	151	16.7	15	Payne and (2008)	63	6.3
16	Hui and Bateson (1991)	451	15.5	16	Rose et al. (2012)	31	5.2
17	Bitner et al. (1997).	340	14.7	17	Grewal et al. (2009)	46	5.1
18	Otto and Ritchie (1996)	348	14.5	18	Berry et al (2002)	80	5.0
19	Tax et al. (2013)	87	12.4	19	Patricio et al. (2011)	35	5.0
20	Gallan et al. (2013)	86	12.2	20	Pine and Gilmore (1998)	93	4.7

and explore the theoretical foundations of customer engagement by drawing on the service-dominant logic (SDL) theory. According to the cited authors, customer engagement is a resulting psychological state due to the virtue of customer experiences with a focal agent/object (e.g. brand) within specific service relationships. Other articles in the 20 most-cited focal articles adopt the SDL perspective as e.g. Refs. [32–35]. Besides those, Verhoef et al. [11] and Lemon and Verhoef [7] take a holistic, multi-dimensional, and dynamic view of the customer experience, trying to conceptualize the CX concept, bring a historical perspective of the roots of customer experience and defining the construct operationalization within a marketing perspective. Lastly, the consistency of customer experience research takes place due to the intensification of a variety of topics. For instance, other most cited focal articles cover CX management in retailing (e.g. [36, 37]; service design (e.g. [38, 39]; tourism (e.g. [40]); quality (e.g. [41]), and in healthcare (e.g. [42]).

Table 2 also shows the most-cited references of the focal articles. They contribute to the intellectual roots of the focal articles and form the basis of the co-citation analysis presented hereafter. At first sight, it is noting the importance of the marketing community for the CX intellectual structure since 11 out the 20 most-cited references were published in marketing journals. In addition, five were in service management journals and four in management/business journals. Furthermore, the analysis of the 20 most-cited references confirms the predominance of references on customer experience topics (15 out the 20). Therefore, CX-related research has rapidly increased over the past 10 years and specialized CX references have occupied a central position in the intellectual network. To delve deeper into the intellectual structure of CX, co-citation results are presented next.

4.1 Co-citation Results in the Customer Experience Field

Reference co-citation analysis was chosen as the main method because it has been increasingly used to comprehend the intellectual structure of research fields (as in Refs. [23, 24, 26]). The intellectual structure represents the foundations upon which current research is being carried out and contains fundamental theories and methodological canons of the field [21]. Since the study sample had over 25,000 references, it is not viable to conduct a co-citation analysis of the whole sample set. Then, some authors suggested that a cut off point can be established to select the most influential papers (e.g. [21, 26]). Therefore, the current study selected the references which had been cited at least 15 times, and thus 176 references were involved in the co-citation analysis, clustered using the VOSviewer software. VOSviewer employs an algorithm to locate nodes (references) in such a way that the distance between any two nodes reflects the relationship between them [25]. The software then assigns the nodes to clusters through a community detection process, in which each cluster of closely related nodes represents a leading theoretical community within a specific research field. In the network, each node is assigned to only one cluster and the size of each node (reference) reflects its relevance (based on the quantity of connections) [25]. Finally, because these software tools provide quantitative and visual information that tends to be descriptive in nature, content analysis was employed to interpret the information and gain insight.

As showed in Fig. 2, the analysis of customer experience co-citation network led to the identification of four clusters that were named based on the main content of most references belonging to them. Each one of the four clusters corresponds to one influent theoretical basis for CX research. The first cluster (65 red nodes) was labelled 'service quality'. The second one (22 yellow nodes) was nominated 'service encounter'. The third cluster (53 green nodes) represents 'service-dominant logic' and the last one (36 blue nodes) is 'customer experience'. These four clusters comprise the intellectual structure of the customer experience research. Upon Helkkula's framework [18], each cluster is detailed in more detail following.

4.1.1 Service Quality

Cluster 1 (red nodes) illustrates the influence of service quality theory on the intel-lectual basis of customer experience research. In a traditional view, service quality is the degree of discrepancy between customer's service perceptions and expectations [43, 44]. It relates to the customer satisfaction concept which is traditionally defined as a customer's judgment of a product or service regarding customer's expectations [45, 46]. In fact, although service quality and customer satisfaction concepts arose sepa-rately around the 80s, as they progressed (over the 90s) researchers began to see parallels between them [47]. Thus, a number of studies demonstrate significant rela-tionships between service quality and customer satisfaction constructs which affect behavioural intentions (such as loyalty) and financial results (such as profitability) as e.g. [48–52].

Service quality theory has provided a fundamental basis for understanding and measuring CX [7]. Indeed, historically customer experience was not dealt with as a separate concept but, rather, as an element related to customer satisfaction and service quality [11]. Only more recently, service quality, satisfaction, and CX could be dis-tinguished as different constructs, however, in some cases, complementary [15]. CX is exactly replacing service quality [53]. Therefore, service quality theory has influenced the customer experience research, primarily the outcome-based view [18].

According to Ref. [18], the outcome-based research is a type of CX research which focuses on the relationships affecting: (i) the outcomes of the CX or (ii) how the CX moderates other relationships. The outcome-based research focus on measure CX in terms of outcome variables as satisfaction, loyalty, and repurchase intentions, or how these outcomes might be influenced by other variables (such as emotions), positing the experience as one element in a model linking a number of variables (or attributes) to

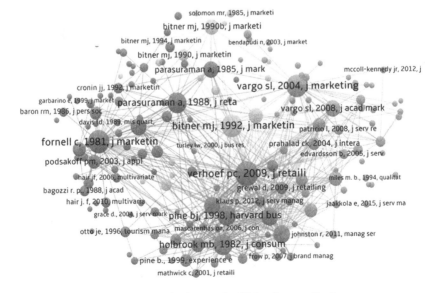

Fig. 2. Co-citation results (Color figure online)

outcomes [18]. For this type of research, service quality provided several models build on a statistical basis, which has sought to measure the influence of specific firm actions on customer's perceptions and behaviours (see Ref. [54] for more details). Thus, Cluster 1 presents several references related to statistical methods, mainly multivariate techniques as e.g. [55].

In a recent review aiming to identify the measures of CX, Bueno et al. [8] demonstrated that SERVQUAL [56] has been utilized by several authors to measure CX in different contexts. In fact, Parasuraman [56] occupies a prominent position in the co-citations network together with Cronin and Taylor [57] which presents SERVPERF as an alternative approach of SERVQUAL. It confirms the assumption that the main contribution of service quality theory remains at the measurement level, where service quality appears as a significant starting point to define and measure CX [7, 8].

4.1.2 Service Encounter

Cluster 2 (yellow nodes) represents the influence of service encounter theory on the intellectual basis of CX research. Service encounter theory was developed from the 90s and it is grounded on the fundamental idea that the customer evaluation of a service depends on the service encounter when the customer interacts directly with the firm [58, 59]. Service encounter encompasses all aspects of the service with which the customer may interact, including its personnel, its physical facilities, and other tangible elements during a consumption ratio [60]. In this domain, the term 'servicescape' refers specifically to the influence of the physical surrounding on customer behaviour [59]. Service encounter theory serves primarily what Helkkula [18] categorized as CX process-based research, whose main focus is on the "architectural elements" of the CX process, which often referred to phases or stages of the service process. Indeed, companies seek to design and execute services that match their brand promise, which implies in manage service encounters at different touch points to deliver a consistent CX [61].

Cluster 2 also encompasses references which deal with the role of employees on the behaviour of customers during the service encounter [59], and how the servicescape can influence employee satisfaction, productivity, and motivation [58–60, 62]. In addition, service encounters often occur in the presence of multiple customers who share the servicescape with each other. Therefore, other customers have an impact on one's customer experience [63]. In this line, servicescape comprises not only objective stimuli but also 'subjective, immeasurable, social, symbolic, and often managerially uncontrollable natural stimuli', which all influence customer decisions and social interactions between the customers [64]. All these architectural elements of the service encounter impact the CX and may to be managed by the companies [61].

4.1.3 Service-Dominant Logic

Cluster 3 (green nodes) portrays the influence of the Service-Dominant Logic (SDL) theory on the intellectual basis of CX research. SDL emerged in the mid-2000s as a 'paradigmatic lens for rethinking the role of services and customers in value creation', replacing: (i) services as the basis of all exchange and (ii) customer as an active participant (co-creator) in relational exchanges with the firm [65, 66]. SDL theory serves primarily the CX process-based research, in terms of Helkkula's framework [18]. In fact, SDL expands the service encounter view to encompass a perspective of providers engaged with a dynamic perspective often referred to as customer journey [7, 67]. In this view, the fundamental premise is that firms must move into the domain of customer experience management in a dynamic way, creating long-term emotional bonds with their customers through the co-creation of memorable experiences potentially involving a network of goods and services [67].

In seeking to translate this fundamental premise in management practices, several authors grounded on SDL theory have dealt with CX trough a service design approach, providing tools to properly manage CX across the different phases and levels of the service [38, 39]. For instance, Bitner et al. [68] describe service blueprints as techniques customer-focused, which allow firms to visualize the service processes, points of customer contact, and the physical evidence associated with their services from their customers' perspective, connecting the underlying support processes throughout the organization that drives and supports CX. In the same line, Patrício et al. [69] introduce the 'service experience blueprint' and Teixeira et al. [70] present the 'customer experience modelling'. Patrício et al. [38] present the 'multilevel service design'. These tools help designers understand how the different levels and stages of customer experience are interrelated and can contribute to design complex service systems.

In addition to providing a theoretical basis for CX process-based research, SDL also supports the phenomenon-based research, which is focused on the individual experiences, usually internal, subjective, event-specific, and context-specific [18]. In this perspective, SDL expands the view of CX beyond hedonic experience by incorporating everyday consumption experiences [71, 72]. Additionally, SDL supports the idea of co-creation of CX, which occurs when interpersonal interaction with other actors within or beyond the service environment influences an actor's subjective response [14, 73]. In this line, SDL allows for a service-ecosystem view of the value co-creation, in which CX is a complex system of different actors and stakeholders [32, 74]. Therefore, experiences are not only subjective but also a relational, social, and inter-subjective phenomenon [32].

4.1.4 Customer Experience

Cluster 4 (blue nodes) represents the customer experience and encompasses what is known as customer experience literature in a specific sense. It provides the seminal works and occupies a central role in the co-citation network of CX research, connecting the different intellectual basis of CX and serving as groundwork. Therefore, it supports phenomenon, process, and outcome-based CX research.

In a phenomenon-based perspective, the seminal article of Holbrook and Hirschman [75] postulates that consumer behaviour links to a subjective state of consciousness regarding 'symbolic meanings', 'hedonic responses', and 'aesthetic criteria' that involves a steady flow of 'fantasies, feelings, and fun'. Later, Pine and Gilmore [76] positioned experience as a distinct entity from goods and services, arguing that consumers want to purchase experiences, which means memorable events that a company stages to engage him/her in an inherently personal way. In a marketing foundation, Schmitt [77, 78] summarizes experiential view as CX-focused, in which consumption is a holistic experience comprised of both the rational and emotional drivers. This holistic approach gained prominence and was adopted by important references to CX research (e.g. [11]).

From a process-based perspective, CX is the process of strategically managing a customers' entire experience with a company to create value both to the customer and the firm [36, 79]. CX refers to the holistic management approach of designing and directing the interactions that customers have with the company and its employees, to improve the total customer experience [11]. Practitioners have to apprise CX management as one of the most promising approaches to address the challenges of consumer markets because CX enhances relationships with customers and builds long-term customer loyalty [79].

In an outcome-based perspective, measuring CX is critical for the firms. Therefore, many companies try to measure and evaluate customer experience through the use of multiple metrics and scales. In this way, the experiential view provides some influential references. For instance, Brakus et al. [80] constructed a brand experience scale that includes four dimensions: 'sensory', 'affective', 'intellectual', and 'behavioural'. Chang and Horng [81] developed a multidimensional scale of experience quality which is composed of five dimensions: 'physical surroundings', 'service providers', 'other customers', 'customers' companions' and the 'customers themselves'. References [9, 53] propose a scale to measure the overall quality of the customer experience, which has four dimensions: 'peace of mind', 'moments of truth', 'focus on output', and 'product experience'.

Finally, due to recent advances in mobile technology and the increasing number of consumers using the Internet to make purchases, the dynamics of e-commerce have changed considerably, making it impressive for companies to re-engineer their interaction and service delivery to gain an ideal CX. In this direction, since the 2000s some studies have focused on the online CX as e.g. in Refs. [82–84]. Next, the following results deals with the growing literature relating technology and CX.

4.2   The Technology Advance at Customer Experience Literature

'Enhancing the customer experience' and 'leveraging technology to advances services' are among the top priorities of contemporary research for advancing the service field [17]. Indeed, the interplay between technology and CX is presented in several recent articles (e.g. Refs. [93–100]), shaping a research stream whose early works emerged in the 2000s based on the service encounter theory. At that time, scholars have already recognized the CX management challenges facing the increasing deployment of technology, since technology infusion changes the essence of service encounters and, therefore, the ways the customer experiences the service [85–87], by enabling service customization, improving service recovery and spontaneously delighting customer [85].

An initial and prolific line of research has focused heavily on self-service technologies, especially on the factors for customer acceptance or lack of acceptance of the technology [86, 88–90]. In addition to the service encounter theory, the service-dominant logic provides a framework for this research line, by applying key themes of value co-creation, co-production, and resource-integration as lens to provide insights into the adoption of self-service technology and CX consequences [91, 92]. Recent studies have dealt with customer's acceptance of emerging technologies such as artificial intelligence (e.g. [93]) and service robots (e.g. Ref. [94]), investigating the effects on the CX and customer's behavioural consequences. Others publications have already dealt with how emerging technologies can support the CX measurement and management, including big data (e.g. Ref. [95]), internet of things (e.g. Ref. [96]), augment reality (e.g. Ref. [97]), and mobile social media (e.g. Ref. [98])

Currently, the concept of service encounter 2.0 has emerged encompassing 'any customer-company interaction that results from a service system that is comprised of interrelated technologies (either company-or customer-owned), human actors (employees and customers), physical/digital environments and company/customer processes' [99]. In this context, the CX literature points to the need for further investigate the interplay among the CX's physical, digital and social realms to create a consistently superior CX [100].

5   Concluding Remarks

This paper identified the most influential published sources and explored the changes that have come about in the intellectual structure of customer experience research using bibliometric techniques. In terms of volume of publication, 2009 reveals a significant increase in the number of publications, with emphasis on the past four years (2015–2018). This increasing interest can be partially explained by the emergence of service-dominant logic and by the fact that companies are now competing mostly on the basis of customer experience. The bibliometric analysis revealed three clusters of research (service quality, service encounter, and service-dominant logic), which constitutes the foundations of the CX research field. Nevertheless, in the past years, the CX literature has evolved to form its own body of knowledge and, consequently, a number of studies specifically dedicated to the experiential aspects of the consumption have started to gain influence in the field.

Based on the current results, some implications for the customer experience research can be summarized:

- The emergence of customer-centric service theories has contributed to the development of the CX research field. With the emergence of service quality theory in the late 70s and early 80s, the customer perspective came to occupy a central focus in the service evaluation. In the 90s, service encounter theory has continued the centrality of the customer, emphasizing the direct interactions between the customer and organization and how service processes should be managed from the customer perspective. Later, in the mid-2000s, the emergence of the service-dominant logic theory deepened the central role of the customer, while broadening the spectrum of influence over the customer to increasingly complex value networks that translate into the service ecosystem. Taken all of them together, these customer-centric theories have allowed the development of a theoretical corpus culminated in the emergence of CX specific literature.
- The initial studies focusing on CX emphasized the hedonic aspects of consumption. In this sense, these seminal articles positioned CX as a phenomenon related to

memorable experiences staged by organizations to customers, which are involved in 'fantasies, feelings, and fun'. Conversely, later studies have changed the conceptualization of customer experience construct. Indeed, the definition of CX has evolved into a holistic view that represents the customer response to multiple interactions with the organization (objective and subjective aspects), including all routine customer-organization encounters.

- Customer experience has evolved from a static to a dynamic view. Initially, the main focus of the CX research was to evaluate the service delivery from the customer perspective, following the tradition of the service quality theory. Overall, these initial studies focused on investigating customer perceptions during the purchase were losing attention from researchers. As the CX research field evolves and other theoretical influences are incorporated into the intellectual structure, the customer experience view expands to integrate other different stages of the purchase, including the moments before and after its realization. This implies the need for organizations to continuously monitor and manage processes in the different purchase phases to engage the customer. Furthermore, the idea that services are co-created with customers gain impulse and highlights the relationship between experience and co-creation.

- As customer experience research field consolidates, the role of technology has gained prominence. Research into how technology can help monitor and improve the CX has emerged in the late 2000s and especially from 2015. It may result from the consolidation of a theoretical body entirely devoted to customer experience, which allowed the field to expand its related topics. In addition, the increasing importance of the internet in consumption relations also contributed to driving the interest of the CX field to technology. More recently, cheapness and advances in emerging technologies such as artificial intelligence, big data, analytics, and robotics have driven research that links customer experience and technology.

To the best of our knowledge, bibliometric studies are not yet well established in the field of customer experience research. Although this study attempts to systematically map the field using bibliometric co-citation analysis, it is not without its own limitations. In particular, special attention should be given to the fact that the data collection was conducted using exclusively the SSCI-WoS database and, naturally, not all issues and areas have been covered. As such, the next steps of this study would be to extend to other databases in order to complement the results obtained in this phase.

References

1. Gartner: Key findings from the gartner customer experience survey (2018). https://www.gartner.com/en/marketing/insights/articles/key-findings-from-the-gartner-customer-experience-survey. Accessed 1 Aug 2019
2. De Keyser, A., Lemon, K.N., Klaus, P., Keiningham, T.L.: A framework for understanding and managing the customer experience. Marketing Science Institute, Working Paper Series Report, 15-21 (2015)

3. Morgan, B.: The 10 best B2B customer experience [Web log post], 4 June 2019. https://www.forbes.com/sites/blakemorgan/2019/06/04/the-10-best-b2b-customer-experiences/#24ce226653f5. Accessed 22 Aug 2019

4. Patel, S.: Five small businesses that get customer experience right [Web log post] (2018). https://www.entrepreneur.com/article/319520. Accessed 31 Aug 2019

5. McColl-Kennedy, J.R., et al.: Fresh perspectives on customer experience. J. Serv. Mark. **29**(6/7), 430–435 (2015)

6. Jain, R., Aagja, J., Bagdare, S.: Customer experience – a review and research agenda. J. Serv. Theory Pract. **27**(3), 642–662 (2017)

7. Lemon, K.N., Verhoef, P.C.: Understanding customer experience throughout the customer journey. J. Mark. **80**(6), 69–96 (2016)

8. Bueno, E.V., Beauchamp Weber, T.B., Bomfim, E.L., Kato, H.T.: Measuring customer experience in service: a systematic review. Serv. Ind. J. **39**(11–12), 779–798 (2019)

9. Klaus, P., Maklan, S.: Towards a better measure of customer experience. Int. J. Market Res. **55**(2), 227–246 (2013)

10. Maklan, S., Klaus, P.: Customer experience: are we measuring the right things? Int. J. Market Res. **53**(6), 771–772 (2011)

11. Verhoef, P.C., Lemon, K.N., Parasuraman, A., Roggeveen, A., Tsiros, M., Schlesinger, L. A.: Customer experience creation: determinants, dynamics and management strategies. J. Retail. **85**(1), 31–41 (2009)

12. Mahr, D., Stead, S., Odekerken-Schröder, G.: Making sense of customer service experiences: a text mining review. J. Serv. Mark. **33**(1), 88–103 (2019)

13. Kranzbühler, A.M., Kleijnen, M.H.P., Morgan, R.E., Teerling, M.: The multilevel nature of customer experience research: an integrative review and research agenda. Int. J. Manag. Rev. **20**(2), 433–456 (2018)

14. Jaakkola, E., Helkkula, A., Aarikka-Stenroos, L.: Service experience co-creation: conceptualization, implications, and future research directions. J. Serv. Manag. **26**(2), 182–205 (2015)

15. Vasconcelos, A.M.D., Barichello, R., Lezana, A., Forcellini, F.A., Ferreira, M.G.G., Cauchick Miguel, P.A.: Conceptualisation of the service experience by means of a literature review. Benchmarking: Int. J. **22**(7), 1301–1314 (2015)

16. Ostrom, A.L., et al.: Moving forward and making a difference: research priorities for the science of service. J. Serv. Res. **13**(1), 4–36 (2010)

17. Ostrom, A.L., Parasuraman, A., Bowen, D.E., Patrício, L., Voss, C.A.: Service research priorities in a rapidly changing context. J. Serv. Res. **18**(2), 127–159 (2015)

18. Helkkula, A.: Characterising the concept of service experience. J. Serv. Manag. **22**(3), 367–389 (2011)

19. Hwang, J., Seo, S.: A critical review of research on customer experience management: theoretical, methodological and cultural perspectives. Int. J. Contemp. Hosp. Manag. **28**(10), 2218–2246 (2016)

20. Lipkin, M.: Customer experience formation in today's service landscape. J. Serv. Manag. **27**(5), 678–703 (2016)

21. Zupic, I., Čater, T.: Bibliometric methods in management and organization. Organ. Res. Methods **18**(3), 429–472 (2015)

22. Bragge, J., Kauppi, K., Ahola, T., Aminoff, A., Kaipia, R., Tanskanen, K.: Unveiling the intellectual structure and evolution of external resource management research: insights from a bibliometric study. J. Bus. Res. **97**(1), 141–159 (2019)

23. Ferreira, F.A.F.: Mapping the field of arts-based management: bibliographic coupling and co-citation analyses. J. Bus. Res. **85**, 348–357 (2018)

24. Leung, X.Y., Sun, J., Bai, B.: Bibliometrics of social media research: a co-citation and co-word analysis. Int. J. Hosp. Manag. **66**, 35–45 (2017)
25. van Eck, N.J., Waltman, L.: Software survey: VOSviewer, a computer program for bibliometric mapping. Scientometrics **84**, 523–538 (2010)
26. Dzikowski, P.: A bibliometric analysis of born global firms. J. Bus. Res. **85**, 281–294 (2018)
27. Gurzki, H., Woisetschläger, D.M.: Mapping the luxury research landscape: a bibliometric citation analysis. J. Bus. Res. **77**, 147–166 (2017)
28. Følstad, A., Kvale, K.: Customer journeys: a systematic literature review. J. Serv. Theory Pract. **28**(2), 196–227 (2018)
29. Braun, V., Clarke, V.: Using thematic analysis in psychology. Qual. Res. Psychol. **3**(2), 77–101 (2006)
30. Nag, R., Hambrick, D.C., Chen, M.-J.: What is strategic management, really? Inductive derivation of a consensus definition of the field. Strateg. Manag. J. **28**(5), 935–955 (2007)
31. Brodie, R.J., Hollebeek, L.D., Jurić, B., Ilić, A.: Customer engagement: conceptual domain, fundamental propositions, and implications for research. J. Serv. Res. **14**(3), 252–271 (2011)
32. Chandler, J.D., Lusch, R.F.: Service systems: a broadened framework and research agenda on value propositions, engagement, and service experience. J. Serv. Res. **18**(1), 6–22 (2015)
33. Heinonen, K., Strandvik, T., Mickelsson, K.J., Edvardsson, B., Sundström, E., Andersson, P.: A customer-dominant logic of service. J. Serv. Manag. **21**(4), 531–548 (2010)
34. Payne, A., Storbacka, K., Frow, P., Knox, S.: Co-creating brands: diagnosing and designing the relationship experience. J. Bus. Res. **62**(3), 379–389 (2009)
35. Wu, P.-L., Yeh, S.-S., Huan, T.C., Woodside, A.G.: Applying complexity theory to deepen service dominant logic: configural analysis of customer experience-and-outcome assessments of professional services for personal transformations. J. Bus. Res. **67**(8), 1647–1670 (2014)
36. Grewal, D., Levy, M., Kumar, V.: Customer experience management in retailing: an organizing framework. J. Retail. **85**(1), 1–14 (2009)
37. Puccinelli, N.M., Goodstein, R.C., Grewal, D., Price, R., Raghubir, P., Stewart, D.: Customer experience management in retailing: understanding the buying process. J. Retail. **85**(1), 15–30 (2009)
38. Patrício, L., Fisk, R.P., FalcãoeCunha, J., Constantine, L.: Multilevel service design: from customer value constellation to service experience blueprinting. J. Serv. Res. **14**(2), 180–200 (2011)
39. Zomerdijk, L.G., Voss, C.A.: Service design for experience-centric services. J. Serv. Res. **13**(1), 67–82 (2010)
40. Otto, J.E., Ritchie, J.R.B.: The service experience in tourism. Tour. Manag. **17**(3), 165–174 (1996)
41. Lemke, F., Clark, M., Wilson, H.: Customer experience quality: an exploration in business and consumer contexts using repertory grid technique. J. Acad. Mark. Sci. **39**(6), 846–869 (2011)
42. Gallan, A.S., Jarvis, C.B., Brown, S.W., Bitner, M.J.: Customer positivity and participation in services: an empirical test in a health care context. J. Acad. Mark. Sci. **41**(3), 338–356 (2013)
43. Grönroos, C.: A service quality model and its marketing implications. Eur. J. Mark. **18**(4), 36–44 (1984)
44. Parasuraman, A., Zeithaml, V.A., Berry, L.L.: A conceptual model of service quality and its implications for future research. J. Mark. **49**(4), 41–50 (1985)

45. Oliver, R.L.: A cognitive model of the antecedents and consequences of satisfaction decisions. J. Mark. Res. **17**(4), 460–469 (1980)
46. Oliver, R.: Cognitive, affective, and attribute bases of the satisfaction response. J. Consum. Res. **20**, 418–430 (1993)
47. Oliver, R., Rust, R., Varki, S.: Customer delight: foundations, findings, and managerial insights. J. Retail. **73**(3), 311–336 (1997)
48. Anderson, E., Fornell, C., Lehman, D.: Customer satisfaction, market share, and profitability: findings from Sweden. J. Mark. **58**, 53–66 (1994)
49. Bolton, R.N., Drew, J.H.: Multistage model of service customers' quality and value assessments. J. Consum. Res. **17**(4), 375–384 (1991)
50. Cronin Jr., J.J., Brady, M.K., Hult, G.T.M.: Assessing effects of quality, value and customer satisfaction. J. Retail. **76**(2), 193–218 (2000)
51. Tam, J.L.M.: Customer satisfaction, service quality and perceived value: an integrative model. J. Mark. Manag. **20**, 897–917 (2004)
52. Zeithaml, V.A., Berry, L.L., Parasuraman, A.: The behavioral consequences of service quality. J. Mark. **60**(2), 31–46 (1996)
53. Klaus, P., Maklan, S.: EXQ: a multiple-item scale for assessing service experience. J. Serv. Manag. **23**(1), 5–33 (2012)
54. Gupta, S., Zeithaml, V.: Customer metrics and their impact on financial performance. Mark. Sci. **25**(6), 718–739 (2006)
55. Hair, J.F., Black, W., Babin, B., Anderson, R.: Multivariate Data Analysis. Pearson New International Edition (2010)
56. Parasuraman, A., Zeithaml, V., Berry, L.: SERVQUAL: a multiple-item scale for measuring consumer perceptions of service quality. J. Retail. **64**, 12–40 (1988)
57. Cronin, J.J., Taylor, S.A.: Measuring service quality: a reexamination and extension. J. Mark. **56**(3), 55–68 (1992)
58. Bitner, M.J.: Evaluating service encounters: the effects of physical surroundings and employee responses. J. Mark. **54**(2), 69–82 (1990)
59. Bitner, M.J.: Servicescapes: the impact of physical surroundings on customers and employess. J. Mark. **56**(2), 57–71 (1992)
60. Shostack, G.L.: Planning the service encounter. In: Czepiel, J.A., Solomon, M.R., Suprenant, C.F. (eds.) The Service Encounter: Managing Employee/Customer Interaction in Service Businesses, pp. 243–254. Lexington Books, New York (1985)
61. Bolton, R.N.: Service timing: designing and executing service in a dynamic environment. In: Maglio, P.P., Kieliszewski, C.A., Spohrer, J.C., Lyons, K., Patrício, L., Sawatani, Y. (eds.) Handbook of Service Science, Volume II. SSRISE, pp. 13–33. Springer, Cham (2019). https://doi.org/10.1007/978-3-319-98512-1_2
62. Hui, M.K., Bateson, J.E.G.: Perceived control and the effects of crowding and consumer choice on the service experience. J. Consum. Res. **18**(2), 174–184 (1991)
63. Grove, S.J., Fisk, R.P.: The impact of other customers on service experiences: a critical incident examination of "getting along". J. Retail. **73**(1), 63–85 (1997)
64. Rosenbaum, M.S., Massiah, C.: An expanded servicescape perspective. J. Serv. Manag. **22**(4), 471–490 (2011)
65. Vargo, S.L., Lusch, R.F.: Evolving to a new dominant logic for marketing. J. Mark. **68**, 1–17 (2004)
66. Vargo, S.L., Lusch, R.F.: Service-dominant logic: continuing the evolution. J. Acad. Mark. Sci. **36**(1), 1–10 (2008)
67. Tax, S.S., McCutcheon, D., Wilkinson, I.F.: The service delivery network (SDN): a customer-centric perspective of the customer journey. J. Serv. Res. **16**(4), 454–470 (2013)

68. Bitner, M.J., Ostrom, A.L., Morgan, F.N.: Service blueprinting: a practical technique for service innovation. Calif. Manag. Rev. **50**(3), 66–94 (2008)
69. Patrício, L., Fisk, R.P., Falcão, E., Cunha, J.: Designing multi-interface service experiences: the service experience blueprint. J. Serv. Res. **10**(4), 318–334 (2008)
70. Teixeira, J., Patricio, L., Nunes, N.J., Nobrega, L., Fisk, R.P., Constantine, L.: Customer experience modeling: from customer experience to service design. J. Serv. Manag. **23**(3), 362–376 (2012)
71. Carù, A., Cova, B.: Marketing theory: a more humble but complete view of the concept. Mark. Theory **3**(2), 267–286 (2003)
72. Edvardsson, B., Enquist, B., Johnston, R.: Cocreating customer value through hyperreality in the prepurchase service experience. J. Serv. Res. **8**(2), 149–161 (2005)
73. Helkkula, A., Kelleher, C., Pihlström, M.: Practices and experiences: challenges and opportunities for value research. J. Serv. Manag. **23**(4), 554–570 (2012)
74. Vargo, S.L., Lusch, R.F.: Institutions and axioms: an extension and update of service-dominant logic. J. Acad. Mark. Sci. **44**(1), 5–23 (2016)
75. Holbrook, M.B., Hirschman, E.C.: The experiential aspects of consumption: consumer fantasies, feelings, and fun. J. Consum. Res. **9**(2), 132–140 (1982)
76. Pine, J., Gilmore, J.: Welcome to the experience economy. Harvard Bus. Rev. **76**(4), 97–105 (1998)
77. Schmitt, B.H.: Experiential marketing. J. Mark. Manag. **215**(1–3), 53–67 (1999)
78. Schmitt, B., Zarantonello, L.: Consumer experience and experiential marketing: a critical review. Rev. Mark. Res. **10**, 25–61 (2013)
79. Homburg, C., Jozić, D., Kuehnl, C.: Customer experience management: toward implementing an evolving marketing concept. J. Acad. Mark. Sci. **45**(3), 377–401 (2017)
80. Brakus, J.J., Schmitt, B.H., Zarantonello, L.: Brand experience: what is it? How is it measured? Does it affect loyalty? J. Mark. **73**(3), 52–68 (2009)
81. Chang, T.Y., Horng, S.C.: Conceptualizing and measuring experience quality: the customer's perspective. Serv. Ind. J. **30**(14), 2401–2419 (2010)
82. Bilgihan, A., Kandampully, J., Zhang, T.: Towards a unified customer experience in online shopping environments: antecedents and outcomes. Int. J. Qual. Serv. Sci. **8**(1), 102–119 (2016)
83. Novak, T.P., Hoffman, D.L., Yung, Y.F.: Measuring the customer experience in online environments: a structural modeling approach. Mark. Sci. **19**(1), 22–42 (2003)
84. Rose, S., Clark, M., Samouel, P., Hair, N.: Online customer experience in e-retailing: an empirical model of antecedents and outcomes. J. Retail. **88**(2), 308–322 (2012)
85. Bitner, M.J., Brown, S.W., Meuter, M.L.: Technology infusion in service encounters. J. Acad. Mark. Sci. **28**(1), 138–149 (2000)
86. Bitner, M.J., Ostrom, A.L., Meuter, M.L.: Implementing successful self-service technologies. Acad. Manag. Perspect. **16**(4), 96–108 (2002)
87. Meuter, M.L., Ostrom, A.L., Roundtree, R.I., Bitner, M.J.: Self-service technologies: understanding customer satisfaction with technology-based service encounters. J. Mark. **64**(3), 50–64 (2000)
88. Davis, F.D.: Perceived usefulness, perceived ease of use, and user acceptance of information technology. MIS Q. **13**(3), 319–340 (1989)
89. Meuter, M.L., Ostrom, A.L., Bitner, M.J., Roundtree, R.: The influence of technology anxiety on consumer use and experiences with self-service technologies. J. Bus. Res. **56**(11), 899–906 (2003)
90. Salomann, H., Kolbe, L., Brenner, W.: Self-services in customer relationships: balancing high-tech and high-touch today and tomorrow. E-Service **4**(2), 65–84 (2006)

91. Hilton, T., Hughes, T., Little, E., Marandi, E.: Adopting self-service technology to do more with less. J. Serv. Mark. **27**(1), 3–12 (2013)

92. Nilsson, E., Ballantyne, D.: Reexamining the place of servicescape in marketing: a service-dominant logic perspective. J. Serv. Mark. **28**(5), 374–379 (2014)

93. Ostrom, A.L., Fotheringham, D., Bitner, M.J.: Customer acceptance of AI in service encounters: understanding antecedents and consequences. In: Maglio, P.P., Kieliszewski, C.A., Spohrer, J.C., Lyons, K., Patrício, L., Sawatani, Y. (eds.) Handbook of Service Science, Volume II. SSRISE, pp. 77–103. Springer, Cham (2019). https://doi.org/10.1007/978-3-319-98512-1_5

94. Wirtz, J., et al.: Brave new world: service robots in the frontline. J. Serv. Manag. **29**(5), 907–931 (2018)

95. Burton, J., Nasr, L., Gruber, T., Bruce, H.L.: Special section: advancing customer experience and big data impact via academic–practitioner collaboration. J. Serv. Mark. **31**(2), 142–147 (2017)

96. Balaji, M.S., Roy, S.K.: Value co-creation with internet of things technology in the retail industry. J. Mark. Manag. **33**(1–2), 7–31 (2017)

97. Hilken, T., de Ruyter, K., Chylinski, M., Mahr, D., Keeling, D.I.: Augmenting the eye of the beholder: exploring the strategic potential of augmented reality to enhance online service experiences. J. Acad. Mark. Sci. **45**(6), 884–905 (2017)

98. Carlson, J., Rahman, M.M., Taylor, A., Voola, R.: Feel the VIBE: examining value-in-the-brand-page-experience and its impact on satisfaction and customer engagement behaviours in mobile social media. J. Retail. Consum. Serv. **46**, 149–162 (2019)

99. Larivière, B., et al.: "Service Encounter 2.0": an investigation into the roles of technology, employees and customers. J. Bus. Res. **79**, 238–246 (2017)

100. Bolton, R.N., et al.: Customer experience challenges: bringing together digital, physical and social realms. J. Serv. Manag. **29**(5), 776–808 (2018)

On Finding the Voice of the Customer in the Digital Traces of the Michelin-Star Gastronomy Experience: Unveiling Insights for Service Design

Jesús Alcoba[1]([⊠]), Susan Mostajo[2], Rowell Paras[2],
Romano Angelico Ebron[2], and Adela Balderas-Cejudo[3,4]

[1] International Graduate School, Centro Superior de Estudios Universitarios La
Salle, Universidad Autónoma de Madrid, Madrid, Spain
jesus@lasallecampus.es
[2] De La Salle University-Dasmariñas, Dasmariñas, Philippines
{stmostajo, rrparas, rtebron}@dlsud.edu.ph
[3] Oxford Institute of Population Ageing, University of Oxford, Oxford, UK
abalderas@bculinary.com
[4] Basque Culinary Center, San Sebastián, Spain

Abstract. Services contribute to create experiences, and the critical question is how the management can systematically orchestrate the service layer that underlies experiences, taking into account that this endeavor, in most cases, require the participation of all departments and systems of a firm. In that context, the first issue is how to find the insights that are the basis for design. In a service economy where co-creation is a trend, it is mandatory to incorporate the voice of the customer from the first moment. By means of sentiment-analysis and statistics, this paper draws insights from opinions on a digital platform of customers of renowned Michelin-starred restaurants for the improvement of a gastronomic experience design. The Michelin-star system leads service provision in the realm of haute cuisine identifying and giving greater visibility to those restaurants whose value proposal stands out from the rest. The basic assumption of this paper is that analyzing sector leaders can be of great help for the rest of the actors in the field.

Keywords: Service science · Service design · Experience design · Voice of the customer · Sentiment analysis · Gastronomy · Tourism

1 Introduction

The basic pillar of economy is that customers exchange their money for that which provides them with value. Recently, the evolution of economic value has experienced major changes that can be explained through service-dominant logic [1–5], service science [6–10] and customer experience [11–13]. From this approach, experiences can help them to construct identity [14–16] and because of that they are of prime importance when delivering value to clients, due to the immense value that this implies for

© Springer Nature Switzerland AG 2020
H. Nóvoa et al. (Eds.): IESS 2020, LNBIP 377, pp. 21–34, 2020.
https://doi.org/10.1007/978-3-030-38724-2_2

them. It is not the objective of this article to describe the history of these paradigms nor to detail their concepts. However, while it is true that the economy is increasingly dominated by the immense value that customers place on experiences, it is equally true that services represent part of the generation of an experience (the rest is provided by the client, as experiences are subjective). In fact, tools like service blue print [17, 18] can be considered in the design of customer journeys [19]. In line with this assumptions, the capture of deep customer insights is of paramount importance in services design, as a basic requirement within the concept of value co-creation that has been emphasized by the service-dominant logic approach.

Building on this framework, to find out the voice of the customer [20] is of prime importance with the purpose of improving service design through gathering insights in the digital traces that, in the form of opinions, they leave in online platforms. This study starts from this approach and focuses on the tourism industry, more specifically on gastronomy.

Tourism, as a new driver of rapid economic growth, has immensely progressed over time. As tourists become more informed, have more disposable income and leisure time, factors affecting their travel decisions and motivations have also changed. An escalating number of tourists are increasingly seeking for new, meaningful experiences; and gastronomy is viewed as an experience. Experiences and attractions related to food have gained a lot of attention, hence, seen as one of the travel motivations and become an essential ingredient of the entire tourist experience whilst the gastronomic tourism has emerged as one of the most dynamic segments of tourism sector [21].

Gastronomic tourism has surfaced as a key player in the tourism market, attracting more and more tourists every year. Gastronomy tourism has emerged as particularly important not only because food and drink is central to any tourist experience, but also "because the concept of gastronomy tourism has evolved to embrace cultural practices and include in its discourse the ethical and sustainable values of the territory, the landscape, the sea, local history, values and culture heritage" [22].

Furthermore, food and beverages are vital and influential tools for a place to become attractive tourist spot. Rural tourism development and areas image empowerment will result from these unique local products, thus, positive contribution to the local and rural economies creating job and other economic opportunities. These would not only stimulate the food and beverage industry but also would have considerable effects on other productive sectors [23].

More particularly, gastronomic tourism is considered as one of the fastest-growing travel motivations in Spain. Over five million out of 56.7 million tourists who chose Spain as a tourism destination were increasingly motivated and enthralled by the Spanish food and wine, which generated remarkable receipts of over five billion euros. Moreover, the food and agriculture sector in Spain represents 7.6% of GDP and employs more than 400,000 individuals [24]. Clearly, as gastronomic tourism develops, economy accelerates in accordance resulting in wealth accumulation and job creation.

2 The Michelin Star System

Initially meant for the court and aristocracy, modern culinary creativity moved to restaurants, with chefs first as employees then, with the arrival of nouvelle cuisine, as restaurant owners [25].

The restaurant industry is driven by two key characteristics: (1) as a typical experience good, it needs signaling devices that consumers may trust in their decision-making process [26]. And (2) Haute-cuisine is also special to the extent that consumers expect from it some sort of "magic of discovery". Thus, experience, excellence and trust seem to be not only important aspects but driven forces.

Haute cuisine has its institutions, convention setters and tradition guardians [25]. The Guide Michelin, created in 1900, was initially designed by the Michelin tire company to help motorists by giving various technical advices [27].

It is based on a standardized system of fixed criteria [28], based on anonymous inspections and independence, featuring a selection of best hotels and restaurants in all comfort and price categories [29]. The guide is published annually with a standard classification assessment by recognizing the cultural differences among 21 countries worldwide [30, 31]. Michelin stars are awarded to restaurants on five criteria: the quality of the products, mastering of flavors and cooking, the personality of the cuisine, value for money and consistency [29, 31]. One star is considered a very good restaurant in its category, two stars reflect excellent cooking, worth a detour, while three stars display exceptional cuisine, worth a special journey [32].

Nowadays, the Michelin star is acknowledged internationally as the most serious [30–32], and its conception of haute cuisine grows influential throughout the world [28]. More specifically, the Michelin stars are the criterion most valued by all field agents [33].

Very first-class chefs around the world have to adapt their supply to luxury criteria to become a compulsory call for both business and leisure luxury tourists [34]. Some studies stated that professionally evaluated awards are the most effective way of boosting the market success of Haute-Cuisine among chefs, restaurateurs, culinary experts [29, 30].

Central to the key features of this system is the "mystique" culture and clandestine nature of Michelin that is an essential element of the Red Guide itself that contributes to preserve creativity in the segment. An aspect of this is that Michelin inspectors who test and rate the restaurants do so anonymously and visits are unannounced, so that Chefs simply do not know when they are serving the inspectors [26]. It is important to highlight that its influence in restaurant choice is both impressive and fearsome [30].

3 Gastronomy in San Sebastian (Spain)

There is a worldwide rise of concern for gastronomy and, specifically, for top gastronomy. It is fashionable to talk about gastronomy and luxury gastronomy has become a mass topic [34]. Ferran Adriá states [35] that all Spanish culinary connoisseurs point to the national cuisine's giant leap over the last twenty years as the birth of what came to be called New Basque Cuisine. The Basque Country has long been recognized for the excellence of its food.

In recent decades, Basque chefs have obtained worldwide recognition, winning numerous international awards. San Sebastian is internationally known for holding one of the highest concentrations of Michelin stars per square kilometer [36]. The reputation of the Basque Country, which has had an enormously positive impact on the region's image, is largely due to the efforts of an initial group of chefs—and others who came in their wake—in bringing innovation to their work and in disseminating Basque culture through their cuisine [37].

A whole generation of Basque chefs has accomplished the impressive feat of bringing together identity and food [36].

It is to a great extent as a result of this collaboration that many Basque chefs, and their restaurants, have now become international reference points, making the Basque Country a pole of attraction and winning the highest national and international plaudits in the world of cuisine [37].

A landmark in the collaboration movement among Basque chefs came in 1976, when they got in touch with other chefs from the area and formed the "gang of twelve" of what came to be known as Basque Nouvelle Cuisine—Juan Mari Arzak, Pedro Subijana, Ricardo Idiáquez, María Jesús Fombellida, Pedro Gómez, Manuel Iza and Jesús Mangas, Patxi Kintana, José Juan Castillo, Karlos Arguiñano, Ramón Roteta, Xabier Zapirain and Luis Irizar [38].

The network of collaboration created around the group of Basque "top chefs" transcends the limits of the industry, extending to relationships with suppliers, producers of foodstuffs and other products, kitchen and dining room material, financial resources, decoration services, consultancy, legal and research firms [29]. In March 2009, seven Basque chefs (Juan Mari Arzak, Pedro Subijana, Karlos Arguiñano, Hilario Arbelaitz, Martín Berasategui, Andoni Luis Aduriz and Eneko Atxa) teamed up with the University of Mondragón to create Basque Culinary Center to transfer and generate knowledge in San Sebastian [37].

4 Gastronomy Experience: Beyond Cooking

The paradigm shift on how organizational assets are viewed can be linked to the evolution of production economy to knowledge economy. Intangible resources were recognized as contributing factor for economic success in the knowledge economy. Intangible resources otherwise considered as the human capital is the knowledge, abilities and skills of people employed in the organization [39, 40]. Thus, the value of human resources for organization's success was given emphasis being the one that create sustainable competitive advantage of the firm – the company's value, long-term strength and most important asset [41].

Recognizing the value of people in organizational performance justifies the reason why companies worldwide strive to hire the best people to work with them. In such case, tourism facilities also aim to acquire the best workforce who possesses the required knowledge, skills and attitude for the job. Literature suggests the necessity of soft skills over hard skills for the hospitality and tourism employees [42, 43]. The importance of teamwork, communication skills, customer service, professional image, comprehension of performance standards, and prioritizing needs of others over personal needs were specifically expressed by the same authors. The ability to work in a team as a vital competency was also supported by other authors [44]. Additionally, various characteristics like professional background, culture and change adaptability, technical skills and personality traits of human resources in the global tourism industry for valued services were recognized [43].

In gastronomy, personnel were mentioned by Kocbek as part of the factors affecting the experience of quality service in addition to the taste and price of served food or beverage, environment, service quality, hygiene, cleaning and atmospheric elements [45]. The Michelin Guide states certain standards in the quality of products/food served, and the competency of the chef and the staff in providing the service is also mentioned a necessity [46]. Thus, it is expected that a Michelin Starred restaurant should maintain the concept of high quality food and service approaches, the chef's creativity and mastery of flavor and cooking techniques as well as the manner how staff delivers the food from the kitchen to the customer for the provision of an exceptional diner's experience.

The necessity to create a meaningful experience is a prime factor in the quality of service in tourism industry [16] which is remarkable to discover in gastronomy. And in a Michelin-starred restaurant, there might be other factors fundamental in the creation of a meaningful gastronomy experience apart from cooking. Perhaps the human resources' contribution in this context can be thought of in order to call the experience worth a detour and/or worth a special journey. As argued by an author, human resources are recognized to play a more important role in the fine dining innovation than other product innovation situations in a Michelin Star [46]. Likewise, social interaction was recognized as one of the drivers of customer service experience in restaurants where employees are friendly, make guests feel treated well and cared, and knowledgeable and skillful yet not too dominant [47].

Having the perspective that the quality of product or service is the worth of every company in the market [31], experts advocated to continuously improve business processes to ensure service quality. Being anchored on this framework, it is the intention of this research to unveil the insights that make up the gastronomic customer experience in a Michelin-starred restaurant. Consequently, and considering also that emotions are related to memorability [48] the findings of this research can serve as basis in service design, in order to create a gastronomy experience that is considered memorable by clients.

5   Methodology

Emanating from the perspective that customers' perception is necessary to improve service quality, the current study used the concept of voice of customer [20], a view on how organizations gather, analyze, and take actions of customers' feedback in order to improve customer experience for organizational benefits. Through this principle applied to gastronomy, and following similar research in comparable topics [14, 16, 45, 49–51] all the available opinions given by Spanish clients from the 9 Michelin-starred restaurants in San Sebastian were downloaded from TripAdvisor, in line with a "no topic is off-limits" approach [20].

There were 3664 opinions, and they were analyzed by a sentiment-analysis engine, a tool that can analyze large amounts of text to extract what are the topics discussed and to calculate the intensity and direction of feelings towards them [52]. The value of this tool lies in the automation of the voice of the customer by extending the narrow focus of classic questionnaires to the analysis of large amounts of data.

The software identified 16367 text lines, highlighting a "sentiment topic", the "sentiment text" where the topic is in, a "sentiment score" and a polarity (positive or negative, depending on the case) for each one. Out of these 16367 text lines, the study eventually covers only 12856, since the sentiment-analysis engine identified no topic in the others.

The goal of this study is to get insights on how the customer describes the gastronomic experience in a Michelin-starred restaurant. The ultimate goal is to provide keys for the design of the service layer that underlies the customer experience. With the aim of analyzing this issue in greater depth, the analysis strategy involved the following steps:

1. Frequency analysis. For the analysis of the most recurrent topics mentioned by customers the sentiment topic was used to identify which words were used the most. The ten sentiment topics that customers more frequently mentioned were selected. In order to analyze which words customers used to describe their experience in each of these most mentioned sentiment topics, a frequency analysis of the "sentiment text" was carried out. These are the text strings associated to each topic. Eventually, all the comments were arranged from higher to lower according to their "sentiment score" to identify the most frequently mentioned "sentiment topics" in both the best and the worst opinions.

2. Statistical analyses. There were two strategies carried out in this step to establish statistical relationship of data. First, the mean of both the "sentiment analysis score" and "numerical score" given by customers to the restaurants were computed, compared and analyzed, similar to the strategy of previous research [40]. Second, the sentiment score of each of the ten most mentioned sentiment topics were analyzed using Tukey's multiple comparison method. An ANOVA analysis was also carried out on the sentiment "score" of each of the ten most mentioned sentiment topics.

6   Results

With the attempt of this study to unveil insights from the voice of customer through online opinions, relevant findings as revealed by the statistically treated data are presented below. These results can serve as significant information for service design in the gastronomy sector.

Frequency Analysis. Frequency analysis revealed that the ten most frequently mentioned topics, the number of times they appeared and their percentage of occurrence were as follows: service (789 −4%), restaurant (777 −4%), food (725 −4%), dish (617 −3%), experience 493 −3%), menu (431 −2%), treatment (370 −2%), kitchen (353 −2%), price (329 −2%), and view (325 −2%).

Since "service" is the most frequently used word by customers, an analysis was then carried out on the most frequent words used to describe this term. After a filter for non-significant words, they were the following: "good", "excellent", "kind", "impeccable", "correct", "professional", "perfect", "pleasant", and "exquisite".

Similarly, as the term "experience" appears among the most frequently mentioned and since experience economy is the last segment in the progression of economic value [9], an analysis was carried out on how customers describe this term. The most frequently used adjectives were: "unforgettable", "great", "best", "good", "wonderful", "fantastic", "enjoy, "excellent", "recommendable".

Furthermore, and due to the fact that price is a stressed variable in this type of restaurants, an analysis was done of the words customers used to describe this term. They are: "expensive", "high", "excessive", "reasonable", "cheap", "good", "affordable" and "exorbitant". It should be noted that using words such as "cheap" or "affordable" does not necessarily mean that the customer perceives these restaurants as such, since there could be words that may significantly modify the meaning, such as 'not cheap' or 'not affordable at all'. Thus, an additional statistical analysis was carried out.

Finally, a further analysis was carried out on the text frequency in the higher and lower section of opinions according to their "sentiment score" (those over the mean plus two standard deviations were chosen together with those below the mean minus two standard deviations, according to the statistical analysis after this section). The most repeatedly used word in higher section restaurants was "exquisite" and the most frequently mentioned in the lower section was "expensive". However, three out of the ten most frequently used words in these restaurants referred to the concept "disappointing", which actually becomes the most recurrently mentioned.

Statistical Analysis. Two descriptive statistical analysis were carried out on the 12856 text lines with an identified sentiment topic. On the one hand, the sentiment analysis scores were analyzed and, on the other, the scores directly given to restaurants by customers on TripAdvisor (ranging from 1 to 5). The results are shown in Table 1.

Table 1. Descriptive statistics

Sentiment analysis scores		Numerical scores	
Mean	2,334	Mean	4,550
Standard error	0,020	Standard error	0,008
Median	3	Median	5
Mode	2	Mode	5
Standard deviation	2,317	Standard deviation	0,901
Kurtosis	1,308	Kurtosis	5,137
Range	30	Range	4
Minimum	−7,6	Minimum	1
Maximum	22,5	Maximum	5
No.	12856	No.	12856

Concerning the ten "sentiment topics" most frequently mentioned, an analysis was carried out to check whether there were significant differences among the means of the groups of text lines corresponding with each topic (including those lines in which that topic was named first), as it can be seen in Table 2. The sentiment topic with the highest mean was "treatment" while "price" got the lowest mean.

Table 2. Descriptive statistics of each category as considered in the analysis

	No.	Mean	St. deviation	St. error	95% mean confident interval		Min	Max
Treatment	318	3.69	1.893	.106	3.48	3.90	−5	14
View	239	3.41	1.587	.103	3.21	3.61	−5	9
Food	597	3.04	2.149	.088	2.87	3.21	−8	8
Service	661	2.99	2.179	.085	2.82	3.15	−5	9
Menu	434	2.76	2.169	.104	2.55	2.96	−5	8
Kitchen	268	2.68	1.842	.113	2.46	2.90	−5	7
Restaurant	619	2.55	2.061	.083	2.39	2.71	−5	10
Experience	420	2.45	1.957	.095	2.26	2.64	−5	8
Dish	438	2.29	2.358	.113	2.07	2.51	−5	10
Price	217	−.05	2.787	.189	−.42	.32	−6	5
Total	4211	2.68	2.245	.035	2.61	2.75	−8	14

Once the previous analyses were done to check the suitability of the test, an ANOVA of one factor was carried out. The results reveal that there is a statistically significant difference between the means, as shown in Table 3.

Table 3. ANOVA results of one factor

	SS	df	RMS	F	Sig.
Inter-groups	2309.120	9	256.569	57.028	.000
Intra-groups	18900.139	4201	4.499		
Total	21209.259	4210			

Next, *a posteriori* tests were carried out through Tukey's multiple comparison method in order to find where the differences among means lay. In Table 4, the key statistically different means pairs can be specifically seen. Specifically, results revealed that there is a significant difference between the mean of "treatment" and the rest of the sentiment topics with the exception of "view". Likewise, statistical result showed that there is a significant difference between the mean of "price" and all the other sentiment topics.

Table 4. Tukey's results for "treatment" and "price"

(I) category	(J) category	Means difference (I-J)	St. error	Sig.	95% confident interval	
					Sup.	Inf.
Treatment	Kitchen	1.006(*)	.176	.000	.45	1.56
	Food	.651(*)	.147	.000	.18	1.12
	Experience	1.240(*)	.158	.000	.74	1.74
	Menu	.933(*)	.157	.000	.44	1.43
	Dish	1.399(*)	.156	.000	.90	1.89
	Price	3.739(*)	.187	.000	3.15	4.33
	Restaurant	1.137(*)	.146	.000	.67	1.60
	Service	.701(*)	.145	.000	.24	1.16
	VIEW	.280	.182	.874	−.29	.86
Price	Kitchen	−2.734(*)	.194	.000	−3.35	−2.12
	Food	−3.089(*)	.168	.000	−3.62	−2.56
	Experience	−2.499(*)	.177	.000	−3.06	−1.94
	Menu	−2.807(*)	.176	.000	−3.36	−2.25
	Dish	−2.341(*)	.176	.000	−2.90	−1.78
	Restaurant	−2.602(*)	.167	.000	−3.13	−2.07
	Service	−3.038(*)	.166	.000	−3.56	−2.51
	Treatment	−3.739(*)	.187	.000	−4.33	−3.15
	View	−3.459(*)	.199	.000	−4.09	−2.83

7 Discussion and Conclusions

The frequency analysis shows some relevant aspects for discussion, such as the following:

1. The aspect that is mentioned the most is "service", with a highly superior frequency than others as "kitchen" or "price".
2. If experience categories ("experience", "restaurant", "service", "treatment") are grouped together they appear with more frequency than product categories ("dish", "kitchen", "menu", "food"), although with less text lines devoted to them.
3. Among the most frequently mentioned terms there are no specific references to ingredients, quality, innovation or creativity. This is remarkable considering that the Michelin star criteria consider these factors necessary.
4. The term "experience" appear among the ten most frequently mentioned by customers.
5. The analysis of restaurants that were assessed more negatively shows that they are perceived as "expensive" and "disappointing".

The further statistical analysis of data provides the following elements for discussion:

1. The numerical scores are higher than the sentiment score ones. This confirms that customers are more positive when expressing opinions with numbers than when using words, as it has been found in a previous research [45].
2. The sentiment topic with the highest mean is "treatment", and the lowest is "price".
3. The means of categories belonging to the area of experience ("experience", "restaurant", "service", "treatment") are, globally considered, slightly higher than the ones regarding the product ("dish", "kitchen", "menu" and "food").
4. The only sentiment topic that shows significant differences with the rest is "price". If we were to exclude "view" (since not all the restaurants have a view), then "treatment", and "price" are the ones placed significantly further apart from the rest.
5. The sentiment topic "experience" is the third one starting from the end.

Based on sentiment and statistical analyses, the study draws insights from opinions representing the voice of customer in online opinions, unveiling insights for the improvement of the service design that contributes to create the gastronomic experience. Specifically, it can be concluded that there are two aspects that make up the customer experience in a Michelin-starred restaurant: one of them is related to food itself (as expressed in concepts such as "dish", "kitchen", "menu" and "food"), and the other one refers to experiential factors ("experience", "restaurant", "service" and "treatment").

The second group of variables seems to be more commented and considered more positively than the first ones on the whole, which may be confirmed since 'treatment' is ranked as a significantly higher value than the rest (confirming the above stated importance of human resources in the gastronomy experience). As a matter of fact, it is the most positively considered, showing a statistically significant difference with respect to "food". That is to say, it seems that, in these restaurants, what customers value above all are the experiential aspects. This outcome is compatible with a report [46] that highlighted that the experiential dimensions of service and atmosphere are the ones that make the difference as regards to reputation and added value. To further support this claim, it can be noted that "treatment" can be linked to "service", and both sentiment topics are associated with human resources. The kind of treatment and quality of service provided by the staff help to define the customer experience, which eventually is described as meaningful or not. This result implies that dining experience could be more meaningful if service is coupled with a pleasing human and environmental atmosphere having "view" as the second in the rank. This could be one of the reasons why "experience" is not the highest in the list. Thus, it is recommended that further research be done to find out why such semantic value was given by the customers to this factor, and to determine the areas for improvement in this issue.

Moreover, it can be observed that price is a very critical variable in these restaurants, being actually the most negatively considered factor. Perhaps because prices in this kind of restaurants are known to be high. Even so, however, the customer penalizes this factor. Despite the limited number of observations in the finding, the one with the most repeated terms in the most negative opinions is "disappointing". Again, this might mean that there could be room for improvement in the customers' expectation to management on behalf of restaurants. Thus, further research can also be directed towards this area for the benefit of service design.

The findings in this research may have some practical implications, as it is the significance and importance of experiential aspects. Besides, the need for creating experiences providing customers with an appropriate environment generated by employees and staff is a key aspect. Customer satisfaction and better experience stems from employee experience itself.

These findings suggest and highlight that employee engagement help drive customer experiences. Thus, it is paramount and crucial for restaurants to create a corporate and company culture of engaged employees by focusing on empowering employees and make them feel valued, respected and trusted.

As mentioned above, in these restaurants, what customers value above all are the experiential aspects. Restaurants need to create a framework that permits customers to achieve the experience that they are pursuing.

The greater the knowledge and deeper understanding of insights of customers, the greater the capacity of improvement of a gastronomic experience design.

Finally, this paper shows several limitations. First, since the whole study is based on digital traces, that is, the opinions written by a certain group of customers on the Internet. Therefore, the opinions of those who did not express themselves by this means are missing. A possible subsequent research line could be drawing comparisons between these results and those that may be obtained by capturing the voice of those customers that do not leave a digital trace. Second is the study's presumption that Michelin-starred restaurants are sufficiently homogeneous as to be considered as a group. It is clear that they all share a position as market leaders, but differences among them, which may exist in numerous aspects, could be altering data. The third limitation is that, even assuming the previous one, conclusions could only be extrapolated to haute cuisine restaurants. That is, it may be possible that these findings may not be applied to standard restaurants. Lastly, a great part of the study is based on analyzing the most frequently used terms. Nevertheless, as it can be noted, the mentioning frequencies are generally low. That is to say, the customers' words to describe their experience show a high degree of dispersal.

To sum up, this research presents a voice of the customer analysis methodology in the realm of haute cuisine to find out how they perceive the gastronomic experience. Through the data presented, and assuming all the limitations, it can be said that the value of human service, like the kind of treatment and service being offered, may provide more meaning to customer experience when compared to the mere cooking or the food. Thus, value proposals connected to human interaction which is perceived by clients to generate favorable experience should be given importance in service design. On the other hand, this presented context should be explored further in future research undertakings in order to establish a more concrete data as regards service design improvements in gastronomy that generate the experience itself.

For example, if the treatment is the most valued aspect by customers, it becomes a priority challenge how to operationalize this factor in the service layer, while incorporating the semantic categories that the client actually uses. In a market driven by digital transformation, the incorporation of these insights into service design becomes a priority concern, given that the attractiveness of the value proposition has to be maintained.

As the world and society evolve within an economy based on experiences, it is increasingly vital to understand in depth how those experiences occur in the client's mind, and how to design the service layer that underlies them. If there has been a substantial change in the market in recent decades, it is that the institutions want to systematically design the services, thus achieving greater value and allowing a more memorable and authentic experience. Within that equation, achieving a deep and precise definition of the voice of the customer is a task that new tools, such as sentiment analysis, can facilitate, just as this work has just shown. However, much remains to be done, since experiences are subjective events that occur into the customer´s mind, a dynamic and complex context. And that is one of the reasons that makes research in this field fascinating.

Acknowledgements. The authors of this paper would like to express their gratitude to Bitext for providing the sentiment analysis technology that made possible the data exploration.

References

1. Spohrer, J., Anderson, L., Pass, N., Ager, T.: Service science and service-dominant logic. Presented at the Otago Forum 2, Otago (2008)
2. Vargo, S.L., Lusch, R.F.: Evolving to a new dominant logic for marketing. J. Mark. **68**, 1–17 (2004). https://doi.org/10.1509/jmkg.68.1.1.24036
3. Vargo, S.L., Lusch, R.F.: Institutions and axioms: an extension and update of service-dominant logic. J. Acad. Mark. Sci. **44**, 5–23 (2015). https://doi.org/10.1007/s11747-015-0456-3
4. Dohmen, P., Kryvinska, N., Strauss, C.: "S-D logic" business model - backward and contemporary perspectives. In: Snene, M. (ed.) IESS 2012. LNBIP, vol. 103, pp. 140–154. Springer, Heidelberg (2012). https://doi.org/10.1007/978-3-642-28227-0_11
5. Vargo, S.L., Lusch, R.F.: Service-dominant logic: continuing the evolution. J. Acad. Mark. Sci. **36**, 1–10 (2007). https://doi.org/10.1007/s11747-007-0069-6
6. IfM and IBM: Succeeding Through Service Innovation: A Service Perspective for Education, Research, Business and Government. University of Cambridge Institute for Manufacturing, Cambridge (2008)
7. Spohrer, J., Maglio, P.P., Bailey, J., Gruhl, D.: Steps toward a science of service systems. Computer **40**, 71–77 (2007). https://doi.org/10.1109/MC.2007.33
8. Spohrer, J., Maglio, P.P.: The emergence of service science: toward systematic service innovations to accelerate co-creation of value. Prod. Oper. Manag. **17**, 238–246 (2008). https://doi.org/10.3401/poms.1080.0027
9. Maglio, P.P., Vargo, S.L., Caswell, N., Spohrer, J.: The service system is the basic abstraction of service science. Inf. Syst. E-Bus. Manag. **7**, 395–406 (2009). https://doi.org/10.1007/s10257-008-0105-1
10. Qiu, R.G.: Service Science: The Foundations of Service Engineering and Management. Wiley, Somerset (2014)
11. Pine, J., Gilmore, J.H.: Welcome to the experience economy. Harv. Bus. Rev., 97–105 (1998)
12. Joseph Pine II, B., Gilmore, J.H.: The Experience Economy, Updated Edition. Harvard Business Review Press, Boston (2011)
13. Gracia, S.S.A., et al.: La experiencia de cliente rentable: manual para directivos y profesionales. Asociación DEC (2017)
14. Alcoba, J., Mostajo, S., Clores, R., Paras, R., Mejia, G.C., Ebron, R.A.: Tourism as a life experience: a service science approach. In: Nóvoa, H., Drăgoicea, M. (eds.) IESS 2015.

LNBIP, vol. 201, pp. 190–203. Springer, Cham (2015). https://doi.org/10.1007/978-3-319-14980-6_15

15. Alcoba, J.: Beyond the paradox of service industrialization: approaches to design meaningful services. In: Management Science, Logistics, and Operations Research. IGI Global, Hershey (2014)

16. Alcoba, J., Mostajo, S., Paras, R., Mejia, G.C., Ebron, R.A.: Framing meaningful experiences toward a service science-based tourism experience design. In: Borangiu, T., Drăgoicea, M., Nóvoa, H. (eds.) IESS 2016. LNBIP, vol. 247, pp. 129–140. Springer, Cham (2016). https://doi.org/10.1007/978-3-319-32689-4_10

17. Bitner, M.J., Ostrom, A.L., Morgan, F.N.: Service blueprinting: a practical technique for service innovation. Calif. Manag. Rev. **50**, 66–94 (2008). https://doi.org/10.2307/41166446

18. Szende, P., Dalton, A.: Service blueprinting: shifting from a storyboard to a scorecard. J. Foodserv. Bus. Res. **18**, 207–225 (2015). https://doi.org/10.1080/15378020.2015.1051429

19. Lemon, K.N., Verhoef, P.C.: Understanding customer experience throughout the customer journey. J. Mark. **80**, 69–96 (2016). https://doi.org/10.1509/jm.15.0420

20. Jaworski, B., Kohli, A.K.: Co-creating the voice of the customer. In: The Service-Dominant Logic of Marketing. Dialog, Debate, and Directions. Routledge, New York (2006). https://doi.org/10.4324/9781315699035-18

21. Gheorghe, G., Tudorache, P., Nistoreanu, P.: Gastronomic tourism, a new trend for contemporary tourism. Cactus Tour. J. **9**, 12–21 (2014)

22. World Tourism Organization: Affiliate Members Report, Volume sixteen – Second Global Report on Gastronomy Tourism. UNWTO, Madrid (2017)

23. Kocaman, M., Kocaman, E.M.: The importance of cultural and gastronomic tourism in local economic development: Zile sample. Int. J. Econ. Finance Issues **4**, 735–744 (2014)

24. Alvarez, M.C.: Tasting Spain: the creation of a product club for gastronomic tourism. UNWTO Global report on food tourism, Madrid (2012)

25. Svejenova, S., Mazza, C., Planellas, M.: Cooking up change in haute cuisine: Ferran Adrià as an institutional entrepreneur. J. Organ. Behav. **28**, 539–561 (2007). https://doi.org/10.1002/job.461

26. Surlemont, B., Johnson, C.: The role of guides in artistic industries. Manag. Serv. Qual. Int. J. **15**, 577–590 (2005). https://doi.org/10.1108/09604520510634032

27. Karpik, L.: Le Guide rouge Michelin. Sociol. Trav. **42**, 369–389 (2000)

28. Bouty, I., Gomez, M.-L.: Creativity in haute cuisine: strategic knowledge and practice in gourmet kitchens. J. Culin. Sci. Technol. **11**, 80–95 (2013). https://doi.org/10.1080/15428052.2012.728979

29. Eren, S., Güldemir, O.: Factors affecting the success of internationally awarded Turkish chefs. J. Hum. Sci. **14**, 2409–2416 (2017)

30. Johnson, C., Surlemont, B., Nicod, P., Revaz, F.: Behind the stars: a concise typology of Michelin restaurants in Europe. Cornell Hotel Restaur. Adm. Q. **46**, 170–187 (2005). https://doi.org/10.1177/0010880405275115

31. Ottenbacher, M., Harrington, R.J.: The innovation development process of Michelin-starred chefs. Int. J. Contemp. Hosp. Manag. **19**, 444–460 (2007). https://doi.org/10.1108/09596110710775110

32. Harrington, R.J., Ottenbacher, M.C.: Culinary tourism—a case study of the gastronomic capital. J. Culin. Sci. Technol. **8**, 14–32 (2010). https://doi.org/10.1080/15428052.2010.490765

33. Durand, R., Rao, H., Monin, P.: Code and conduct in French cuisine: Impact of code changes on external evaluations. Strateg. Manag. J. **28**, 455–472 (2007). https://doi.org/10.1002/smj.583

34. Barrère, C., Bonnard, Q., Chossat, V.: Luxury gastronomy as an attractive activity for luxury tourism. Presented at the Lisbon Advances in Tourism Economics Conference (2009)
35. AAVV: Basque: Creative Territory. Agapea, Málaga (2016)
36. Leizaola, A.: Matching national stereotypes?: eating and drinking in the Basque borderland. Anthropol. Noteb. **12**, 79–94 (2006)
37. Aguirre García, M.S., Aldamiz-Echevarría, C., Aparicio de Castro, M.G.: Keys to success in an example of inter-competitor cooperation: the case of the "Big Seven" basque chefs. Innobasque, Zamudio (2011)
38. Hess, A.: The social bonds of cooking: gastronomic societies in the Basque country. Cult. Sociol. **1**, 383–407 (2007). https://doi.org/10.1177/1749975507082056
39. Noe, R.A., Hollenbeck, J.R., Gerhart, B.A., Wright, P.M.: Human Resource Management: Gaining a Competitive Advantage, 9th edn. McGraw-Hill Education, New York (2015)
40. Armstrong, M., Taylor, S.: Armstrong's Handbook of Human Resource Management Practice, 12th edn. Kogan Page, London (2012)
41. Burma, Z.A.: Human resource management and its importance to today's organizations. Int. J. Educ. Soc. Sci. **1**, 85–94 (2014)
42. Shariff, N.M., Kayat, K., Abidin, A.Z.: Tourism and hospitality graduates competencies: industry perceptions and expectations in the Malaysian perspectives. World Appl. Sci. J. **31**, 1992–2000 (2014)
43. Alcoba, J., Mostajo, S.T., Ebron, R.A.T., Paras, R.: Balancing value co-creation: culture, ecology, and human resources in tourism industry. In: Handbook of Research on Strategic Alliances and Value Co-Creation in the Service Industry, pp. 285–304 (2017). https://doi.org/10.4018/978-1-5225-2084-9.ch014
44. Bucak, T., Kose, Z.C.: The application of Michelin star standards in restaurant business: Hamburg Le Canard Sample. J. Tour. Hosp. Manag. **2**, 21–35 (2014)
45. Alcoba, J., Mostajo, S., Paras, R., Ebron, R.A.: Beyond quality of service: exploring what tourists really value. In: Za, S., Drăgoicea, M., Cavallari, M. (eds.) IESS 2017. LNBIP, vol. 279, pp. 261–271. Springer, Cham (2017). https://doi.org/10.1007/978-3-319-56925-3_21
46. Serrano, D., Jodar, A., González, R.: Estudio de Reputación Online de los restaurantes de Donostia-San Sebastián 2017, San Sebastián (2017)
47. Walter, U., Edvardsson, B., Öström, Å.: Drivers of customer's service experiences: a study in the restaurant industry. Manag. Serv. Qual.: Int. J. **20**(3), 236–258 (2010). https://doi.org/10.1108/09604521011041961
48. Tambini, A., Rimmele, U., Phelps, E.A., Davachi, L.: Emotional brain states carry over and enhance future memory formation. Nat. Neurosci. **20**, 271–278 (2017). https://doi.org/10.1038/nn.4468
49. Miguéis, V.L., Nóvoa, H.: Using user-generated content to explore hotel service quality dimensions. In: Borangiu, T., Drăgoicea, M., Nóvoa, H. (eds.) IESS 2016. LNBIP, vol. 247, pp. 155–169. Springer, Cham (2016). https://doi.org/10.1007/978-3-319-32689-4_12
50. Thelwall, M.: Sentiment analysis for tourism. In: Sigala, M., Rahimi, R., Thelwall, M. (eds.) Big Data and Innovation in Tourism, Travel, and Hospitality, pp. 87–104. Springer, Singapore (2019). https://doi.org/10.1007/978-981-13-6339-9_6
51. Molnar, E., Moraru, R.: Content analysis of customer reviews to identify sources of value creation in the hotel environment. In: Za, S., Drăgoicea, M., Cavallari, M. (eds.) IESS 2017. LNBIP, vol. 279, pp. 251–260. Springer, Cham (2017). https://doi.org/10.1007/978-3-319-56925-3_20
52. Benjamins, V.R., Cadenas, D., Alonso, P., Valderrabanos, A., Gomez, J.: The voice of the customer for digital telcos. Presented at the 13th International Semantic Web Conference, ISWC (2014). https://doi.org/10.13140/2.1.1982.2087

Quality and Efficiency Evaluation of Airlines Services

Agnese Rapposelli[(✉)] and Stefano Za

University "G. d'Annunzio" of Chieti-Pescara, Pescara, Italy
{agnese.rapposelli,stefano.za}@unich.it

Abstract. Frontier analysis methods are able to investigate the technical efficiency of service productive systems. In our opinion, in the service sector field the analysis of service productivity must be also linked to service quality. Hence, the aim of this work is to outlines a new efficiency assessment based on these two productivity components (technical efficiency and service quality). More specifically, we adapt efficiency measurement techniques to airline industry by also considering an indicator of service quality represented by the average delay. We then evaluate the operational performance of an Italian airline by applying a Principal Component Analysis (PCA) - Data Envelopment Analysis (DEA) model and we verify which airline routes are ranked among the most efficient ones by also including, in the proposed model, the presence of this undesirable output.

Keywords: Airlines services · Service evaluation · Data Envelopment Analysis (DEA) · Principal Component Analysis (PCA) · Flight delays

1 Introduction

Frontier analysis methods, such as Data Envelopment Analysis (DEA), are able to investigate the technical efficiency of service productive systems which typically perform the same function, by using the same set of inputs to produce the same sets of outputs [1]. The role of input/output systems in assessing performance is very important. Apart of their nature, it must be remembered that questions can also be raised concerning the appropriate number of inputs and outputs for describing an activity process [2]. Introduction of too many, and especially redundant, variables will result in a large number of units with high efficiency scores [3].

DEA method measures technical efficiency relative to a deterministic best practice frontier (also called the efficiency frontier), which is built empirically from observed inputs and outputs using linear programming techniques. DEA is a performance management approach with more than forty years of history [4], created as an accounting tool in the analysis of non-profit organizations [5]. It is usually adopted for identifying the most efficient interaction routines in a service system resolving trade-off decisions [6]. There are several examples of DEA application, such as: a benchmark of microcomputers [5], or a comparison of the restaurants performances [7] or the identification of the firms with the highest IT productivity [8]. Moreover, it is possible to recognize

© Springer Nature Switzerland AG 2020
H. Nóvoa et al. (Eds.): IESS 2020, LNBIP 377, pp. 35–46, 2020.
https://doi.org/10.1007/978-3-030-38724-2_3

several examples in the healthcare sector, concerning organizational performance [9], patience's quality of life [10], and diagnosis classification [11] among others.

DEA is a nonparametric approach. It uses a mathematical programming model for constructing an efficient frontier over the data. Using this approach, it is possible to identify each data point's efficiency respect to the efficiency frontier. Every data point represents a decision-making unit, in which the main aim is to efficiently convert inputs into outputs [12]. For instance, the DEA efficiency score could be used as a measure of the productivity factor [13].

Its main advantage is that it allows several inputs and several outputs to be considered at the same time. However, it is well known that if all inputs and outputs are included in assessing the efficiency of units under analysis, then they will all be fully efficient. As DEA allows flexibility in the choice of inputs and outputs weights, the greater the number of factors included the lesser discriminatory the method appears to be. More specifically, the number of inputs and outputs included in a DEA model should be as small as possible in relation to the number of units being assessed. Discrimination can be increased, therefore, by being parsimonious in the number of factors. This can be achieved by using Principal Component Analysis (PCA), which is able to reduce the data to a few principal components whilst minimising the loss of information [2].

Following on from the above discussion, the main objective of this paper is to adapt DEA technique to a specific field of service sector, represented by airline industry. The PCA-DEA approach has been already used and validate in the airline context. In particular, Rapposelli [2] has performed the efficiency analysis of airline routes by means of this DEA formulation. However, previous studies have focused only on productivity service and have omitted service quality variables that could address the different results registered among routes. In this work, we want to fill this gap by focusing also on service quality and by examining its impact on the operational performance of this airline.

To this purpose, this work outlines a new efficiency assessment based on both service productivity and service quality. More specifically, this study aims to analyze the operational performance of an Italian airline by also considering an indicator of service quality. In the airline production process under analysis we have identified a very high number of inputs to be included in the performance evaluation. Moreover, the application of efficiency techniques to the context of air transportation has motivated the inclusion of a special kind of output, an undesirable output [14, 15] represented by the average delay for each route, which represents a negative factor for an airline company both in terms of costs and passenger demand [5].

Many studies on flight delays have focused on their impact in terms of cost to airlines and airports. By focusing on the airline context, flight delays have been treated as service failures [16], that reduce passenger demand and raise airfares [17]. Suzuki [18], in fact, states that passengers who have experienced flight delays are more likely to switch airlines for the subsequent flights than those who have not experienced delays [18]. Also Kim and Park [19] find that flight delays negatively influence repurchase intention [19].

Hence, the efficiency analysis is carried out by including the average delay in the proposed PCA-DEA model to measure the technical efficiency of an Italian airline, by verifying which domestic routes are ranked among the most efficient ones.

The article is organized as follows. Section 2 reviews the PCA-DEA methodology, Sect. 3 presents the data used, Sect. 4 describes the analysis results and Sect. 5 provides a set of preliminary conclusions.

2   Method

This study involves two main steps:

1. application of Principal Component Analysis (PCA) in order to aggregate the original input variables into principal components that summarize the information contained in the original data [20];
2. application of a specific formulation of Data Envelopment Analysis (DEA) in order to obtain an efficiency value for each route. This new DEA model includes:
   (a) principal components identified in the previous phase (that replace the original input variables);
   (b) output variables;
   (c) an undesirable output, i.e. the average delay for each route, which is an indicator of airline service quality.

2.1   The PCA-DEA Formulation

The basic DEA models measure the technical efficiency of one of the sets of n Decision Making Units, DMU j_0, in terms of maximal radial contraction to its input levels (input orientation) or expansion to its output levels feasible under efficient operation (output orientation).

In this study we calculate the technical efficiency of DMUs under analysis by implementing a VRS model (which assumes variable returns to scale of activities), that is also in line with the findings of Pastor [21], and we use the following output-oriented formulation of DEA method [22]:

$$e_0 = \max \phi_0 \tag{1}$$

subject to

$$\sum_{j=1}^{n} \lambda_j x_{ij} \leq x_{ij0} \tag{2}$$

$$\phi_0 y_{rj0} - \sum_{j=1}^{n} \lambda_j y_{rj} \leq 0 \tag{3}$$

$$\sum_{i=1}^{n} \lambda_j = 1 \tag{4}$$

$$\lambda_j \geq 0, \forall j \tag{5}$$

where y_{rj} is the amount of the r-th output ($r = 1, \ldots, s$) produced by unit $j(j = 1, \ldots, n)$, x_{ij} is the amount of the i-th input ($i = 1, \ldots, m$) consumed by unit j, λ_j are the weights of unit j and ϕ_0 is the scalar expansion factor for DMU j_0 under evaluation. The efficiency score obtained for each DMU is bounded between zero and one: a technical efficient DMU will have a score of unity.

As introduced in Sect. 1, in order to overcome the difficulties that classical DEA models encounter when there is an excessive number of inputs or outputs in relation to the number of DMUs to be evaluated, DEA can be combined with PCA to aggregate and, therefore, to reduce input and output data.

Principal Component Analysis (PCA) is a multivariate statistical method devised for dimensionality reduction of multivariate data with correlated variables. This technique accounts for the maximum amount of the variance of a data matrix by using a few linear combinations (Principal Components) of the original variables. The aim is to take p variables $X_1, X_2, \ldots, X_p$ and find linear combinations of them to produce principal components $X_{PC1}, X_{PC2}, \ldots, X_{PCp}$ that are uncorrelated. The principal components are also ordered in descending order of their variances so that X_{PC1} accounts for the largest amount of variance, X_{PC2} accounts for the second largest amount of variance, and so on: that is, $\text{var}(X_{PC1}) \geq \text{var}(X_{PC2}) \geq \ldots \geq \text{var}(X_{PCp})$. Often much of the total system variability can be accounted for by a small number k of the principal components, which can then replace the initial p variables without much loss of information [23].

Hence, principal component scores can be used to replace either all the original inputs and/or outputs simultaneously or alternatively groups of variables [24]. To this purpose we modify the above DEA formulation by incorporating principal components (PCs) directly into the linear programming problem.

In particular, constraint (2) is replaced with the following ones:

$$\sum_{j=1}^{n} \lambda_j x_{PCij} \leq x_{PCij0} \tag{6a}$$

$$\sum_{j=1}^{n} \lambda_j x_{Oij} \leq x_{Oij0} \tag{6b}$$

and constraint (3) is replaced with the following ones:

$$\phi_0 y_{PCrj0} - \sum_{j=1}^{n} \lambda_j y_{PCrj} \leq 0 \tag{7a}$$

$$\phi_0 y_{Orj0} - \sum_{j=1}^{n} \lambda_j y_{Orj} \leq 0 \tag{7b}$$

where:

x_{Oij} denote original input variables;
y_{Orj} denote original output variable;
x_{PCij} denote principal component input variables;
y_{PCij} denote principal component output variables.

Hence, the output-oriented DEA model (1–5) refers to both the original data and PCs and presents the following generalized formulation:

$$e_0 = \max \phi_0 \tag{2}$$

subject to

$$\sum_{j=1}^{n} \lambda_j x_{PCij} \leq x_{PCij0} \tag{6a}$$

$$\sum_{j=1}^{n} \lambda_j x_{Oij} \leq x_{Oij0} \tag{6b}$$

$$\phi_0 y_{PCrj0} - \sum_{j=1}^{n} \lambda_j y_{PCrj} \leq 0 \tag{7a}$$

$$\phi_0 y_{Orj0} - \sum_{j=1}^{n} \lambda_j y_{Orj} \leq 0 \tag{7b}$$

$$\sum_{i=1}^{n} \lambda_j = 1 \tag{4}$$

$$\lambda_j \geq 0, \forall j \tag{5}$$

2.2   The PCA-DEA Model with Undesirable Outputs

The efficiency measures analysed above are not suitable in contexts where at least one of the variables that have to be radially contracted or expanded is not a "good".

It was already mentioned in the seminal work of Koopmans that the production process may also generate a third kind of factor [25], i.e. undesirable outputs (or "bads") that should be treated differently than traditional variables (inputs and outputs) [15, 26]: they should be reduced in order to improve the performance.

The current study provides an alternative method in dealing with desirable and undesirable outputs in a different way. Undesirable outputs will be included directly into the linear programming problem, just like inputs that have to be radially reduced [27].

Hence, the PCA- DEA formulation (1–5) introduced in Sect. 2.1 is made more specific for incorporating the undesirable output by adding the following constraint:

$$\sum_{j=1}^{n} \lambda_j h_{tj} \leq h_{tj0} \tag{8}$$

where h_{tj} is the amount of the t-th undesirable output ($t = 1, \ldots, z$) produced by unit j.

3 Data

In this study the PCA-DEA approach is used in order to evaluate the performance of 30 domestic routes. In order to respect homogeneity assumptions about the DMUs under assessment, we have not included international routes and seasonal destinations. As mentioned above, the selected variables accurately reflect the production activity of the analysed DMUs (in our case airline routes). The production process, in our specific case, involves several inputs (seven) to produce both three outputs and an undesirable output, the average delay for each route.

The domestic airline industry provides a particularly rich setting for this empirical study. In order to assess airline routes, the inputs and the outputs of the function they perform must be identified. However, there is no definitive study to guide the selection of inputs and outputs in airline applications of efficiency measurement. In the production process under analysis we have identified seven inputs and three outputs to be included in the performance evaluation. The input selected are the number of seats available for sale, block time hours and several airline costs categories such as total variable direct operating costs (DOCs), total fixed direct operating costs (FOCs), commercial expenses, overhead costs and financial costs [2].

The number of seats available for sale reflects aircraft capacity. Block time hours is the time for each flight sector, measured from when the aircraft leaves the airport gate to when it arrives on the gate at the destination airport. With regard to the cost categories considered, according to the ICAO (International Civil Aviation Organization) classification, variable or "flying" costs are costs which are directly escapable in the short run, such as fuel, handling, variable flight and cabin crew expenses, landing charges, passenger meals, variable maintenance costs. These costs are related to the amount of flying airline actually does, hence they could be avoided if a flight was cancelled [28]. Fixed or "standing" costs are costs which are not escapable in the short or medium term, such as lease rentals, aircraft insurance, fixed flight and cabin crew salaries, engineering overheads. These costs are unrelated to amount of flying done, hence they do not vary with particular flights in the short run; they may be escapable but only after a year of two, depending on airlines [28]. Both DOCs and FOCs are dependent on the type of aircraft being flown [29]. Commercial expenses, such as reservations systems, commissions, passengers reprotection, lost and found, and overhead costs, such as certain general and administrative costs which do not vary with output (legal expenses, buildings, office equipment, advertising, etc.), are not directly dependent on aircraft operations [29]. Finally, we have included financial costs, such as interests, depreciation and amortisation.

With regard to the output side of the model, the output variables are the number of passengers carried, passenger scheduled revenue and other revenues. Passenger scheduled revenue is the main output for a passenger focused airline, whilst other revenues includes charter revenue and a wide variety of non-airline businesses (incidental services) such as ground handling, aircraft maintenance for other airlines and advertising and sponsor. Even if incidental services are not airline's core business, they are considered in the production process under evaluation because they utilise part of the inputs included in the analysis [30].

Finally, we consider a third kind of variable, the average delay (measured in minutes) for each route, which may reflect the service quality of the airline network.

All data have been developed from Financial Statements and from internal Reports (Table 1).

Table 1. Description of variables

Variables	Meaning
Inputs	
Seats	The number of seats available for sale
Block time hours	The time for each flight sector
DOCs	Total variable direct operating costs (such as fuel, handling, variable flight and cabin crew expenses, landing charges, passenger meals, variable maintenance costs)
FOCs	Total fixed direct operating costs (such as lease rentals, aircraft insurance, fixed flight and cabin crew salaries, engineering overheads)
Commercial expenses	It includes costs for reservations systems, commissions, passengers reprotection, lost and found
Overhead costs	It includes certain general and administrative costs, such as legal expenses, costs for buildings, office equipment, advertising
Financial costs	It includes interests, depreciation and amortisation
Outputs	
Number of passengers	The number of passengers carried in the period analyzed
Passenger scheduled revenue	The revenue produced from scheduled flights
Other revenues	The revenue produced from non-scheduled flights (such as charter flights) and from incidental services (such as ground handling, aircraft maintenance for other airlines and advertising and sponsor)
Undesirable output	
Average delay	It reflects the service quality for each route

4 Results

In this section we discuss the results we have obtained by applying the modified PCA-DEA model introduced in Sect. 2.1 to the selected set of variables.

The airline production process defined in this empirical study is characterized by a higher number of inputs, that are very highly correlated. This is therefore good material for using PCA to produce a reduced number of inputs by removing redundant information [2].

Table 2 gives the eigenanalysis of the correlation matrix of data set. The first principal component X_{PC1} explains 98.22% of the total variance of the data vector, so the input variables will be included in the DEA model via the first principal component.

Table 2 Eigenvalues and total variance explained

Component	Eigenvalue	Proportion (%)	Cumulative (%)
1	6.876	98.22	98.22
2	7.420E−02	1.06	99.28
3	3.509E−02	0.50	99.78
4	1.410E−02	0.20	99.98
5	6.151E−04	8.788E−03	99.99
6	3.485E−04	4.978E−03	100.00
7	3.428E−13	4.897E−12	100.00

It should be noted that principal components used here are computed based on the correlation matrix rather than on covariance, as the variables are quantified in different units of measure. Generally inputs and outputs of DEA models need to be strictly positive, but the results of a PCA can have negative values [31]. It has been argued [21] that the BCC output-oriented model used in the current study is input translation invariant and vice versa. Hence the efficiency classification of DMUs is preserved if the values of principal component X_{PC1} are translated by adding a sufficiently large scalar β (1 in this case) such that the resulting values are positive for each DMU j.

We apply therefore DEA on the translated first component and not on the whole set of the original input variables, as we have identified a single principal component that explains 98.22% of the total variance. In order to incorporate X_{PC1} directly into the linear programming problem the general DEA formulation has to be modified, as specified in Sect. 2.1. With regard to the output variables (in terms of both "goods" and "bads"), they are not included in terms of principal components.

Hence, the PCA-DEA model with undesirable outputs to solve is the following one:

$$e_0 = \max \phi_0 \tag{9}$$

subject to

$$\sum_{j=1}^{n} \lambda_j x_{PCij} \leq x_{PCij0} \tag{10}$$

$$\sum_{j=1}^{n} \lambda_j h_{tj} \leq h_{tj0} \tag{11}$$

$$\phi_0 y_{Orj0} - \sum_{j=1}^{n} \lambda_j y_{Orj} \leq 0 \tag{12}$$

$$\sum_{i=1}^{n} \lambda_j = 1 \tag{13}$$

$$\lambda_j \geq 0, \forall j \tag{14}$$

where:

x_{PCij} denote the translated first component;

h_{tj} denote the t-th undesirable output;

y_{Orj} denote the original r-th output.

The above linear programming model must be solved n times, once for each route in the sample. The linear program associated with the modified PCA-DEA model has been performed by using DEA-Solver, a software developed by Kaoru Tone [32]. For each route evaluated, the output-orientated VRS DEA efficiency scores obtained from the above model, in descending order of efficiency, are presented in Table 3.

Table 3 DEA efficiency scores by domestic routes

DMU	DEA score	DMU	DEA score	DMU	DEA score
A	1	EE	0,94622	H	0,737039
E	1	L	0,841838	N	0,729675
J	1	T	0,827512	QQ	0,72945
V	1	FF	0,823246	G	0,712585
Z	1	X	0,802699	B	0,683339
CC	1	F	0,7912	JJ	0,680024
DD	1	U	0,780837	OO	0,660772
TT	1	W	0,771784	SS	0,643053
VV	1	Q	0,754694	PP	0,629553
S	0,967441	AA	0,754097	MM	0,619797

Two different remarks can be made. The average level of technical efficiency is 0.8295. Nine routes are fully efficient, and two more routes are quite close to the best practice frontier. On the other hand, the remaining DMUs are sub-efficient but they do not show very low ratings. These results suggest that these routes are operating at a high level of efficiency, although there is room for improvement in several routes.

With respect to poorly performing units it is also possible to identify their relatively efficient DMUs (peer units), because DEA method is able to give information on the extent to which an efficient unit is used as an efficient peer for other DMUs.

Focusing on efficient units, the number of citations in peer groups can be interpreted as a measure of the "robustness" of best practice units. Table 4 displays the frequency with which efficient routes appear in the peer group of the inefficient ones. We may note that route V and CC appear very frequently in the reference sets (19 and 17 times, respectively). The remaining efficient units (A, E, DD and T) are not likely to be a better role models for less efficient units to emulate.

Table 4 Reference sets

Peer set	Frequency to other DMUs
V	19
CC	17
VV	13
Z	8
J	6

5 Conclusion

In this paper we have evaluated the operational performance of an Italian airline with respect to the average delay of each domestic route by means of the non-parametric approach to efficiency measurement, represented by DEA. To this purpose, we have obtained measures of technical efficiency from VRS production frontiers.

It is well known that the discriminatory power of DEA often fails when there is an excessive number of inputs and outputs in relation to the number of DMUs [33]. We have applied a new model formulation within DEA framework that can be used in efficiency measurement when there is a large number of inputs and outputs variables that can be omitted with least loss of information. The use of a combined PCA-DEA model has provided a more parsimonious description of a relatively large multivariate data set. Moreover, this approach has been applied by also considering a special kind of factor, an undesirable output, i.e. the average delay, which is an indicator of airline quality.

The inclusion of this service indicator in our analysis has made more understandable the efficiency results registered among routes. The results provided by the applied model show that out of the 30 units analysed nine are on the efficient frontier and that several routes are operating at a high level of efficiency. These results show an improvement in the identification of efficient routes that could have implications for managers both for the identification of sub-efficient routes and for the actions to pursue in order to improve their level of efficiency.

However, these results can be improved. This study suggests some main avenues for future research. First of all, the usefulness of the method could be explored for large data sets: for example, in further application studies we could add international routes or other air carriers. Besides, the efficiency analysis applied in this work can be improved by developing a performance analysis over time. Finally, this method could be integrated or combined with simulation approaches in order to explore new concepts, ideas, boundaries, and limitations and to build predictive and prescriptive theories [34].

References

1. Pang, M.-S., Tafti, A., Krishnan, M.S.: Information technology and administrative efficiency in U.S. state governments: a stochastic frontier approach. MIS Q. **38**, 1079–1101 (2014). https://doi.org/10.25300/misq/2014/38.4.07
2. Rapposelli, A.: Route-based performance evaluation using data envelopment analysis combined with principal component analysis. In: Di Ciaccio, A., Coli, M., Ibanez, J.M.A. (eds.) Advanced Statistical Methods for the Analysis of Large Data-Sets. Studies in Theoretical and Applied Statistics, pp. 351–360. Springer, Berlin (2012). https://doi.org/10.1007/978-3-642-21037-2_32
3. Golany, B., Roll, Y.: An application procedure for DEA. Omega **17**, 237–250 (1989)
4. Cook, W.D., Seiford, L.M.: Data envelopment analysis (DEA) - Thirty years on. Eur. J. Oper. Res. **192**, 1–17 (2009). https://doi.org/10.1016/j.ejor.2008.01.032
5. Doyle, J., Green, R.: Strategic choice and data envelopment analysis: comparing computers across many attributes. J. Inf. Technol. **9**, 61–69 (1994). https://doi.org/10.1057/jit.1994.7
6. Becker, J., Beverungen, D., Breuker, D., Dietrich, H.A., Rauer, H.P.: Performance benchmarking for designing interaction routines -managing trade-offs in service co-creation with the data envelopment analysis. In: Proceedings of the 21st European Conference on Information Systems, ECIS 2013 (2013)
7. Banker, R.D., Kauffman, R.J., Morey, R.C.: Measuring gains in operational efficiency from information technology: a study of the Positran deployment at Hardee's Inc. J. Manag. Inf. Syst. **7**, 29–54 (1990). https://doi.org/10.1080/07421222.1990.11517888
8. Nevo, S., Wade, M.R., Cook, W.D.: An examination of the trade-off between internal and external IT capabilities. J. Strateg. Inf. Syst. **16**, 5–23 (2007). https://doi.org/10.1016/j.jsis.2006.10.002
9. Lin, F., Deng, Y.J., Lu, W.M., Kweh, Q.L.: Impulse response function analysis of the impacts of hospital accreditations on hospital efficiency. Health Care Manag. Sci. 394–409 (2019). https://doi.org/10.1007/s10729-019-09472-6
10. Borg, S., Gerdtham, U.-G., Eeg-Olofsson, K., Palaszewski, B., Gudbjörnsdottir, S.: Quality of life in chronic conditions using patient-reported measures and biomarkers: a DEA analysis in type 1 diabetes. Health Econ. Rev. **9**, 31 (2019). https://doi.org/10.1186/s13561-019-0248-4
11. Ji, A., Qiao, Y., Liu, C.: Fuzzy DEA-based classifier and its applications in healthcare management. Health Care Manag. Sci. **22**, 560–568 (2019). https://doi.org/10.1007/s10729-019-09477-1
12. Ayabakan, S., Bardhan, I.R., Zheng, Z.: (Eric): a data envelopment analysis approach to estimate IT-enabled production capability. MIS Q. **41**, 189–205 (2017). https://doi.org/10.25300/MISQ/2017/41.1.09

13. Mehra, A., Langer, N., Bapna, R., Gopal, R.: Estimating returns to training in the knowledge economy: a firm-level analysis of small and medium enterprises. MIS Q. **38**, 757–771 (2014). https://doi.org/10.25300/MISQ/2014/38.3.06

14. Scheel, H.: Undesirable outputs in efficiency valuations. Eur. J. Oper. Res. **132**, 400–410 (2001)

15. Seiford, L.M., Zhu, J.: A response to comments on modeling undesirable factors in efficiency evaluation. Eur. J. Oper. Res. **161**, 579–581 (2005)

16. Ferrer, J.C., Rocha e Oliveira, P., Parasuraman, A.: The behavioral consequences of repeated flight delays. J. Air Transp. Manag. **20**, 35–38 (2012)

17. Britto, R., Dresner, M., Voltes, A.: The impact of flight delays on passenger demand and societal welfare. Transp. Res. Part E **48**, 460–469 (2012)

18. Suzuki, Y.: The relationship between on-time performance and airline market share: a new approach. Transp. Res. Part E **36**, 139–154 (2000)

19. Kim, N.Y., Park, J.W.: A study on the impact of airline service delays on emotional reactions and customer behavior. J. Air Transp. Manag. **57**, 19–25 (2016)

20. Agovino, M., Rapposelli, A.: Inclusion of disabled people in the Italian labour market: an efficiency analysis of law 68/1999 at regional level. Qual. Quant. **47**, 1577–1588 (2013). https://doi.org/10.1007/s11135-011-9610-2

21. Pastor, J.T.: Translation invariance in data envelopment analysis: a generalization. Ann. Oper. Res. **66**, 93–102 (1996)

22. Banker, R.D., Charnes, A., Cooper, W.W.: Some models for estimating technical and scale inefficiencies in data envelopment analysis. Manag. Sci. **30**, 1078–1092 (1984)

23. Johnson, R.A., Wichern, D.W.: Applied Multivariate Statistical Analysis, Upper Saddle River (2002)

24. Adler, N., Yazhemsky, E.: Improving discrimination in data envelopment analysis: PCA-DEA or variable reduction. Eur. J. Oper. Res. **202**, 273–284 (2010)

25. Koopmans, T.C.: An analysis of production as an efficient combination of activities. In: Activity Analysis of Production and Allocation, Cowles Commission for Research in Economics, Monograph n.13. Wiley, New York (1951)

26. Seiford, L.M., Zhu, J.: Modeling undesirable factors in efficiency evaluation. Eur. J. Oper. Res. **142**, 16–20 (2002)

27. Coli, M., Nissi, E., Rapposelli, A.: Monitoring environmental efficiency: an application to Italian provinces. Environ. Model Softw. **26**, 38–43 (2011)

28. Doganis, R.: Flying Off Course: The Economics of International Airlines. Routledge, London (2002)

29. Holloway, S.: Straight and Level. Practical Airline Economics, Aldershot (1997)

30. Oum, T.H., Yu, C.: Winning Airlines: Productivity and Cost Competitiveness of the World's Major Airlines. Springer, Heidelberg (1998). https://doi.org/10.1007/978-1-4615-5481-3

31. Adler, N., Berechman, J.: Measuring airport quality from the airlines' viewpoint: an application of data envelopment analysis. Transp. Policy **8**, 171–181 (2001)

32. Cooper, W.W., Seiford, L.M., Tone, K.: Data Envelopment Analysis, A Comprehensive Test with Models, Applications, References and DEA-Solver Software, Boston (2000)

33. Adler, N., Golany, B.: Including principal component weights to improve discrimination in data envelopment analysis. J. Oper. Res. Soc. **53**, 985–991 (2002)

34. Za, S., Spagnoletti, P., Winter, R., Mettler, T.: Exploring foundations for using simulations in IS research. Commun. Assoc. Inf. Syst. **42**, 268–300 (2018). https://doi.org/10.17705/1CAIS.04210

Reducing the Expectation-Performance Gap in EV Fast Charging by Managing Service Performance

Stephanie Halbrügge[1][(✉)], Lars Wederhake[2], and Linda Wolf[1]

[1] FIM Research Center, University of Augsburg, Augsburg, Germany
{stephanie.halbruegge,linda.wolf}@fim-rc.de
[2] Project Group Business and Information Systems Engineering
of the Fraunhofer FIT, Augsburg, Germany
lars.wederhake@fit.fraunhofer.de

Abstract. Electric mobility is considered pivotal to decarbonising transport. The operation of fast charging services has become a mobility business model. Its value proposition rests on the promise that fast chargers re-empower drivers to fulfil their mobility needs within acceptable servicing times. This is in particular important when levels for tolerance are low like on long-distance journeys. That value proposition might set inflated customer expectations. Due to economic considerations and operational restrictions, charging park operators might not live up to these expectations. This leads to an expectation-performance gap, which has received little scientific attention, to date. This paper presents an information system (IS) design, which aims at reducing that gap by managing performance. Our findings indicate significant benefits by the IS and highlights further opportunities for the IS discipline. Also, this article invites researchers from service science to discover opportunities for better expectation management and further reduction of the identified gap.

Keywords: Electric mobility · Fast charging · Customer expectation · Service performance · Information system

1 Introduction

There is overwhelming certainty about humankind's detrimental influence on global warming stemming from the steep increase of emitted greenhouse gas emissions. Today, the transport sector emits at least a fourth of global emissions and projections indicate some 50% by 2050 [1]. By then, transport will be the largest pollutant sector and continues to be a focal point of decarbonisation efforts. These efforts currently centre on electric mobility (e-mobility) [2]. Still, the adoption of battery electric vehicles (BEV) and consequently market penetration rates lag behind targets [3]. Major barriers to BEV adoption result from today's limited driving ranges in comparison to conventional vehicles, as well as, similarly, long servicing times for charging [4].

The fast charging service, especially along highways, sets out to reduce such barriers. Unsurprisingly, among others Neamieh et al. [5] find evidence for the importance of fast chargers for BEVs [6, 7]. The operation of fast chargers has become

© Springer Nature Switzerland AG 2020
H. Nóvoa et al. (Eds.): IESS 2020, LNBIP 377, pp. 47–61, 2020.
https://doi.org/10.1007/978-3-030-38724-2_4

an e-mobility business model [8]. The value proposition rests on the promise that fast chargers re-empower BEV drivers to fulfil their mobility needs within acceptable servicing times. This is in particular important when levels for tolerance are low like on long-distance journeys or on business trips [6]. This contrasts sharply with e.g. charging at home [9] or at parking facilities [10]. Carmakers advertise shortest technically feasible servicing times. These advertisements can set inflated customer expectations [11] as charging park operators do not or cannot always live up to them. Reasons are mainly of economic nature such as reducing expensive peak demand and/or limited technical constraints. As a result, customers might experience longer servicing times than expected. Churchill and Surprenant [12] as well as Palawatte [13] support the hypothesis of a strong negative correlation between customer satisfaction and the gap between expected and actual servicing times. Worse, adding to this, Lin et al. [14] applied prospect theory to validate the seemingly intuitive assumption that losing one minute of time ten times is preferred over losing ten minutes once. Thus, this paper identifies the gap between expected and actual servicing time as a challenge to the value proposition of the fast charging service. We want to improve customer servicing in the context of fast charging service business models. Previous research is vast on financial aspects (revenues and cost structure) [4, 6, 15] and on technical simulation/optimisation [10, 16] but scarce on customer behaviour as Motoaki and Shirk [17] find. Motoaki and Shirk [17] provide first empirical insights but lack a design perspective. To the best of our knowledge there is no prior work directly targeting at reducing the expectation-performance gap in the context of fast charging service business models. This paper therefore sets out to contribute to the scientific discourse in the following ways:

– The paper identifies the gap between expected and actual servicing time as an indicator for customer satisfaction in fast charging services derived from satisfaction theory.
– The paper proposes an information system (IS) design, which embeds an optimisation model to reduce that gap by improving information use for managing the service performance in on-site operations.
– The paper provides a basis to stimulate discussion on how service science can further address the expectation-performance gap in the fast charging service beyond what this conference paper can methodically deliver.

2 Paper Structure and Research Process

Similar to Wagner et al. [10] and Brandt et al. [18], who follow a design science approach [19], the centrepiece of this paper's contributions is an IS design and a resultant artefact for sustainability [20]. In line with those solution-oriented sustainability approaches in the IS discipline, we structure the remainder of the paper as follows. The following section sets out to characterise the problem environment and the theoretical foundations based on related scientific contributions. Next, we formalise and summarise the relevant concepts from the problem environment in the setting, before we construct the IS artefact. The artefact adjusts actual charging performance for individual customers as to reduce the gap between the customer's expectation and actual performance. The kernel of the artefact is represented by a deterministic optimisation model relying on theory from service science (satisfaction theory) [12–14]. Using available information from e.g. BEV data, the model optimises power allocations to on-site charging stations for a single point in time, where nominal charging power cannot be delivered to all BEVs, i.e. that the IS sets the allocation once for the entire charging process. In the IS design evaluation we study the usefulness of the IS artefact when it is challenged with different charging park setups, i.e. vehicles, stations, and available power. We simulate fictive but realistic charging park setups as to test the IS artefact and its optimisation, because we cannot evaluate our IS artefact by a physical prototype system in a field study. This is so because, by now, only a few fast charging parks exist, charging happens at only a very limited scale, and charging park operators (CPOs) consider intervening in operations hardly feasible (for the time being). In fact, fast charging parks, which allocate power intelligently, are not yet available in series production. The form of the IS artefact in a physical prototype may also diverge from the version implemented in this paper. Such implementations will certainly contain many further technical restrictions. Note, however, that further restrictions can only extend actual servicing times. That, in turn, results in a widening gap because customer expectations do not change as a consequence of more (to the customer intransparent) technical restrictions. Therefore, we willingly choose to limit technical constraints to the necessary minimum as to demonstrate the benefits of the IS design. In times when BEVs compete for the scarce available charging power, CPOs for fast charging risk to forfeit their value proposition. This holds for two or more charging stations. In this conference paper, we choose to analyse setups consisting of two charging stations not only for comparability's sake but also because it is the most prevalent setup (also along highways) today [21] and larger setups share similar limitations. Also, demonstrating benefits for the dual setup provides a basis to hypothesise that analysing larger setups, in future research, will yield presumably better results. In this context, our minimal setup represents a lower benchmark. We study the effects based on the present distribution of BEVs derived from real world data. For benchmarking purposes, we consider the status quo and use it as a reference case. Nowadays the power is allocated uniformly across all charging stations. We compare the results of the status quo case to our optimised case. The optimised case uses information provided by the BEVs and the charging stations more intelligently. We eventually outline the results by comparing the status quo case to the optimised case. Also, we provide a variation analysis with regard to the power available, which might be technically or economically constraint in order to reduce peak demand and corresponding charges. We outline the parametrisation and the procedure, which generates the setups. Finally, we discuss the results in the larger fast charging service business model context as to conclude by giving implications for decision-makers.

3   Literature and Theoretical Background

Information and communication technology (ICT) has enabled entirely new classes of service business models [22], e.g. charge point directories and networks in e-mobility [23]. Also, considering human and organisational aspects, information systems aim at aligning ICT to an organisation's objectives. In fact, IS are known for both their aligning and enabling roles [24]. In the domain of electric vehicle charging, studying this dual role requires to link research on (e-mobility) business models (1), service science (2), and power systems (3). In this section, we thus shape the problem context building upon existing research from all three fields and highlight their important intersections as to make the facets of the paper's contributions accessible.

(1) Charging services, in general, are considered as one type of e-mobility business models [8]. Fast charging, in specific, is a variant of this type. The importance of the fast charging service business models is about to overcome debate as large-scale initiatives in Europe (ionity) and America (ElectrifyAmerica) follow up roll-out plans. Characterising the fast charging service business models on charging parks, we apply the established canvas approach by Osterwalder and Pigneur [25]: at the heart of the business model rests its value proposition which addresses a specific customer need [26]. In expectation of the fulfilment of his/her needs, the customer commits to a value exchange [27]. In the case of fast charging, the customer exchanges a payment for the promise of fulfilling his/her mobility needs within acceptable servicing times, i.e. a negligible fraction of the total trip duration. The fast charging service sets out to relieve vehicle drivers of range anxiety, i.e. "a stressful experience of a present or anticipated range situation, where [...] resources [...] are perceived to be insufficient" [28]. Supporting the value proposition, the charging process itself constitutes the key activity. Key resources of the business model are charging stations capable of fast charging and a grid coupling point connecting a set of charging stations to the electricity grid. The grid coupling point consists of a transformer reducing the grid's voltage and inverters to change the grid's alternating current to direct current. Further key resources are the parking space, and an IS coordinating power and information flows in order to facilitate the key activity. We refer to this set of resources as the charging park, which is coordinated by the CPO. In order to carry out the fast charging service, electricity providers, charge point directories, maintenance services, and parking space operators partner to deliver the business model's value proposition [4]. Because the market for BEVs is comparatively immature, segmenting customers with regard to fast charging cannot be observed at this time. It addresses occasional customers by channels such as charge point directories and/or charging networks [23]. Capital costs and depreciation, as well as operating expenses [15] such as delivered energy and costs for peak demand [29] characterise the dominant parts of the business model's cost structure [15, 29]. The by far most important revenue stream stems from billing customers for the fast charging service [6, 15].

(2) Service science suggests that after a customer commits to a value exchange, his/her evaluation leads to a degree of (dis)satisfaction [30]. The gap of prior

expectation and actual performance plays a central role in satisfaction as it is defined as "an outcome of purchase and use resulting from the buyer's comparison of the rewards and costs of the purchase in relation to the anticipated consequences" [12]. The relationship between the gap (of prior expectation and actual performance) and the resulting satisfaction seems to be disproportionately strong, especially on the downside, as previous research by Lin et al. [14] suggests referring to prospect theory [31]. In addition, mobility is a human basic need and this presumably holds in particular for already started road trips. The model by Kano et al. [32] proposes a convex functional relationship for basic (must-be) requirements of a service and satisfaction. Thus, both, the theory and the model, corroborate the seemingly intuitive hypothesis that the longer a charging process deviates from a customer's expectation, the increasingly less satisfied the customer will be. On top of this, research found that volatility of service performance affects satisfaction negatively, as well [33]. Therefore, this paper sets out to apply satisfaction theory in the context of fast charging services by considering the gap of prior expectation and actual performance.

(3)   There are two non-exclusive options for operators to influence the expectation-performance gap: adjust expectations or actual performance. Altering expectations is hard for CPOs because of e.g. carmakers interests. Carmakers are known for influential information dissemination [34] and advertise servicing times only possible under optimal circumstances, e.g. Tesla [35]. Additionally, research also predominantly refers to nominal charging power [5, 36]. Charge point directories might reduce information asymmetries with regard to pricing and help to align expectations. However, for the time being they list nominal power rather than actual power, which leaves customers subjectively anticipating servicing times using the information at hand. Research suggests that this generates inflated expectations [11]. Additionally, expectations in the context of fast charging are hard to adjust based on prior experiences. This is because of non-transparent bottlenecks, e.g. peak reduction and peak charges [29], and endogenous factors, e.g. the other BEV(s) and their technical specifications as well as their battery level at arrival, that are opaque for the customer. All this influences the available power for each customer. Consequently, it is difficult for customers to form context-specific expectations even for the very same charging park.

Turning to actual performances, a CPO wants to operate economically. In fear of customer queues, CPOs will place a sufficient number of charging stations on the park [7]. CPOs are aware of the fact that the aggregated power demand of all charging stations might surpass economic or technical constraints. Thus, they would rather risk customers having to charge somewhat slower than forfeiting them. Additionally, volatility of daily traffic volume and the corresponding charging demand leads to uneven demand patterns i.e. peaks. The CPO wants to manage peak demand as to lower operational costs. Fast charging parks typically have power tariffs similar to industrial consumers [29]. They pay for used energy and peak demand separately. The latter typically represents the majority of costs for electricity [29]. Even in cases when peak demand is not billed, CPOs will not be able to surpass technically feasible limits as represented by batteries, inverters, and transformers. When the economic or technical limit is reached, charging performance will drop below nominal charging power [5, 36]. This affects at least one customer. In order to generate the most efficient use of power, CPOs should manage the allocation of power. This paper, therefore, sets out to manage performance in existing charging parks to address an issue most prevalent in the current BEV ramp up phase. Enhancing charging parks' capabilities by performance management avoids investments in infrastructure and additional technical equipment.

As an implication, reducing the expectation-performance gap by allocating power and thereby managing performance is crucial in fast charging services. Consequently, the objective of the paper is to manage and improve service performance.

4   Information System Design

As a deterministic optimisation model represents the kernel of the IS, we next derive and develop the model and relate it to the theoretical background stated above.

4.1   Model Setting

In order to formulate the optimisation model, we first describe the actors and components in the model setting before we follow up on the information and power flows between them.

The model setting represents an existing publicly accessible charging park with fast charging technology to allow acceptable servicing times for charging on long-distance journeys, e.g. in a motorway service area. The charging park comprises actors (a CPO and vehicle drivers) and system components (a grid coupling point, charging stations, BEVs, and a power management system). The CPO is a business entity operating all system components on the charging park that are permanently installed including the grid coupling point, the charging stations, and the power management system. A CPO is concerned with satisfying all customers, i.e. each vehicle driver $d \in D$. Every vehicle driver d belongs to one BEV which is connected to one of the charging stations $s \in S$. Charging stations may differ in their maximum charging power but are accessible to all types of BEVs. A charging station is either occupied by a BEV or vacant. Our model considers a situation where at least two stations are occupied ($|S| \geq |D| \geq 2$), i.e. the charging BEVs compete for the shared power. It is important to note that the vehicles do not need to arrive simultaneously. The system component grid coupling point connects all charging stations by a single point to the electricity grid. This represents a physical part of the power bottleneck, which is also constrained by economic considerations, e.g., peak reduction. The power management system is the IS artefact which we design to optimise the power allocation as to manage performance.

All components provide information about their technical specifications and charging states. We refer to the act of passing on information as information flow. Each charging station informs about its occupation status, its maximum (i.e. nominal) power deliverable to vehicle driver d's BEV ($CSPowMax_d$), and the efficiency of the charging process (η^{act}). The power management system receives information from each BEV. These are the battery's initial and final state of charge $\left(SoC_d^{\text{init}}, SoC_d^{\text{fin}}\right)$, the maximum power the battery is able to be charged with ($BEVPowMax_d$), and its battery capacity (BC_d). The grid coupling point transmits information on its maximum available power to the power management system, which then determines the overall available power ($PowMax$). Furthermore, the power system sets a parameter for the efficiency expected by the vehicle driver d (η^{exp}). Obviously, this is an optional parameter, which is set different from 1 only if the CPO can draw on empirical data. The power management system receives and processes the information to allocate the optimal amount of power (Pow_d) to the BEV of each vehicle driver d. The power management system sends control signals to the power bottleneck via information flow and as a consequence initiates the power flow.

Figure 1 reflects the model including cardinalities.

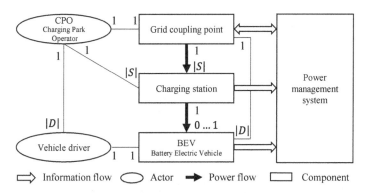

Fig. 1. Model setting

We refer to the time between initiation and termination of vehicle driver d's charging process as the actual servicing time ($TimeAct_d$). Each vehicle driver forms expectations about the servicing time based on the nominal charging power of the charging station and his/her BEV's technical specifications and charging state. We refer to a vehicle driver d's expected servicing time as $TimeExp_d$. We model an energy-quantity-based pricing for the charging service as applied by most CPOs today [21]. These thus do not have monetary incentives to speed up or slow down charging processes, which would distort the relationship between servicing time and satisfaction.

Table 1 summarises the indices, parameters, and variables.

Table 1. Indices, parameters, and variables used in the optimisation model.

Indices	
d	Index for a vehicle driver with his/her corresponding BEV, $d \in D$
s	Index for a charging station, $s \in S$
Parameters	
BC_d	Vehicle driver d's BEV's battery capacity
$BEVPowMax_d$	The maximum power the vehicle driver d's BEV can be charged with
$PowMax$	The power bottleneck
$CSPowMax_d$	The maximum power the vehicle driver d's charging station is able to provide
$SoC_d^{\mathrm{init}}, SoC_d^{\mathrm{fin}}$	The initial and final state of charge of vehicle driver d's BEV
$\eta^{\mathrm{act}}, \eta^{\mathrm{exp}}$	The efficiency of the charging process
Decision variable	
Pow_d	The power transferred via each charging station to the vehicle driver d's BEV
Variables	
$TimeAct_d$	The vehicle driver d's actual servicing time
$TimeExp_d$	The vehicle driver d's expected servicing time

4.2 Optimisation Model

In the following, we develop the deterministic optimisation model as part of the power management system. We first formally derive the objective function and afterwards the relevant constraints. The power management system can only exert control on the decision variable (Pow_d), i.e. the power the system allocates via the power bottleneck and the charging station to each vehicle driver d's BEV.

By the objective function, we strive to reduce the expectation-performance gap and target at a substitute measure derived from satisfaction literature: the gap between a customer's actual and expected servicing time, i.e. $TimeAct_d - TimeExp_d$. A vehicle driver is neutral, if actual and expected servicing time match, i.e. $TimeAct_d - TimeExp_d = 0$. Note that the actual servicing time is longer than the expected servicing time, i.e. $TimeAct_d \geq TimeExp_d$, if the vehicle driver expects to charge at nominal charging power. Nominal charging power is only theoretically feasible. On exceeding expected servicing time, a vehicle driver becomes increasingly less satisfied. Based on Lin et al. [14] we stick with the concept that customer satisfaction decreases disproportionately on larger deviations. Formally, this maps to a convex relationship between satisfaction as the dependent variable and the gap as the independent variable. Theoretically, all power functions with exponents greater 1 are conceivable. In analogy to Kobayashi and Salam [37] and for the sake of tractability, we square the deviation as to comply with convexity requirements, i.e. $(TimeAct_d - TimeExp_d)^2$. The CPO treats vehicle drivers equally such that the CPO will thus weigh all vehicle-driver-individual deviations equally. The CPO thus does not optimise individual but aggregated deviations between expectation and performance. Hence, the objective function minimises the sum of the squared deviations by identifying the optimal allocation of power, i.e. setting the decision variables $\left(\overrightarrow{Pow}\right)$:

$$Min\, f\left(\overrightarrow{Pow}\right) = \sum_{d=1}^{D}\left(TimeAct_d - TimeExp_d\right)^2 \tag{1}$$

We model the actual servicing time of vehicle driver d ($TimeAct_d$) which depends on the BEV's state of charge (SoC), the battery capacity, the allocated charging power Pow_d, and its efficiency η^{act}, as represented in Formular 2 [36]. The expected servicing time by vehicle driver d ($TimeExp_d$) is based on information available to the vehicle drivers. We consider that the vehicle drivers know their BEV's maximum power and the maximum power of the charging station as listed in chargepoint directories [36]. Consequently, we model the expected servicing time as represented in Formular 3 which implies that vehicle drivers perform simple calculations on a rule of proportion basis [36], i.e. proportional to the energy outstanding. Vehicle drivers will consequently estimate the servicing time proportional to the quantity of energy to be charged. This estimate grounds on adjustments of expectations by observable limitations, i.e. minimum of the maximum BEV power and maximum charging station power. Also, theoretically, vehicle drivers could learn the charging process' efficiency η^{exp}.

$$TimeAct_d = \frac{\left(SoC_d^{fin} - SoC_d^{init}\right) \cdot BC_d}{Pow_d \cdot \eta^{act}} \qquad \forall d \in D. \tag{2}$$

$$TimeExp_d = \frac{\left(SoC_d^{fin} - SoC_d^{init}\right) \cdot BC_d}{\min(BEVPowMax_d; CSPowMax_d) \cdot \eta^{exp}} \qquad \forall d \in D. \tag{3}$$

By the constraints, we ensure that optimal solutions suffice both technical and service limits. The maximum power the battery is able to be charged with limits the optimised power for the vehicle driver d's BEV. Additionally, the power allocated to vehicle driver d cannot surpass the charging stations' maximum power (Formular 4).

$$Pow_d \le \min(BEVPowMax_d; CSPowMax_d) \qquad \forall d \in D. \tag{4}$$

Also, the sum of allocated power must not surpass the limits of the power bottleneck (Formular 5).

$$PowMax \ge \sum_{d=1}^{D} Pow_d \qquad \forall d \in D. \tag{5}$$

5   IS Design Evaluation

We evaluate the IS design by fictive but realistic charging setups. The evaluation aims at challenging the IS design under different charging park setups. We present a scenario for dual charging station setups, i.e. two stations compete for a single scarce resource. We choose to do so for two reasons: first, charging setups with two BEVs and corresponding stations competing for a shared and constraint power is the most prevalent setup today - also along motorways [21]. Second, larger setups like Tesla's Superchargers frequently featuring double-digit numbers of stations come with limitations as characterised in this paper because they pair two stations sharing the same power electronic capacity [35]. Also, demonstrating benefits for the dual setup provides a basis to hypothesise that analysing larger setups, in future research, will yield similar or presumably better results.

5.1   Parameterisation and Data Input

Predominantly targeting existing and current charging park setups, we gather and derive information for BEVs and charging stations for a scenario parameterisation from publicly available resources (databases and scientific studies). We briefly outline the essential information on the fictive setups herein. We first introduce BEV-related data and then the charging-station-related data.

We obtained fast-charging-capable BEV characteristics from a database aggregating comprehensive information on currently registered vehicle types in Germany [21]. The assignment of different battery capacities to a maximum charging power ensures that we generate only setups with a combination of the two parameters that represent existing BEV types. Since we generate fictive but realistic setups, we need to consider the likelihood of a vehicle appearing at the charging park. Therefore, we use available information on the number of registered BEVs capable of fast charging in Germany in the year of 2018 [38]. Based on the number of registrations we determine market shares for BEVs and assign each BEV a likelihood of appearance in terms of a discrete distribution function. The code generates a setup including exactly two BEVs from that distribution and assigns them to the two charging stations. The vehicle drivers initiate the charging process at different SoCs. We calibrate the initial and final SoC referring to empirical data from 18,000 charging events at public charging stations in a large field test [39], as conditional probability distributions. In terms of the charging process efficiency, we refer to Zhang et al. [40] setting it to 85% (η^{act}) as to calculate the actual servicing time. Since the vehicle drivers either might not be aware of the underlying physics and/or refrain from the computational hassle, we parameterise an efficiency of 1 (η^{exp}) for calculating the expected servicing time in absence of empirical data. According to Longo [41] we set the maximum power of each charging station to 50 kW (Table 2).

Table 2. Parametrisation

Parameter	Data	Source
SoC^{init}, SoC^{fin}	Conditional discrete probability function over [0.05; 0.90] and [0.20; 1.00] respectively	[39]
η^{act}	0.85	[40]
η^{exp}	1.00	
BC	[14.5 kWh; 100 kWh]	[21, 38]
$BEVPowMax$	Discrete probability distribution function over [40 kW; 135 kW]	[21, 38]
$CSPowMax$	50 kW	[41]

Regarding the either economic or technically constraint power bottleneck, we conduct a variation analysis. By doing so, we take into account that the power bottleneck may vary among the existing charging parks. A variation analysis of the bottleneck allows us to quantify the benefit of the IS artefact on reducing the gap between actual and expected servicing time for a wide breadth of fast charging park setups. We vary the power bottleneck ($PowMax$) in interval steps of 1 kW. We set the minimum variation limit to be capable satisfying two BEVs of the slowest charging type regarding our data (80 kW). As the charging station is the upper limiting factor, the maximum variation is set to two times the maximum power of the charging stations (100 kW).

5.2 Scenario Analysis with Power Bottleneck Variation

In this section, we describe how we measure the performance of our IS design and present the results with regard to a power bottleneck variation. Because the parameterisation involves stochastics, we run all variation steps a thousand times having outputs converge to below 0.1% accuracy. Finally, we measure and evaluate results for each variation step based on two different metrics: first, we measure the average gap over all runs (a) and second, we measure the standard deviation of the gap over all runs (b). The first metric is in line with theory on customer satisfaction as introduced in Sect. 3. As theory suggests that this uncertainty additionally negatively affects satisfaction [33]. The second metric captures the uncertainty of the gap.

Looking at the results, we find that the artefact either improves or at least yields the same results (i.e. the metrics (a) and (b) are smaller or equal) as the status quo case.

Figure 2(i) demonstrates that improvements exist in relative terms for both metrics and all variation steps - except for the power bottleneck 80 kW and 100 kW. In these setups, the power management system cannot allocate more intelligently as technical restrictions kick in. At a power bottleneck of 90 kW improvements reach a maximum. In this case, the power management system has maximum flexibility to optimise compared to the status quo case, as BEV types of 40 kW and 50 kW can be satisfied in the status quo case as well. We find that the average deviation shrinks by up to 14.1%, and the standard deviation declines by up to 31.9% depending on the variation step.

Figure 2(ii) presents the improvements in absolute values for both metrics and all variation steps. Similar to Fig. 2(i) improvements for both metrics and all variation steps exist except for the technical restriction, i.e. 80 kW and 100 kW. Again, the graph reaches its maximum at a power bottleneck of 90 kW. Concerning the absolute improvements, we find that the average deviation shrinks by up to 0.65 min and the standard deviation reduces by up to 1.43 min.

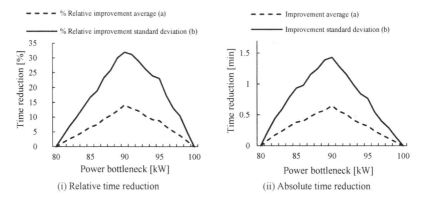

Fig. 2. Improvement corresponding to a variation of the power bottleneck

The main finding is that the IS design on average reduces the gap between the expected and actual servicing times illustrated by the reduction in the average deviations over all customers when comparing the status quo and optimised cases (a). Furthermore, the IS design significantly reduces the uncertainties, i.e. the average of standard deviations (b). We find that a CPO is generally better off with the proposed IS design than without. Therefore, we rest confident that the IS design can align power allocation in a way that it can contribute to customer satisfaction. Finally, this paper presents a suitable approach for CPOs to address customer satisfaction without large investments in infrastructure and additional technical equipment or high costs for peak demand and correspondingly high charges.

6   Discussion and Conclusion

A power bottleneck caused by economic or technical restrictions along with uncontrolled charging causes volatility of servicing times in fast charging services. Customers perceive deviations from their expected servicing times as uncertainties. Uncertainties stimulate range anxiety [28], which is known as a principal barrier to the acceptance of e-mobility and thus long-term decarbonisation goals. In this vein, the proposed IS design demonstrates potential to reduce the expectation-performance gap regarding fast charging. Contributing to fast charging operations, our IS design helps to relieve the pain from range anxiety by reducing the volatility of servicing times. Regarding fast charging service business models, our work and its results directly support the value proposition. The IS design improves the charging process – the business model's key activity – by a more efficient use of the key resources – the charging stations, the power bottleneck, and information managed by the power management system. To the best of our knowledge, this is the first (scientific) contribution reducing the expectation-performance gap in this field. Since it is important to apply this approach in larger charging parks, as well, we want to analyse this setup in our future research. The results of this study are subject to pricing based on the quantity of energy deployed, which is most common to date [15]. When considering time-based discounts, however, pricing might become a mediating variable on satisfaction. It remains to research to discover the relationship between these concepts. Prioritising charging processes by service differentiation might increase the volatility of the servicing time further. Policymakers should be aware of potential side effects of service differentiation, if market penetration of e-mobility is considered pivotal [4]. Because there is dispute on erecting and financing additional infrastructure, there is merit in looking into additional ways to improve customer satisfaction. This paper's contributions also stem from the identification of additional ways to address the expectation-performance gap. Our analysis suggests to further explore advanced IS designs, as well. E.g. it is interesting to reduce information asymmetries arising from opaque charging park capacities. Additionally, IS designs might help to form more realistic expectations. These designs influence the expectation (customer) side of the gap. This is likewise to reducing the gap by managing performance. Concepts from behavioural economic such as nudging are a possible approach we consider in future research. When reviews and ratings in charge point directories affect customer relationships, applying IS theories to improve these directories is another interesting field. In summary, there are various potential roles for IS in the field of fast charging business models – especially with regard to reducing the expectation-performance gap.

References

1. International Energy Agency: CO2 Emissions from Fuel Combustion. OECD Publishing, Paris (2017)
2. Hall, D., Moultak, M., Lutsey, N.: Electric vehicle capitals of the world: demonstrating the path to electric drive. International Council on Clean Transportation (2017)
3. Fetene, G.M., Hirte, G., Kaplan, S., Prato, C.G., Tscharaktschiew, S.: The economics of workplace charging. Transp. Res. Part B: Methodol. **88**, 93–118 (2016)
4. Kley, F., Lerch, C., Dallinger, D.: New business models for electric cars—a holistic approach. Energy Policy **39**, 3392–3403 (2011)
5. Neaimeh, M., Salisbury, S.D., Hill, G.A., Blythe, P.T., Scoffield, D.R., Francfort, J.E.: Analysing the usage and evidencing the importance of fast chargers for the adoption of battery electric vehicles. Energy Policy **108**, 474–486 (2017)
6. Schroeder, A., Traber, T.: The economics of fast charging infrastructure for electric vehicles. Energy Policy **43**, 136–144 (2012)
7. Jabbari, P., MacKenzie, D.: EV everywhere or EV anytime? Co-locating multiple DC fast chargers improves both operator cost and access reliability. In: EVS29 International Battery, Hybrid and Fuel Cell Electric Vehicle Symposium, Montreal, Canada (2016)

8. Stryja, C., Fromm, H., Ried, S., Jochem, P., Fichtner, W.: On the necessity and nature of e-mobility services – towards a service description framework. In: Nóvoa, H., Drăgoicea, M. (eds.) IESS 2015. LNBIP, vol. 201, pp. 109–122. Springer, Cham (2015)

9. Fridgen, G., Mette, P., Thimmel, M.: The value of information exchange in electric vehicle charging. In: Proceedings of the 35th International Conference on Information Systems (2014)

10. Wagner, S., Brandt, T., Neumann, D.: IS-centric business models for a sustainable economy - the case of electric vehicles as energy storage. In: Proceedings of the 12th International Conference on Wirtschaftsinformatik (2015)

11. Rosenberg, L.J., Czepiel, J.A.: A marketing approach for customer retention. J. Consumer Market. 1, 45–51 (1984)

12. Churchill, G.A., Surprenant, C.: An investigation into the determinants of customer satisfaction. J. Market. Res. 19, 491–504 (1982)

13. Palawatte, T.M.: Waiting times and defining customer satisfaction. Vidyodaya J. Manag. 1, 15–24 (2015)

14. Lin, Y.-T., Xia, K.-N., Bei, L.-T.: Customer's perceived value of waiting time for service events. J. Consumer Behav. 14, 28–40 (2015)

15. Madina, C., Zamora, I., Zabala, E.: Methodology for assessing electric vehicle charging infrastructure business models. Energy Policy 89, 284–293 (2016)

16. Kahlen, M., Ketter, W., van Dalen, J.: Balancing with electric vehicles: a profitable business model. In: Proceedings of the 22nd European Conference on Information Systems (2014)

17. Motoaki, Y., Shirk, M.G.: Consumer behavioral adaption in EV fast charging through pricing. Energy Policy 108, 178–183 (2017)

18. Brandt, T., Feuerriegel, S., Neumann, D.: Shaping a sustainable society: how information systems utilize hidden synergies between green technologies. In: Proceedings of the 34th International Conference on Information Systems (2013)

19. Hevner, A.R., March, S.T., Park, J., Ram, S.: Design science in information systems research. MIS Q. 28, 75–105 (2004)

20. Gholami, R., Watson, R., Hasan, H., Molla, A., Bjorn-Andersen, N.: Information systems solutions for environmental sustainability: how can we do more? J. Assoc. Inf. Syst. 17, 521–536 (2016)

21. Bach, M.: Elektromobilität entdecken. https://www.e-stations.de/

22. Brendel, A.B., Lichtenberg, S., Nastjuk, I., Kolbe, L.M.: Adapting carsharing vehicle relocation strategies for shared autonomous electric vehicle services. In: Proceedings of the 38th International Conference on Information Systems (2017)

23. EVgo Services LLC: New Low Pricing. Charge Faster, Drive Farther, Pay Less. https://www.evgo.com/

24. Henderson, J.C., Venkatraman, H.: Strategic alignment: leveraging information technology for transforming organizations. IBM Syst. J. 32, 472–484 (1993)

25. Osterwalder, A., Pigneur, Y.: Business Model Generation. A Handbook for Visionaries, Game Changers and Challengers. Wiley, Hoboken (2010)

26. Osterwalder, A., Pigneur, Y., Bernarda, G., Smith, A.: Value Proposition Design. How to Create Products and Services Customers Want. Wiley, Hoboken (2014)

27. Wilson, D.T.: Value exchange as the foundation stone of relationship marketing. Market. Theory 3, 175–177 (2003)

28. Rauh, N., Franke, T., Krems, J.F.: Understanding the impact of electric vehicle driving experience on range anxiety. Hum. Factors 57, 177–187 (2015)

29. Knupfer, S., Noffsinger, J., Sahdev, S.: How battery storage can help charge the electric-vehicle market. http://bit.ly/2BIgiXT

30. Sirgy, M.J.: A social cognition model of consumer satisfaction/dissatisfaction an experiment. Psychol. Market. **1**, 27–44 (1984)
31. Kahneman, D., Tversky, A.: Prospect theory. an analysis of decision under risk. Econometrica: J. Econometric Soc. **47**, 263–291 (1979)
32. Kano, N., Seraku, N., Takahashi, F., Tsuji, S.: Attractive quality and must-be quality. Hinshitsu: J. Jpn. Soc. Qual. Control **14**, 39–48 (1984)
33. Meyer, R.J.: A model of multiattribute judgments under attribute uncertainty and informational constraint. J. Market. Res. **18**, 428–441 (1981)
34. Knight, B.: Study: Most EU carmakers report false CO2 emissions. http://www.dw.com/en/study-most-eu-carmakers-report-false-co2-emissions/a-19250251
35. Tesla: Supercharger. https://www.tesla.com/supercharger?redirect=no
36. Gnann, T., Funke, S., Jakobsson, N., Plötz, P., Sprei, F., Bennehag, A.: Fast charging infrastructure for electric vehicles: today's situation and future needs. Transp. Res. Part D: Transp. Environ. **62**, 314–329 (2018)
37. Kobayashi, K., Salam, M.U.: Comparing simulated and measured values using mean squared deviation and its components. Agron. J. **92**, 345–352 (2000)
38. Zinke, E.: Kraftfahrt-Bundesamt. Wir punkten mit Verkehrssicherheit! https://www.kba.de/DE/Home/home_node.html
39. Francfort, J.E.: What Use Patterns Were Observed for Plug-In Electric Vehicle Drivers at Publicly Accessible Alternating Current Level 2 Electric Vehicle Supply Equipment Sites? United States (2015)
40. Zhang, L., Brown, T., Samuelsen, S.: Evaluation of charging infrastructure requirements and operating costs for plug-in electric vehicles. J. Power Sources **240**, 515–524 (2013)
41. Longo, L.: Optimal design of an EV fast charging station coupled with storage in Stockholm. KTH Industrial Engineering and Management (2017)

Data Analytics in Service

Collaborative Recommendations with Deep Feed-Forward Networks: An Approach to Service Personalization

Giovanni Luca Cascio Rizzo[1], Marco De Marco[2], Pasquale De Rosa[1], and Luigi Laura[2(✉)]

[1] LUISS University of Rome, Rome, Italy
glcasciorizzo@luiss.it, pasquale.derosa@alumni.luiss.it
[2] International Telematic University UNINETTUNO, Rome, Italy
{marco.demarco,luigi.laura}@uninettunouniversity.net

Abstract. The aim of this article is to discuss an advanced approach to recommendation systems, based on the adoption of Deep Feed-Forward Neural Networks. Recommendation engines are data-driven infrastructures designed to help customers in their decision-making process, and nowadays represent the "state of the art" in designing smart and personalized services, in accordance with the new customer-centric perspective. For this purpose, we followed a quantitative methodological approach, comparing the predictive ability of traditional "Collaborative" recommendation algorithms, like the k-Nearest Neighbors (k-NN) and the Singular Value Decomposition (SVD), with Feed-Forward Neural Networks; given these assumptions, we finally demonstrated that a "Deep" Neural architecture could achieve better results in terms of "loss" generated by the model, laying the foundations for a new, innovative paradigm in service recommendation science.

Keywords: Recommendation systems · Collaborative Filtering · Deep feed-forward networks · Smart services · Service personalization · Service innovation

1 Introduction

This paper discusses the application of a Deep-Learning-based paradigm for recommendation engines, that are smart service infrastructures designed to help customers in their decision-making process [1,2].

Smart services are closely related to big data and analytics, that provide accurate measures of a dynamic world and insights on dynamic customers, allowing the rise of more instrumented, interconnected and intelligent businesses and societal systems. The role of data science in designing advanced and personalized services determined the development of research studies which aimed to outline the potential contribution of advanced analytics to service value creation [3–7].

© Springer Nature Switzerland AG 2020
H. Nóvoa et al. (Eds.): IESS 2020, LNBIP 377, pp. 65–78, 2020.
https://doi.org/10.1007/978-3-030-38724-2_5

For example, a recent study conducted by Blöcher and Alt [1] discussed the concept of "Customer-Centered Smart Services", a different, marketing-based service perspective that uses Customer-Dominant Logic (CDL) and focuses mainly on analyzing the consumer behaviour instead of the interactions between service providers and their clients. In this perspective, the role of recommendation systems as smart service architectures that make use of customer data to assist them with relevant suggestions is surely prominent.

In this paper we explore an approach to recommendation systems, based on the adoption of Deep Feed-Forward Neural Networks; to this aim, we followed a quantitative methodological approach, comparing the predictive ability of traditional "Collaborative" recommendation algorithms, like the k-Nearest Neighbors (k-NN) and the Singular Value Decomposition (SVD), with Feed-Forward Neural Networks; given these assumptions, we finally demonstrated that a "Deep" Neural architecture could achieve better results in terms of "loss" generated by the model, laying the foundations for a new, innovative paradigm in service recommendation science.

Our research goal is to show that, with a carefully designed ad-hoc neural network, in particular a deep forward one, it is possible to develop a state of the art recommendation system, that can be to deploy a dedicated service, or it can be employed to suggest services in a multi-user, multi-service environment.

Structure of this Paper. This paper is organized as follows: preliminary notions are recalled in the next section. In Sect. 3 we detail our experimental findings, whilst concluding remarks are addressed in Sect. 4.

2   Preliminaries

In this section we provide the necessary background; in particular, we first provide an overview of Recommendation Systems and the *collaborative filtering* approach, and then we focus on the specific techniques we will compare in our research: k-Nearest Neighbors (Sect. 2.2.1), Singular Value Decomposition (Sect. 2.2.2), and Artificial Neural Networks (Sect. 2.2.3).

2.1   Recommendation Systems and Collaborative Filtering

In the introductory paragraph, we briefly described the adoption of big data analysis and customer-centric approaches in nowadays business transformation processes (with a relevant effect on service innovation). One noticeable example of this new paradigm, based on service personalization and one-to-one marketing, is represented by the increasingly common adoption of recommendation systems for almost all companies, with a significant impact on their revenues and profits [8]. A recommendation system allows to simplify the decision-making process for a customer, providing him/her with effective and relevant purchase suggestions. Thereby, companies can adopt more focused marketing policies, in order to increase their transaction volumes by accurately matching the characteristics of their products with needs and desires of actual and potential customers. Recommendation engines are based on a complex system of data and information regarding customers, products to recommend and transactions; the volume and type of information collected may vary according to the complexity of the system itself. For example, it is possible to collect complex and detailed information on customers, like demographics, psychographics and social data; on the other hand, data and insights on products and transactions can both be stored in a structured format, like user ratings, or in a more unstructured scheme, like the content of textual reviews [9].

In literature, several approaches to recommendation systems emerged, among which it is worth mentioning Collaborative Filtering, Content-based Filtering, and Hybrid paradigms between the first two [10–14]; however, for our research purposes we will focus only on the first perspective. Collaborative Filtering is an approach to recommendation systems based on the convergence between the preferences of different users: indeed, each customer will be provided with recommendations of services, that had already been appreciated by other users with a similar and consistent behaviour in the past [9,13]. The main reason why both service and non-service businesses commonly adopt this approach to recommendation, instead of Content-based and Hybrid paradigms, is that Collaborative Filtering systems can extend a customer's purchase horizon to other service types and categories he may never have considered before [15]. As a result, each user can "discover" a new range of service types, acquiring new interests and allowing companies to determine new personalized and data-driven value propositions for their customers.

On the other hand, the adoption of Collaborative Filtering systems could present certain peculiar disadvantages, among which we can mention [15]:

- The "cold-start problem", common to new services, that generally present an insufficient number of reviews/ratings by customers, not allowing then the creation of "collaborative" recommendation systems;
- The "sparse" nature of user ratings, that are mostly made up by missing values;
- The need for huge volumes of data related to each customer in order to build consistent Collaborative Filtering systems.

In order to build effective Collaborative Filtering engines, in literature is attested the use of the user/rating matrix, an analysis tool in which the customers' preferences are represented by a $m \times n$ matrix, where m is the overall number of users in the company's portfolio and n the total number of services (or items) for which the users previously expressed a synthetic judgment, in the form of a rating [13].

$$R = \begin{pmatrix} r_{11} & r_{12} & r_{13} & \cdots & r_{1n} \\ r_{21} & r_{22} & r_{23} & \cdots & r_{2n} \\ \vdots & \vdots & \vdots & \cdots & \vdots \\ r_{m1} & r_{m2} & r_{m3} & \cdots & r_{mn} \end{pmatrix}$$

The user/rating matrix is sparse by definition, i.e. constituted mostly by zeros and missing values, as it would be extremely complex for each user to express a rating for every single service provided by the company [13].

The user/rating matrix allows the creation of sophisticated algorithms that, starting from a known structure of preferences of different users, are capable of generating predictions about the ratings that the same customers would attribute to unknown and unreviewed services. Furthermore, those predictions could be used to define the basis of subsequential recommendations.

The range of Collaborative Filtering algorithms, that are commonly used in current business applications, can be mainly sub-divided in two different families of techniques: Neighborhood-based and Model-based [13]. The Neighborhood-based algorithms originate from the nearest neighbors concept [9]: those neighbors are basically a subset consisting of the k most similar users to a specific customer, whose ratings will further be calculated as a weighted combination of the ratings expressed on the same services by his nearest neighbors in the past [13]. This family of methods is based on the k-Nearest Neighbors supervised learning algorithm, for which, once defined a similarity metric ω_{au} between a certain user a and every other customer $u \neq a$, the predicted rating for a related to the service s will be equal to [9]:

$$\hat{r_{as}} = \frac{\sum_{u \in k_s} \omega_{au} \cdot r_{us}}{\sum_{u \in k_s} \omega_{au}} \tag{1}$$

Where k_s is the overall number of users u that are nearest neighbors of the customer a and rated the service s, r_{us} is the rating attributed by each user u to the service s, and ω_{au}, as we previously anticipated, is the chosen similarity metric. The expression $\sum_{u \in k_s} \omega_{au}$ also appears in the denominator, in order to normalize the weights [9].

From a marketing perspective, the adoption of Neighborhood-based methods for Collaborative Filtering is linked to the word-of-mouth principle, that describes the customer decision journey as an evaluation process influenced by reviews and opinions expressed by other people with similar tastes and preferences.

Despite their remarkable conceptual simplicity, the Neighborhood-based techniques are widely used to create effective Collaborative Filtering engines. This is likely due to some noticeable advantages that those methods present, among which we can mention [9]:

- The computational efficiency, that makes those techniques appropriate when dealing with large databases;
- Proven stability and consistency faced with variations in data structure;
- "Serendipity", that is the capability of arousing the customer's interest in new and relatively unknown services.

The "Model-based" methodologies, instead, make use of statistical techniques and models to estimate and predict the ratings provided by a particular user [13]; among these, we can mention the classification and regression algorithms, often adopted in combination with some dimensionality reduction techniques (like the Principal Component Analysis), in order to pre-process the user/rating matrix and to decrease its sparsity [13]. However, the standard Machine Learning algorithms do not exhaust the entire set of statistical methodologies that can be used to create effective Collaborative Filtering models. It is certainly worth mentioning the Latent Factors Models, a family of techniques that allows to "extrapolate" from the explicit rating attributed by a generic customer the underlying "latent factors" that characterized his decision [9].

The most common examples of Latent Factors Models are represented by the Matrix Factorization (or Decomposition) techniques, like the Singular Value Decomposition (SVD) algorithm, based on the assumption that the similarity between two or more users is determined by the presence of latent and hidden structures in the data [13]; for this reason, both users and services can be represented by two vectors $\overline{u}$ e $\overline{s} \in R^k$ (where k are the "latent" dimensions), and ratings will be calculated through the dot product $\overline{u}^T \cdot \overline{p}$ [13]. This operation will reproduce a user/rating matrix, where each element corresponds to the dot product between the user vector and the service/product vector.

Also Artificial Neural Networks (ANNs), as we discuss further in the following paragraphs, can be used to explain latent structures in the data, and consequently should be regarded as another example of Latent Factors Model [9].

2.2 Approaches to Collaborative Recommendations

Nowadays, Machine Learning-based techniques constitute the "state of the art" in the development of advanced Collaborative recommendation systems; more specifically, some algorithms like the "k-Nearest Neighbors" (k-NN) [9] and the Singular Value Decomposition (SVD) [16] traditionally represent real milestones in this field. Nevertheless, some recently emerged approaches, including the one proposed in this article, focus on complex, Neural-based systems, in order to generate more accurate and effective service recommendations to the customers [17]. This paragraph discusses in detail the characteristics of these algorithms, in order to further describe the findings and the managerial implications of our personal research.

2.2.1 K-Nearest Neighbors

The k-Nearest Neighbors (k-NN) is a well-known classification and regression algorithm, widely used in several supervised learning applications [18], which, as anticipated in the previous paragraph, categorizes and predicts the testing values of interest (in this case, user ratings) by taking into account the first k most "similar" observations in the training dataset [18].

In order to compute the similarity between different users, several metrics are widely used in literature, among which we can certainly mention [19] (Fig. 1):

– The "Mean-Squared Difference" (MSD), which can be defined as:

$$MSD(x,y) = \frac{1}{|S(x,y)|} \cdot \sum_{s \in S(x,y)} [r(x,s) - r(y,s)]^2 \tag{2}$$

Where x and y are two generic users, $r(x,s)$ and $r(y,s)$ are the ratings attributed by x and y on the service s, and $S(x,y)$ represents the set of total services rated both by x and y [19].
The MSD can only be ≥ 0, and follows a diminishing trend if the similarity between the variables x and y increases;
– The "Cosine Similarity", which can be computed as:

$$cos(x,y) = \frac{\sum_{s \in S(x,y)} r(x,s) \cdot r(y,s)}{\sqrt{\sum_{s \in S(x,y)} r(x,s)^2} \cdot \sqrt{\sum_{s \in S(x,y)} r(y,s)^2}} \tag{3}$$

– The "Pearson Coefficient", which can be computed as:

$$p(x,y) = \frac{\sum_{s \in S(x,y)} [r(x,s) - \bar{x}] \cdot [r(y,s) - \bar{y}]}{\sqrt{\sum_{s \in S(x,y)} [r(x,s) - \bar{x}]^2} \cdot \sqrt{\sum_{s \in S(x,y)} [r(y,s) - \bar{y}]^2}} \tag{4}$$

Both the Cosine Similarity and the Pearson Coefficient can, by definition, only take values between -1 and 1, so comparing the variables x e y, their similarity will increase for higher values of $cos(x,y)$ and $p(x,y)$

Once determined the similarity measure for the observations in the dataset, the next step is to select the optimal value for the k parameter, that represents the number of "neighbors", so that the model does not underfit nor overfit, with respect to the data used to train the algorithm [18].

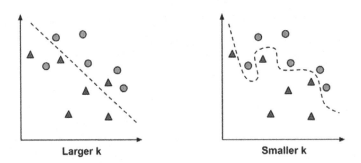

Larger k   **Smaller k**

Fig. 1. The "bias-variance tradeoff" when selecting the k parameter [18].

2.2.2 Singular Value Decomposition

The Matrix Decomposition (or Factorization) techniques are one of the main families of algorithms traditionally adopted for the creation of Collaborative Filtering systems [20]; more specifically, these techniques aim to reduce the complexity of a matrix, that describes the original architecture of input data, by decomposing it in the product of more matrices, each one representing a certain number of "latent factors" that are able to explain its hidden structures.

The main Matrix Factorization technique adopted for Collaborative Filtering applications is certainly the Singular Value Decomposition (SVD), for which, given a certain matrix A whose dimensions are $m \times n$, its decomposition results in the following product:

$$A = U \cdot \Sigma \cdot V^T \qquad (5)$$

Where, respectively:

- U is an orthogonal $m \times m$ matrix;
- Σ is a diagonal matrix, whose elements of the main diagonal, $\sigma_1, \sigma_2, \sigma_3, ..., \sigma_n$, are defined the "singular values" of A;
- V^T is the transpose of an orthogonal $n \times n$ matrix V.

An alternative version of the model was proposed by Simon Funk, who described the rating prediction $\hat{r}_{us}$ given by an user u on a service s as the dot product of two matrices, as it follows [20–22]:

$$\hat{r}_{us} = q_s^T \cdot p_u \qquad (6)$$

Where q_i^T and p_u are embeddings representing the latent factors of the service s and the user u that underlie the rating $\hat{r}_{us}$.

2.2.3 Artificial Neural Networks

Artificial neural networks (ANNs) are Machine Learning algorithms whose structure substantially replies the aspect of biological Neural Networks [23]. Just like the human neuron, in fact, the artificial neuron is composed by three fundamental elements [24]:

- The Dendrites, which receive incoming pulses from other elements in the network;
- The Cellular body, which represents the fulcrum of the brain activity and elaborates input and output electric pulses;
- The Axons, which propagate the output pulses towards the successive layers in the network.

An early example of Artificial Neural Network was proposed, in 1958, by the American psychologist Frank Rosenblatt, which named it "Perceptron" [25,26]: it was a sophisticated classification algorithm, which, starting from a number n of inputs $x_1, x_2, ..., x_n$, returns a single binary output (0 if the observation does not belong to a specific category, or 1 otherwise). More specifically, in the model proposed by Rosenblatt, the variables are classified comparing the sum of the n inputs, weighted for an equal number of elements $\omega_1, \omega_2, ..., \omega_n$, with a threshold value t, so that the following equation applies:

$$\text{Output} = \begin{cases} 0, & \text{if } \sum_n x_n \cdot \omega_n \leq t \\ 1, & \text{if } \sum_n x_n \cdot \omega_n > t \end{cases} \tag{7}$$

The previous expression describes also the "activation" conditions of the Artificial Neuron, and for this reason it is generally defined in literature as the "activation function" of the Neural Network. More recent applications proposed "expanded versions" of the original Perceptron, introducing one or more intermediate levels between the input and the output layers of the Network; those levels, known as "hidden layers", contain additional neurons that apply changes and transformations on the original input data, allowing thus an efficient propagation of the information towards the output layer. In literature, those Neural Network architectures are also defined "Multi-layer Perceptrons" (MLP).

In addition to the above mentioned models, several other complex Neural structures have been introduced over time, among which it is certainly worth to cite:

- The Convolutional Neural Networks, which, due to their characteristic architecture, are particularly suitable for the elaboration of graphic, tridimensional input data, thus avoiding the risk of an information loss [26–28];
- The Recurrent Neural Networks, which are also defined as Feed-Back Networks because of their capability to model sequential data, keeping their temporal dynamics unaltered [26, 29, 30].

3   Experimental Results

We conducted the present study in order to determine the more effective approach to Collaborative Filtering, among the techniques that were previously discussed. More specifically, we trained our models on the "Movielens 100K" dataset [31], that is particularly suitable for Collaborative Filtering implementations [31].

The following sections discuss in detail the results of our research, comparing the effectiveness of early approaches to recommendations (k-NN and SVD) with that of Neural-based algorithms; for this purpose, to evaluate the performance of each technique, we used the "Root-Mean Squared Error" (RMSE), whose mathematical representation is the following [32]:

$$\text{RMSE} = \sqrt{\text{MSE}} = \sqrt{\frac{1}{n} \cdot \sum_x [\hat{r}(x) - r]^2} \tag{8}$$

Where the expression $[\hat{r}(x) - r]$ represents the difference between the predicted rating value and the observed one [32].

3.1   k-Nearest Neighbors and Singular Value Decomposition

We analyzed the performances of the k-Nearest Neighbors algorithm for three values of the k parameter: 32, 50 and 100; moreover, in order to identify the "nearest neighbors" we adopted all the three similarity metrics that were discussed in the previous sections (MSD, Cosine Similarity and Pearson Coefficient). Given these assumptions, the research findings can be summarized in Table 1.

Table 1. Summary of the results for different values of k and similarity metrics.

k	Similarity	RMSE
32	Cosine	1.0226
50	Cosine	1.0219
100	Cosine	1.0225
32	Pearson	1.0205
50	Pearson	1.0156
100	Pearson	1.0158
32	MSD	0.9813
50	MSD	0.9862
100	MSD	0.9973

The best performances were achieved using the Mean-Squared Difference as the similarity metric; in general, the minor loss was obtained with $k = 32$, and there appeared to be a positive correlation between the number of neighbors k and the RMSE. As a result, we decided to limit the selection of k to "low" values, containing the loss associated to the model (Fig. 2).

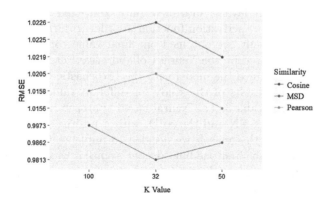

Fig. 2. A graphical representation of the research findings (k-NN).

Regarding the SVD, we used the "Simon Funk" variant of the model [22], fixing the number of latent factors composing the "embeddings", q_i^T and p_u, to 32. Moreover, we adopted the "Regularized-Squared Error" as the cost function and trained the model with the "Stochastic Gradient Descent" (SGD) algorithm for 50 "epochs". Table 2 summarizes the results of the study for values of the Learning Rate η equal to, respectively, 0.001, 0.01 and 0.1.

Table 2. SVD: summary of the results for different values of η.

η	Latent factors	Epochs	Optimization algorithm	RMSE
0.001	32	50	SGD	0.9809
0.01	32	50	SGD	1.0291
0.1	32	50	SGD	1.3016

Similarly to what we observed for the k-NN, also for the SVD there appeared to be a positive correlation between the Learning Rate η and the RMSE of the model; in fact, the best result in terms of performance (RMSE = 0.9809) was obtained with the lowest value of η (0.001).

3.2 The Neural-Based Approach

We implemented a "Feed-Forward" Neural Network model with two "embedding layers", representing the "latent factors" of users and services [17], whose output was a bi-dimensional vector (1,32); furthermore, we included two "flatten layers" to reduce the dimensionality of the previously generated embeddings, in order to "speed up" the entire training process (Fig. 3).

Before the last transformations, we also included a "dropout layer" to prevent a potential overfitting [33], setting the percentage of dropouts to 0.2; finally, we merged the "dropout layers" into one "fully connected layer", applying a "Rectified Linear" (ReLu) activation function in order to turn all negative data into zeros [26]. As a result, we obtained one final value corresponding to the rating prediction generated by the Neural Collaborative Filtering model.

Subsequently, we trained the model with the "Stochastic Gradient Descent" (SGD) algorithm; the results of the study for different values of the Learning Rate η are reported in Table 3. The Neural-based approach resulted to be more effective than both the k-NN and the SVD, with a positive peak for $\eta = 0.01$ (RMSE = 0.9624). For this reason we decided to implement a further Neural architecture, similar to the first one but "deeper", since it was made up by three "fully connected layers" before the final "activation layer".

The Deep Neural Network model registered higher performances than the "Shallow" architecture with a single hidden layer; furthermore, for $\eta = 0.01$, the RMSE of the model was 0.9593, overcoming in effectiveness all the other techniques that we previously analyzed (Fig. 4 and Table 4).

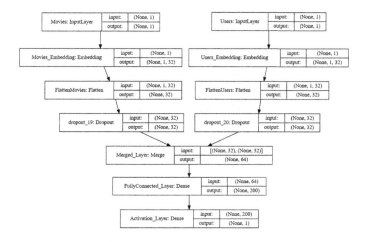

Fig. 3. A graphical representation of the Neural Network model.

Table 3. (Shallow) Neural Network: summary of the results for different values of η.

η	Latent factors	Epochs	Dropout	Optimization algorithm	RMSE
0.001	32	50	0.2	SGD	0.9816
0.01	32	50	0.2	SGD	0.9624
0.1	32	50	0.2	SGD	0.9911

Table 4. (Deep) Neural Network: summary of the results for different values of η.

η	Latent factors	Epochs	Dropout	Optimization algorithm	RMSE
0.001	32	50	0.2	SGD	0.98
0.01	32	50	0.2	SGD	0.9593
0.1	32	50	0.2	SGD	1.01

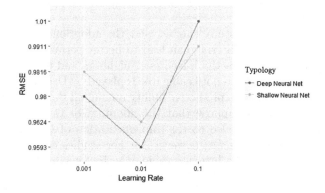

Fig. 4. A graphical comparison between the Neural Network models.

3.3   Model Comparison and Research Findings

The research outcomes highlighted the noticeable performance of the Neural-based Collaborative Filtering approach, also when considered in comparison with traditional algorithms like k-NN and SVD, as seen in Table 5.

The Neural Network models, in fact, registered the lowest values of RMSE, thus resulting more efficient than the other tested approaches; moreover, the Deep Neural architecture was the top-performing algorithm among the proposed techniques, laying the foundations for a possible introduction of a new, innovative and Deep-Learning-based methodological approach to service recommendations.

Table 5. Summary of the research outcomes.

Model	RMSE
k-NN	0.9813
SVD	0.9809
Shallow Neural Network	0.9624
Deep Neural Network	0.9593

4   Conclusions and Future Research

In this paper we explored an approach to recommendation systems, based on the adoption of Deep Feed-Forward Neural Networks; for this purpose, we followed a quantitative methodological approach, comparing the predictive ability of traditional "Collaborative" recommendation algorithms, like the k-Nearest Neighbors (k-NN) and the Singular Value Decomposition (SVD), with Feed-Forward Neural Networks; given these assumptions, we finally demonstrated that a "Deep" Neural architecture could achieve better results in terms of "loss" generated by the model, laying the foundations for a new, innovative paradigm in service recommendation science.

The final results of this study showed that the adoption of a Neural-based approach to Collaborative Filtering can lead to better performances than more traditional methodologies like the k-Nearest Neighbors and the Singular Value Decomposition; furthermore, a noticeable role is played by Deep Neural architectures, that presented the best predictive ability among all the discussed models. These findings lead us to suppose that the adoption of Deep-Learning-based models should be extended also to the implementation of Content-based and Hybrid recommendation systems; however, further studies could be conducted to analyze the suitability of a broader range of techniques to our research purposes [34].

References

1. Blöcher, K., Alt, R.: An approach for customer-centered smart service innovation based on customer data management. In: Satzger, G., Patrício, L., Zaki, M., Kühl, N., Hottum, P. (eds.) IESS 2018. LNBIP, vol. 331, pp. 45–58. Springer, Cham (2018). https://doi.org/10.1007/978-3-030-00713-3_4
2. Dermikan, H., Bess, C., Spohrer, J., Rayes, A., Allen, D., Moghaddam, Y.: Innovations with smart service systems: analytics, big data, cognitive assistance, and the internet of everything. CAIS **37**, 35 (2015)
3. Meierhofer, J., Herrmann, A.: End-to-end methodological approach for the data-driven design of customer-centered digital services. In: Satzger, G., Patrício, L., Zaki, M., Kühl, N., Hottum, P. (eds.) IESS 2018. LNBIP, vol. 331, pp. 208–218. Springer, Cham (2018). https://doi.org/10.1007/978-3-030-00713-3_16
4. Zardini, A., Rossignoli, C., Mola, L., De Marco, M.: Developing municipal e-Government in Italy: the city of Alfa case. In: Snene, M., Leonard, M. (eds.) IESS 2014. LNBIP, vol. 169, pp. 124–137. Springer, Cham (2014). https://doi.org/10.1007/978-3-319-04810-9_10
5. Za, S., Marzo, F., De Marco, M., Cavallari, M.: Agent based simulation of trust dynamics in dependence networks. In: Nóvoa, H., Drăgoicea, M. (eds.) IESS 2015. LNBIP, vol. 201, pp. 243–252. Springer, Cham (2015). https://doi.org/10.1007/978-3-319-14980-6_19
6. Meierhofer, J., Meier, K.: From data science to value creation. In: Za, S., Drăgoicea, M., Cavallari, M. (eds.) IESS 2017. LNBIP, vol. 279, pp. 173–181. Springer, Cham (2017). https://doi.org/10.1007/978-3-319-56925-3_14
7. Meleo, L., Romolini, A., De Marco, M.: The sharing economy revolution and peer-to-peer online platforms. The case of Airbnb. In: Borangiu, T., Drăgoicea, M., Nóvoa, H. (eds.) IESS 2016. LNBIP, vol. 247, pp. 561–570. Springer, Cham (2016). https://doi.org/10.1007/978-3-319-32689-4_43
8. Mackenzie, I., Meyer, C., Noble, S.: How Retailers Can Keep Up with Consumers. McKinsey & Company, New York (2013)
9. Ricci, F., Rokach, L., Shapira, B., Kantor, P.B.: Recommender Systems Handbook, p. 2011. Springer, Heidelberg (2011). https://doi.org/10.1007/978-0-387-85820-3
10. Basu, C., Hirsh, H., Cohen, W.: Recommendation as classification: using social and content-based information in recommendation. In: Proceedings of the Fifteenth National Conference on Artificial Intelligence, pp. 714–720 (1998)
11. Cotter, P., Smyth, B.: PTV: intelligent personalized TV guides. In: Twelfth Conference on Innovative Applications of Artificial Intelligence, pp. 957–964 (2000)
12. Melville, P., Mooney, R.J., Nagarajan, R.: Content-boosted collaborative filtering for improved recommendations. In: Proceedings of the Eighteenth National Conference on Artificial Intelligence, Edmonton, pp. 187–192 (2002)
13. Melville, P., Sindhwani, V.: Recommender systems. In: Sammut, C., Webb, G.I. (eds.) Encyclopedia of Machine Learning and Data Mining, p. 2017. Springer, Boston (2017)
14. Pazzani, M.J.: A framework for collaborative, content-based and demographic filtering. Artif. Intell. Rev. **13**(5–6), 393–408 (1999)
15. Bhatnagar, V.: Collaborative Filtering Using Data Mining and Analysis. IGI Global (2016)
16. Bokde, D., Girase, S., Mukhopadhyay, D.: Matrix factorization model in collaborative filtering algorithms: a survey. Procedia Comput. Sci. **49**, 136–146 (2015)

17. He, X., Liao, L., Zhang, H., Nie, L., Hu, X., Chua, T.-S.: Neural Collaborative Filtering. arXiv:1708.05031v2, August 2017
18. Lantz, B.: Machine Learning with R, 2nd edn, p. 2015. Packt Publishing Ltd., Birmingham (2015)
19. Furht, B.: Handbook of Social Network Technologies and Applications. Springer, Heidelberg (2010). https://doi.org/10.1007/978-1-4419-7142-5
20. Koren, Y., Bell, R., Volinsky, C.: Matrix factorization techniques for recommender systems. Computer **42**(8), 30–37 (2009)
21. Hug, N.: Surprise, a Python library for recommender systems (2017). http://surpriselib.com
22. Funk, S.: Description of the SVD algorithm for Recommendation Systems (2006). http://sifter.org/simon/journal/20061211.html
23. Haykin, S.: Neural Networks and Learning Machines, 3rd edn. Pearson Education, Upper Saddle River (2009)
24. Mesa, J.A.S., Galán, C., Hervás, C.: The use of discriminant analysis and neural networks to forecast with a typical Mediterranean climate. Int. J. Biometeorol. **49**, 355-362 (2005)
25. Rosenblatt, F.: The Perceptron: A Probabilistic Model for Information Storage and Organization in the Brain. Psychol. Rev. **65**(6), 386–408 (1958)
26. Nielsen, M.A.: Neural Networks and Deep Learning. Determination Press, San Francisco (2015)
27. O'Shea, K., Nash, R.: An Introduction to Convolutional Neural Networks, ArXiv e-prints (2015)
28. Krizhevsky, A., Sutskever, I., Hinton, G.: ImageNet classification with deep convolutional neural networks. In: Advances in Neural Information Processing Systems, vol. 25, pp. 1097–1105 (2012)
29. Goodfellow, I., Bengio, Y., Courville, A.: Deep Learning. MIT Press (2016). http://www.deeplearningbook.org
30. Jozefowicz, R., Zaremba, W., Sutskever, I.: An empirical exploration of recurrent network architectures. In: Proceedings of the 32nd International Conference on International Conference on Machine Learning, Lille, vol. 37, pp. 2342–2350 (2015)
31. Harper, M.F., Konstan, J.A.: The MovieLens datasets: history and context. ACM Trans. Interact. Intell. Syst. **5**, 20 (2015)
32. Chai, T., Draxler, R.R.: Root mean square error (RMSE) or mean absolute error (MAE)?- arguments against avoiding RMSE in the literature. Geoscientific Model Dev. **7**, 1247–1250 (2014)
33. Srivastava, N., Hinton, G., Krizhevsky, A., Sutskever, I., Salakhutdinov, R.: Dropout: a simple way to prevent neural networks from overfitting. J. Mach. Learn. Res. **15**(1), 1929–1958 (2014)
34. Hidasi, B., Karatzoglou, A., Baltrunas, L., Tikk, D.: Session-based Recommendations with Recurrent Neural Networks, CoRR abs/1511.06939 (2015)

Empirical Analysis of Call Center Load & Service Level for Shift Planning

Yuval Cohen[1(✉)], Joao Reis[2], and Marlene Amorim[2]

[1] Afeka Tel-Aviv College of Engineering, Tel-Aviv, Israel
yuvalc@afeka.ac.il
[2] University of Aveiro, 3810-193 Aveiro, Portugal
{reis.joao,mamorim}@ua.pt

Abstract. This study characterizes call-center workload as a daily profile, and tests the effects of the "day-of-the-week" on the daily load profile. It then, tests the hypothesis that the service load of incoming calls is correlated with the number of abandoned calls. The analysis main conclusions are: (1) Regular workdays share similar hourly call profile, and same rush hours. (2) Weekend hourly profile is substantially different than the workdays' profiles. (3) Regular workdays share similar profile of hourly abandoned calls. (4) There is a strong correlation between load profiles and profiles of abandoned calls. The discussion which follows summarizes the findings, reports on the same findings at another call center, and suggest a technique for curve fitting to the demand profile, and for computing the required workforce. The analysis process and the workforce planning could be applied on other call centers, and used as a basis for comparison.

Keywords: Frontline employees · Call center · Incoming calls · Abandoned calls · Service level · Demand forecast

1 Introduction

Call centers have evolved immensely over the past few decades [1]. Call centers focus today on service quality and customer satisfaction due to the fact that the service quality is what differentiate one organization from another, making it competitive [2]. The rapid development of such call centers came alongside globalization processes and the rapid development of IT technologies, allowing high availability through means that didn't exist before [3]. A high amount of uncertainty lies with call processes due to the fact that they vary from one day\week\month to another. Based on these processes, managers are challenged with operational and managerial decision making, hence – their importance [4]. A literature [5] as well as our review shows the lack of empirical public information in regards to incoming calls and abandonment processes. Many researches have concentrated on static queueing models and their results bare little similarity to the complex dynamic state of call centers [6]. Thus, the task of decision making in these terms is proven to be a great challenge for managers. As this field of study evolved, grew the understanding that using statistical analysis tools, is a better way of processing data and to derive effective conclusions that match the reality better,

© Springer Nature Switzerland AG 2020
H. Nóvoa et al. (Eds.): IESS 2020, LNBIP 377, pp. 79–91, 2020.
https://doi.org/10.1007/978-3-030-38724-2_6

in terms of modeling the incoming calls and the abandonment processes. Statistical results are also better at reducing the uncertainty of predicting such processes [7]. Therefore, they are more useful for decision making.

This study examines the processes of incoming calls and abandoned calls in a major call center located in Israel. The study presents daily load profiles for incoming calls and abandoned calls throughout the different days of the week.

One of the main challenges in managing CC is the balance between operational efficiency and the quality of service provided by the CC [8]. On one hand, there's the need to grow the resources necessary for handling higher load of calls, and on the other hand more resources mean more costs and decreased profits. As a result, CC managers are often faced with contradictions while making a decision related to the general processes of the CC – incoming calls and abandonment [9]. In general, the only decision parameter managers of CCs must determine is the amount of operators to allocate to each shift over a period of time. Call centers managers are faced with short and long term decision-making every day. Decisions like employee allocation, or how to free time for training and breaks, are a short term kind of decisions; while considering the overall amount of operators, quality and performance measures, are a long-term decisions [10]. Most of these decisions are based on profiling of incoming calls as well as abandonment profiling. The current situation at the observed call center allows its managers to make such decisions based on predictions the managers are able to draw based on call averages and past data collected (similar call behavior of these days in the past). They also use the abandonment rates and past service distribution. The high uncertainty enclosed in running a call center is a great source for concern for these managers. Thus, they are the driving force behind the development of tools to better facilitate decision-making.

These study uses statistical tools to compile a confidence interval around the expected profile of incoming and abandoned calls. The paper contributes additional empirical reference for daily queuing profile of various callers of regular and abandoned calls.

2   Literature Review

Call centers handles different requests originating from customer's incoming calls. Their operations are characterized by three main processes: incoming calls, service duration and abandonment [11]. Thus, in order to operate better there's a significant importance of understanding these processes.

2.1   Incoming Calls

The only parameter managers of call centers can influence is the operator's allocation through each shift. Up to 80% of the call center's budge is comprised of worker's salary. Over staffing is viewed as spending while shortage staffing can lead to long queues and high abandonment. Thus, allocating staff per shift is a tricky task [12]. It is also the reason for many researches to focus on staff allocation methods. Studies shows that the key to optimal allocation is the ability to predict the incoming calls load per call center [13]. Comprising a call center model is quite complicated since incoming calls processes are stochastic, time dependable and are affected by different factors [14]. Empiric evidence shows that incoming calls processes often vary throughout the week and over a period of time as well [15]. Daily incoming calls processes are comprised of different patterns of each day and week, as well as the different operator's skill (customer needs). Many studies have researched the uncertainty of incoming calls that is the cause of such incompatibility between the predictions to the reality [16]. However, most call center's managers base their decisions on predictions alone, resulting in poor performance and incompatibility with their expectations [16]. Most models used in previous studies are based on queueing theory, but one of their errors is the 'known error probability' it is better to analyze and profile the hours of each day rather than inspecting each day of the week in order to be more accurate [17]. Therefore, the prediction distribution is often more compatible to use [14].

Looking through known studies, the ones that publish statistical analysis of empirical data are rare [11]. Since the high uncertainty and variance is a huge obstacle in such processes, some studies focused on inspecting the predictions models weeks ahead, to simulate the challenges facing the call center's managers [15]. Other studies have focused on seasonal effect on call processes and the ability to predict them in a flexible way [15]. In addition, some studies have focused on the prediction per-operator-skill [12]. Taylor claimed that such predictions are the most effective way to reduce manager's uncertainty.

2.2 Service Duration

Setting quality of service levels for the call center is another different challenge managers have to face each day. Literature on the subject at hand aligns with known queueing theory models [11]. In his study, Aktekin [11] examined the service duration distribution of different customer's segmentation in order to detect anomalies in such distributions. Other researches [18] have studied queuing theory models and service duration distribution of complex call centers and presented a new method to allocate staff for a multi- skilled call center.

2.3 Abandoned Calls

Different studies describe abandoning calls in different ways, for example - "abandonment is the opposite of patience which is defined as the time a customer is willing to wait before abandoning the queue" [14, 17]. Abandonment is viewed the same as customer dissatisfaction of the service inquired and as such the call centers will do their best to avoid high abandonment rates. Customers balance between their waiting time and the use of calling the call center in order to determine their level of patience. The customers tend to expect high quality of service and little waiting time, therefore – if their expectations are not met the probability of abandonment is higher [9]. Since most of the literature reviews call centers processes and focuses on queuing theory models, abandonment issues are of little importance. Thus, the lack of studies on abandonment. Having that said, there is a progress in studying that subject over the past few years [9]. These studies have examined the psychological aspects of abandonment processes based on customer profiling. Other studies have examined the reflection of 'time left to wait' on the customer's satisfaction [19]. While some researches have modeled abandonment in queueing theory settings (e.g. [20, 21] only few considered empirical approach (e.g. [22, 23]).

2.4 Service Dominant Logic

Service Dominant Logic (SDL) is a framework that is centered around co-creation of value the customer experience and revolves around [24–26]. SDL provides the ground for better understanding the importance of responding to demand changes during the day and during the week. SDL endorsed the concept that all human economic activities are elements of service. Thus, the demand level of a relevant customers would be directly related to customers' and call-center operators' activities.

It is therefore that demand profile is bundled with SDL. As opposed to the traditional economics theory, SDL does not differentiate between goods and services, and when value is concerned, it is co-created as a common activity between customers and providers. Traditional economics dominant is a logic based on the exchange of goods, which usually are manufactured output, or tangible resources. Over the past several decades, new perspectives have emerged that have a revised the logic focused on intangible resources, the co-creation of value, and relationships. The new perspectives lead to the assertion that service provision rather than goods is fundamental to economic exchange [27].

Customers, and frontline employees alike make value creation an all-encompassing process, without any distinctions between the role of frontline employees and the role of the customers and actions in that process [28, 29]. It is essential to note that, according to SDL framework, both the frontline employee and the customer are considered co-creators of value [30].

Furthermore, the relationship between the customer and suppliers is of strategic value, and would be much more competitive if done as a mutual endeavor. In service science, the role of customer experience is pivotal. However, it is the notion of co-creation that enables putting this experience at the center of the economic arena.

3 Methodology

This study is based on analyzing real data. Therefore, the methodology section starts with a description of data collection. After describing the data-collection procedures a set of hypotheses is presented. Accordingly, a subsection is dedicated to both data collection and hypotheses testing.

3.1 Data Collection

Conducting this study we collected data from a real call center of a major bank operating at Israel. The call center operates between 7AM to 11PM during the week days and on Fridays between 7AM to 2PM, serving all of the customers.

The call center is comprised of different teams. We analyzed data for the busiest team over a period of six months. We also examined abandonment calls processes over a period of 1 month by sampling 12,068 calls. The calls raw data was retrieved from the company's PBX2 system comprising of hourly aggregated data per day and raw data per extension (PBX-is a system that connects telephone extensions to the Public Switched Telephone Network and provides internal communication for a business.) Overall, we sampled 139 days. Analysis was based on hypothesis testing, confidence interval, goodness of fit and ANOVA program, as well as statistical descriptive tools.

We singled out special days and excluded them from the rest of the data in order to avoid anomalies and offsets. Special days mean unusual days such as days for which the call center was closed (Saturdays) or worked shorter shifts (holidays, memorial days). Some of the special days presented unusual high amount of calls due to the proximity to special dates such as – salaries payment, upcoming holidays, unusual trends of the stock market etc.

3.2   Hypotheses

Phone-call arrival and service processes are stochastic, time-dependent, period's dependent and skill dependent. Thus, they often effected by many factors [14]. This study teste the following hypothesis in order to attain insights while characterizing the load and the service level:

H_0-1: There will be a difference in rush hours periods during the week days in terms of incoming calls.

H_0-2: The workdays have the same hourly profile of number of calls.

H_0-3: There will be no difference between regular *work days* and *weekends* in terms of incoming calls.

H_0-4: There is no correlation between daily-call profile and abandoned calls profile during workdays.

H_0-5: The workdays have the same hourly profile of abandoned calls.

These hypotheses were tested according to data collected for validating or rejecting these hypotheses. The hypothesis testing were based on, confidence interval, goodness of fit and ANOVA program, and sometimes were accompanied by graphical descriptive tools.

4   Hypothesis Testing Based on Empirical Data

This section describes the statistical tests for the hypotheses of the methodology section (Sect. 3). As a prior testing we used the F-test ($F(15,64) = 170.17$, with $p < 0.005$) to show that the hourly call rate changes during day so that daily profiles of hourly rates are relevant. The same test for the had similar results for the number of abandoned calls ($F(15,64) = 134.15$, $p < 0.005$). These results are aligned with the literature (e.g. Gans et al. 2015).

4.1 Daily Rush Hours

The data shows that for all workdays including Fridays, rush hours of the bank's calls were between 9–11AM with 10AM as the peak. In addition, there was additional distinct peak of calls between 3–6PM (see Fig. 1).

Thus. no empirical difference at all was present for any difference in rush hours between the different days of the week. We therefore concluded that the rush hours are deterministic and H_0-1 hypothesis can be rejected.

Conclusion: the rush hours are the same for all workdays.

4.2 Workday Profiles for the Number of Hourly Calls

To test whether the workday profiles are similar we computed an average hourly profile presented in Fig. 1. We also computed the corresponding 2σ confidence interval shown in Fig. 1. The mean profile of each workday lies between these boundaries. This effectively proves that workdays have the same hourly profile of number of calls. A goodness of fit test for the profile of each workday to the mean profile of all workdays confirmed that the profile deviations are too small to show a significant deviations for any standard confidence level. Thus, H_0-2 can be accepted.

Conclusion: hourly workday profile of number of calls is not significantly different from the average workday profile.

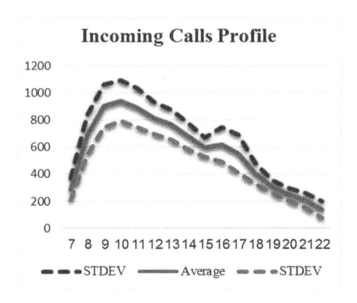

Fig. 1. Profiles of hourly means of calls, and confidence margins in an average workday.

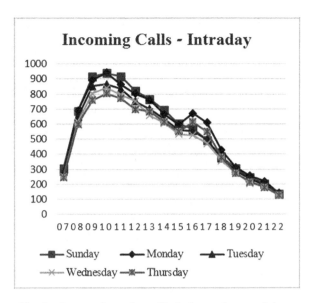

Fig. 2. Average incoming calls during various workdays.

4.3 Comparing Hourly Call Profiles: Workdays vs. Weekends

Weekend hourly call profile is depicted in Fig. 3 and by observation alone could be seen as totally different than the call profiles in Figs. 1 and 2. The difference is huge and visible, as mentioned above. A test of goodness of fit (Table 1) was used to *reject* the third hypothesis *(H_0-3)* that there will be no difference between regular *work days and weekends* in terms of incoming calls. *Conclusion: Reject H_0-3, weekend have different hourly calling pattern.*

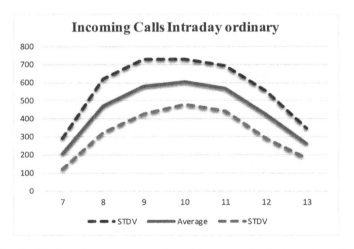

Fig. 3. Weekend hourly means and confidence margins for incoming calls.

4.4   The Correlation Between Daily Call Profile and Abandoned Calls Profile

The correlation between workday's hourly-call profile and abandoned call profile is shown in Table 2. Table 2 is used to *reject H_0-4. Conclusion is that a strong positive correlation exist between the daily profiles of calls and the profiles of abandoned calls.*

The similarity in daily pattern between incoming calls and abandoned calls could be seen visibly by comparing Fig. 5 to Figs. 1 and 2.

Table 1. Goodness of fit test of Sundays vs. mean workdays.

Comparison		Distributions		
x^2	df			
		Week	Sunday	Day
		271	306	7
		646	685	8
		844	917	9
		877	939	10
		834	916	11
		765	822	12
		713	762	13
		642	693	14
		568	601	15
		585	568	16
		523	492	17
		383	383	18
		291	304	19
		230	244	20
		194	202	21
		133	139	22

* $p < 0.05$

Table 2. Correlations matrix between incoming and abandoned calls

		Abandoned Calls				
Weekdays	Sunday	Monday	Tuesday	Wednesday	Thursday	Friday
Sunday	0.9					
Monday		0.81				
Tuesday			0.8			
Wednesday	Inbound calls			0.81		
Thursday					0.78	
Friday						0.61

* $p < 0.05$

4.5   Workday Daily Profiles for the Abandoned Calls

To test whether the workday profiles of abandoned calls are similar we computed an average hourly profile presented in Fig. 4. We also computed the corresponding 2σ confidence interval shown in Fig. 4. The mean abandoned calls profile of each workday lies between these boundaries. This effectively proves that workdays have the same hourly profile of abandoned calls. *Thus, we Accept H_0-5:*

Conclusion: The workdays have the same hourly profile of abandoned calls.

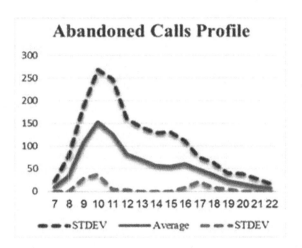

Fig. 4. Profiles of mean hourly abandoned calls, and 2σ margins for mean workday.

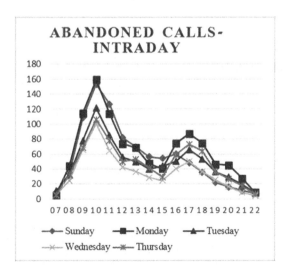

Fig. 5. Average incoming calls during various workdays.

5   Discussion

5.1   Summary of Hypotheses Testing

The following lists the results of the hypotheses testing of Sect. 3:

Reject H_0-1 The rush hours occur at the same daily hours each day for all workdays.
Accept H_0-2: The average workday profile of hourly calls is not significantly different than any workday profile; and thus can represent workdays.
Reject H_0-3, Weekend have different hourly calling pattern.
Reject H_0-4. A strong positive correlation exist between the daily profiles of calls and the profiles of abandoned calls.
Accept H_0-5: The workdays have the same hourly profile of abandoned calls.

5.2   Comparing Results with Another Call-Center

To give better validity to our results and to check that they are not abnormal for call centers, we repeated the same hypothesis testing for a second call-center with very different characteristics. The results of the testing gave the same results as the model presented above. Figure 6 depicts the average daily profiles during weekdays and during the weekend.

5.3   Curve Fitting Model for the Workday Demand Profile

To analyze the overall arrival distribution pattern, we expressed the mean hourly arrival rate as a percentage of total daily arrival, and ran a curve fitting optimization with a plethora of models and parameters. The polynomial approximations yielded better fitting R^2 results, and therefore we chose them as the appropriate approximation alternative. As we increased the polynomial rank from 2 to 6, the improvements in R^2 were always significant, and beyond the rank of 6 the improvements in R^2 were minor to negligible. Therefore, we chose the power of six as our recommended rank for polynomial approximations of arrival rate.

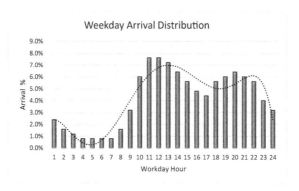

Fig. 6. Comparison profiles of hourly calls of another call-center: workdays

We defined the arrival percentage function of t as AP, and the time in the day (hours) as t. The results are presented in Fig. 6.

In Fig. 6, the weekday's curve-fitting yields the formula of Eq. 1. Using this formula, the results predict the mean hourly arrival rate as with correlation of $R^2 = 0.8942$.

$$AP_{weekday}(t) = -10^{-7}(t^6) + 9 \cdot 10^{-6}(t^5)$$
$$- 10^{-4}[2(t^4) + 28 \cdot (t^3) - 137 \cdot (t^2)] + 0.02(t) + 0.0149 \quad (1)$$

In Fig. 6, the weekend's curve-fitting yields the formula of Eq. 2. Using this formula the results predict the mean hourly arrival rate as with correlation of $R^2 = 0.9341$.

$$AP_{weekend}(t) = -10^{-7}(t^6) + 7 \cdot 10^{-6}(t^5)$$
$$- 10^{-4}t^4[2(t^4) + 22 \cdot (t^3) - 108 \cdot (t^2)] + 0.0139(t) + 0.0295 \quad (2)$$

5.4 Workforce and Service-Level Computations

Based on Figs. 1 and 2 it is clear that the peak hours are between 9 to 11AM invariably. Moreover, Figs. 4 and 5 show the same hours have the peak of abandoned calls. The percentage of abandoned calls varies with the load. For busy hours abandoned calls are shown to reach above 15% of total calls while in other hours they are less than 10% of total calls. Thus, it is clearly that range of hours that should be the basis for workforce planning. We propose the following model for computing the workforce level:

$C(i\%)$ – i^{th} percentile of the number of calls per hour during rush-hour.
$L(j\%)$ – j^{th} percentile of the call duration - during rush hour (random variable)
A – percent of personnel absenteeism
W – total number of operators to recruit for determined levels of i, j
SL – Service level

$$W = [C(i\%) * L(j\%)]/[1 - A] \quad (3)$$

If the standard deviation of the number-of-calls is close to the standard deviation of call-duration - a crude estimation for the service level of W is: $SL = (i + j)/2$.

Otherwise, the i and j may be given weights according to the standard deviations associated with number-of-calls and call-duration

Since this paper is mostly concerned with the number of calls per hour, the following discussion will focus on $C(i\%)$. Figure 1, shows that the peak mean number of calls per hour is 900. However, at this mean level, so close to 50% of the days the call amount is higher. However, choosing the $\mu + 2\sigma$ line ensures that 97.72% of days the call amount will be lower (using the Normal distribution). In Fig. 1, the $\mu + 2\sigma$ peak value is 1100 calls per hour. Thus, for planning according to the median call duration the crude service level estimation for $\mu + 2\sigma$ calls will be: $(0.9772 + 0.5)/2 = 0.738 = 73.8\%$.

6   Conclusion

In this paper we present an analysis of empirical data of a call center. Based on the data and statistical analysis we could test a set of hypotheses and reach the following conclusions: (1) rush hours occur between 9:00 to 11:0 in all workdays; (2) workdays (for different weekdays) share similar profile of hourly calls; (3) weekend profile of hourly calls is different than the profile for workdays; (4) there is strong correlation between the number of calls per hour and number of abandoned calls per hour; (5) workdays (for different weekdays) share similar profile of abandoned calls. A comparison to a different call center reached the same findings. Additional insights from the data analysis show that AR grows non-linearly with the number of calls (reflecting the workload), also the weekend AR is very different than workdays. We hypothesize that this occur due to more patience during weekend. The discussion section reports on the same findings at another call center, and suggest two techniques: (1) Curve fitting technique to model the demand profile, and (2) a technique for computing the required workforce. Future research can use the results of this study as a basis for comparison. The daily profiles could be used as a basis for new forecasting models, for the projection of both load and abandoned calls.

References

1. Robbins, T.R., Harrison, T.P.: New project staffing for outsourced call centers with global service level agreements. Serv. Sci. **3**(1), 41–66 (2011)
2. Tandon, U., Kiran, R., Sah, A.N.: Customer satisfaction as mediator between website service quality and repurchase intention: an emerging economy case. Serv. Sci. **9**(2), 106–120 (2017)
3. Baseman, J., et al.: Impact of new technologies on stress, attrition and well-being in emergency call centers: the Next Generation 9–1-1 study protocol. BMC Public Health **18**(1), 597–606 (2018)
4. Robbins, T.R.: Experience-based routing in call center environments. Serv. Sci. **7**(2), 132–148 (2015)
5. Vergara, F.H., Sullivan, N.J., Sheridan, D.J., Davis, J.E.: The best practice for increasing telephone outreach: an integrative review. Prof. Case Manag. **23**(6), 307–317 (2018)
6. Legros, B., Jouini, O., Koole, G.: Blended call center with idling times during the call service. IISE Trans. **50**(4), 279–297 (2018)
7. Mukherjee, A., Malhotra, N.: Call centre services: the good, the bad, and the ugly. J. Serv. Mark. **23**(5) (2009). https://doi.org/10.1108/jsm.2009.07523eaa.001
8. Song, C., Jang, S., Wiggins, J., Nowlin, E.: Does haste always make waste? Service quantity, service quality, and incentives in speed-intensive service firms. Serv. Bus. **13**(2), 289–304 (2019)
9. Aktekin, T., Soyer, R.: Bayesian analysis of abandonment in call center operations. Appl. Stoch. Models Bus. Ind. **30**(2), 141–156 (2014)
10. Ilk, N., Brusco, M., Goes, P.: Workforce management in omnichannel service centers with heterogeneous channel response urgencies. Decis. Supp. Syst. **105**, 13–23 (2018)
11. Aktekin, T.: Call center service process analysis: Bayesian parametric and semi-parametric mixture modeling. Eur. J. Oper. Res. **234**(3), 709–719 (2014)

12. Taylor, J.W.: Density forecasting of intraday call center arrivals using models based on exponential smoothing. Manag. Sci. **58**(3), 534–549 (2012)
13. Jalal, M.E., Hosseini, M., Karlsson, S.: Forecasting incoming call volumes in call centers with recurrent neural networks. J. Bus. Res. **69**(11), 4811–4814 (2016)
14. Ibrahim, R., Ye, H., L'Ecuyer, P., Shen, H.: Modeling and forecasting call center arrivals: a literature survey and a case study. Int. J. Forecast. **32**(3), 865–874 (2016)
15. Ibrahim, R., L'Ecuyer, P.: Forecasting call center arrivals: fixed-effects, mixed-effects, and bivariate models. Manuf. Serv. Oper. Manag. **15**(1), 72–85 (2013)
16. Li, S., Wang, Q., Koole, G.: Predicting call center performance with machine learning. In: Yang, H., Qiu, R. (eds.) INFORMS-CSS 2018. SPBE, pp. 193–199. Springer, Cham (2019). https://doi.org/10.1007/978-3-030-04726-9_19
17. Brezavšček, A., Baggia, A.: Optimization of a call centre performance using the stochastic queueing models. Bus. Syst. Res. J. **5**(3), 6–18 (2014)
18. Li, C., Yue, D.: A queuing model of the N-design multi-skill call center with impatient customers. Int. J. u- e-Serv. Sci. Technol. **8**(4), 51–60 (2015)
19. Gong, J., Li, M.: Queuing time decision model with the consideration on call center customer abandonment behavior. J. Netw. **9**(9), 2441 (2014)
20. Takagi, H.: Times until service completion and abandonment in an M/M/m preemptive-resume LCFS queue with impatient customers. J. Ind. Manag. Optim. **14**(4), 1701–1726 (2018)
21. Kuzu, K.: Comparisons of perceptions and behavior in ticket queues and physical queues. Serv. Sci. **7**(4), 294–314 (2015)
22. Akşin, Z., Ata, B., Emadi, S.M., Su, C.L.: Impact of delay announcements in call centers: an empirical approach. Oper. Res. **65**(1), 242–265 (2016)
23. Reis, J., Amorim, M., Melão, N.: Omni-channel service architectures in a technology-based business network: an empirical insight. In: Satzger, G., Patrício, L., Zaki, M., Kühl, N., Hottum, P. (eds.) IESS 2018. LNCS, vol. 331, pp. 31–44. Springer, Cham (2018). https://doi.org/10.1007/978-3-030-00713-3_3
24. Vargo, S.L., Lusch, R.F.: Service-dominant logic: continuing the evolution. J. Acad. Mark. Sci. **36**(1), 1–10 (2008)
25. Lusch, R.F., Vargo, S.L.: Service-dominant logic: a necessary step. Eur. J. Mark. **45**(7/8), 1298–1309 (2011)
26. Lusch, R.F., Vargo, S.L.: The Service-Dominant Logic of Marketing: Dialog, Debate, and Directions. Routledge, Abingdon (2014)
27. Vargo, S.L., Lusch, R.F.: Institutions and axioms: an extension and update of service-dominant logic. J. Acad. Mark. Sci. **44**(1), 5–23 (2016)
28. Grönroos, C.: Service logic revisited: who creates value? And who co-creates? Eur. Bus. Rev. **20**(4), 298–314 (2008)
29. Grönroos, C.: Value co-creation in service logic: a critical analysis. Mark. Theory **11**(3), 279–301 (2011)
30. Grönroos, C., Voima, P.: Critical service logic: making sense of value creation and co-creation. J. Acad. Mark. Sci. **41**(2), 133–150 (2013)

Enabling System-Oriented Service Delivery in Industrial Maintenance: A Meta-method for Predicting Industrial Costs of Downtime

Clemens Wolff[(✉)], Niklas Kühl, and Gerhard Satzger

Karlsruhe Service Research Institute, Karlsruhe Institute of Technology,
Karlsruhe, Germany
{clemens.wolff,niklas.kuehl,gerhard.satzger}@kit.edu

Abstract. Nowadays, companies often outsource activities not directly related to their core business to service providers. Service providers—especially in physical services—heavily rely on complex information and optimization systems to increase their efficiency. As of now, most of those systems implement the concept of provider-oriented service delivery (POSD). Whilst POSD optimizes from a provider perspective, it neglects optimization potential on the customer side. Addressing this issue, scholars recently proposed the concept of system-oriented service delivery (SOSD). Although SOSD allows for significant cost reduction compared to POSD, it also requires additional customer cost data. In industrial maintenance, this additional cost data refers to the customer's costs of downtime. Despite some pioneer work in this field, current knowledge does not suffice for the successful application of SOSD in the domain of industrial maintenance. Consequently, the objective of this work is to develop a method to determine a manufacturer's costs of downtime.

Keywords: Industrial cost of downtime · Industrial maintenance · System-oriented service delivery

1 Introduction

Driven by trends such as outsourcing and the servitization of the manufacturing business [25], service providers often offer similar, if not identical, services to different customers. A typical example of such services is industrial maintenance, which in itself is not only a physical, but also the most prominent industrial service [11]. Given such an environment, complex information and optimization systems continue to expand their presence to increase operational efficiency [8].

Given today's practice, however, those complex information and optimization systems implement the concept of *provider-oriented service delivery (POSD)*, as their optimization objectives are provider-oriented (e.g., minimize provider costs, maximize resource utilization) [35]. Whilst this approach is reasonable from a

© Springer Nature Switzerland AG 2020
H. Nóvoa et al. (Eds.): IESS 2020, LNBIP 377, pp. 92–106, 2020.
https://doi.org/10.1007/978-3-030-38724-2_7

provider's perspective, it does not trade-off additional costs on the provider with cost savings on the customer side. Consequently, POSD results in inefficient resource allocation from a system's viewpoint [35]. Addressing this issue, Wolff et al. [35] propose the concept of *system-oriented service delivery (SOSD)*. Through system-cost optimization and an intelligent compensation and benefit distribution mechanism, the concept of SOSD greatly reduces overall system costs [36] whilst ensuring no additional costs on a participant level. Therefore, the concept of SOSD is a Pareto improvement compared to POSD.

In order to enable service delivery according to SOSD, the service provider requires—in addition to the already utilized provider costs—information about the customers' delivery-dependent costs. Applied to industrial maintenance, the concept of SOSD has the following implications: First, upon a machine failure, the customer reports—in addition to his service demand—his delivery-dependent costs for different response times in form of a *cost of downtime curve*. The costs of downtime curve depict the customer's delivery-dependent costs as a function of the response time, i.e., the time until the customer's machine is being repaired. Second, based on the reported cost of downtime curve of all maintenance customers, overall system-cost optimal service delivery is computed. Finally, based on the individual participants' as well as the system-wide cost differences between the provider- and system-optimal resource allocation, the mechanism computes monetary payments to accommodate the compensation and benefit distribution mechanism that ensures no additional costs on a participant level compared to a POSD solution.

With this background, the objective of this work is to develop and present an approach that allows industrial maintenance customers (i.e., industrial manufacturers) to determine their cost of downtime curve. Given this objective, this work contributes to the field of industrial costs of downtime and the development and theorizing process of SOSD itself, which addresses and contributes to the more general field of service delivery, a top research priority in service science [26,27]. In order to develop the required approach, we follow the well-accepted guidelines of *Design Science Research* and build on the kernel theory of *integrated production and distribution scheduling*.

The remainder of this work is structured as follows: In Sect. 2, we present this work's research design. In Sect. 3, we elaborate on related work before highlighting domain-specific requirements in Sect. 4. In Sect. 5, we design and evaluate the meta-method. In Sect. 6, we discuss the developed meta-method. Finally, we conclude this work in Sect. 7.

2   Methodology and Research Design

As an overall research design, we follow a *Design Science Research (DSR)* approach, as it allows the consideration of practical components [23] and has proven itself to be an important and legitimate research paradigm [12]. Following the guidelines of Hevner and Chaterjee [13], this work is characterized by three inherent cycles, namely the rigor, relevance, and design cycle. Within the design

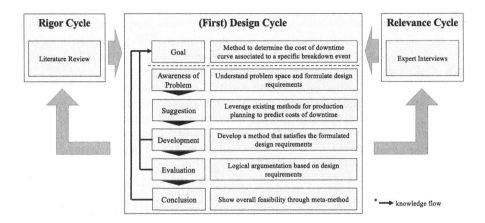

Fig. 1. Overall DSR approach with a first design cycle

cycle, we follow the DSR framework initially proposed by Kuechler and Vaishnavi [19], which extends previous own work [18]. The overall DSR approach of this work is depicted in Fig. 1. The artifact is an *improvement*, since it serves as a new solution for a known problem [12], and of the type *method*, as it consists of "actionable instructions that are conceptual" [29].

3   Rigor Cycle: Fundamentals and Related Work

In this section, we introduce fundamentals and related work as part of the rigor cycle of our DSR project. In detail, we introduce related work on industrial costs of downtime as well as production and distribution planning.

3.1   Industrial Costs of Downtime

In this section, we review literature in the field of industrial costs of downtime, which provides guidelines and examples on how to calculate the costs associated to the downtime of a production asset. Therefore, we do not consider work that uses such information for other decision making, as, for example, supply chain management or maintenance scheduling, as relevant.

Overall, research on the determination of industrial costs of downtime can be separated into two streams of research: Whilst the first stream centers around structuring the field of research and provides guidelines on cost components to consider (e.g., [5,10,24,32]), the other stream focuses on actually determining those. For example, Lincoln [20] presents a method for predicting the costs of downtime associated with downtime of a serial multistage manufacturing line. In his work, he builds on previous work by Chang et al. [2] and Liu et al. [21]—who present a method for determining production capacity loss due to a downtime of duration d in a serial production line—and on activity-based costing (ABC) methods (e.g., [14,17,39]), which have so far mostly been used for the analysis of overhead costs during production. Furthermore, Fox et al. [9] determine the financial impact of cumulative machine downtime on the Australian Post letters sorting process for a historic time frame. They conclude that the loss of capacity has the highest financial impact. Another study of Edwards et al. [6] present a model based on historic data to predict future costs of downtime. Their model, however, is based on average costs per time unit. Similar, Wolff and Schmitz [37] present a method to determine production losses due to downtime for simple production systems and homogeneous production cycles.

Finally, there is work determining the costs of downtime from a contracting perspective (e.g [15,16,38]). Given the contracting perspective, however, such work focuses on deriving costs of downtime associated to a given service level and not to a single downtime incident.

Whilst this shows that there is ongoing research in the determination of industrial costs of downtime, it also shows that current models are not suitable for the determination of a cost of downtime curve as required for the implementation of SOSD in industrial maintenance. In detail, current models are either use-case specific, assume some sort of average costs of downtime per time unit, take an ex-post perspective, or have strong limitations to its applicability. Therefore, this shows that previous research does not allow for the determination of a failure-event specific cost of downtime curve. Previous findings, however, will be considered during the course of this work where possible.

3.2   Production and Distribution Planning

Production planning deals with the timely assignment of production capacity to production demand with the overall objective to "fulfill customer demand at minimum total [...] costs'" [22]. Usually, researchers differentiate between production planning on a short-, medium-, and long-term basis. Whilst long-term decision making usually refers to supply-chain related questions, as, for example, facility location, medium-term planning deals with tactical-related questions, as, for example, production targets for individual facilities as well as transportation partners. Finally, short-term planning deals with the precise assignment of production tasks to different production units and is usually referred to as machine scheduling.

In its core, machine scheduling is an optimization problem dealing with the timely assignment of tasks (orders) to a limited production capacity, which itself is determined through the production assets (resources). Since the optimization problem for individual problems depends on the problem-specific setting, it is evident that there is a large number of publications addressing different problem characteristics. Given the large number of publications related to machine scheduling, Abedinnia et al. [1] provide a taxonomic overview of machine scheduling characteristics based on 129 papers reviewing existing literature on machine scheduling. Their work demonstrates the variety of work on machine scheduling and shows that most machine scheduling problems have a penalty-based optimization objective (i.e., minimize penalties due to delayed delivery) and that machine scheduling models exist for a variety of production systems, as, for example, a single machine, a flow shop, job shop, or parallel machines.

Driven through market trends like *make-to-order*, *build-to-order*, or *assemble-to-order* production strategies as well as increasing computing power, production planning is increasingly performed in an integrated manner, as, for example, production and distribution decisions [28]. This implies that on-site production is not planned independently of order delivery, but instead, on-site production and product distribution are planned jointly. Chen [3] and Fahimnia et al. [7] provide an excellent overviews of integrated production and distribution scheduling.

For the remainder of this work, it is sufficient to understand that—regardless of doing production planning, machine scheduling, or integrated production and distribution scheduling—optimization models exist that assign limited production capacity to the production orders in a timely fashion such that overall costs are minimized. Furthermore, those optimization models can be adopted to account for different problem settings, as, for example, the production system, product delivery, or more sophisticated optimization objectives considering contractual penalties due to delayed delivery.

4 Relevance Cycle: Exploratory Interview Study on Determining Industrial Costs of Downtime

In order to understand the problem further and to verify its practical relevance, we conducted an exploratory interview study as part of our relevance cycle. In the following, we present our methodology and summarize our main findings. To reduce complexity for the interviewed experts, the interviews focused on an approach to determine the costs of downtime upon a machine failure and did not address SOSD as a service delivery approach or the cost of downtime curve.

4.1 Methodology

In order to further understand the application domain and to ensure the necessity of a method to predict the costs of downtime, we conducted several semi-structured expert interviews followed by a qualitative analysis [30]. In order to support a broad applicability of the method and to compare, to contrast, and to learn about specific features from the cases, we use comparison focused sampling [33]. Overall, we conducted four interview cases: The expert of the first case, C1, is responsible for maintenance, repair, and overhaul planning in a management role for a supplier in the automotive industry. The expert of the second case, C2, is a production planner and works for a large automotive company. Third, the expert of case C3 holds a management position in the field of production process development within a large automotive supplier. Finally, the expert of the fourth case C4 is a quality inspector in a management role within the defense industry. Based on the interview transcripts, a qualitative analysis was performed to extract knowledge and derive a sound understanding of the need and requirements for a method to predict the costs of downtime [31].

Following common practice, the interviews were summarized and categorized before the analysis [33].

4.2 Findings

The main findings of the exploratory interview study can be summarized as follows: First, all interview partners stressed the importance of short response times upon an unexpected machine failure and reported similar priorities during production: First, the production output must satisfy production quality, i.e., a manufacturer's main interest lies within producing in a quality as expected. Second, if production quality is met, they aim at delivering the product in-time. Finally, if both previous objectives are achieved, manufacturers aim at achieving those at minimal costs. These results align with reported objectives of a supply chain in academia [4]. Second, all interview partners confirmed the need for a method to determine the costs of downtime associated to a certain downtime. All experts emphasized that this model should be applicable from both, an ex-post as well as an ex-ante perspective. Interestingly, interview partner C1 mentioned that they could—given their highly automated and cycle-time optimized production line—theoretically already calculate their costs of downtime, however, do not currently do it. Third, the experts highlighted that costs increase with downtime duration, but that there are two critical points in time that drastically increase downtime-dependent costs: The first point refers to the point in time from which on normal shipping will not deliver the item in time. Accordingly, if possible, they rely on more expensive express shipping to ensure on-time delivery. The second point refers to the point in time upon which item delivery to the customer will be delayed, regardless of what type of shipping utilized. Hence, current orders, their contracted due dates as well as alternative shipping methods should be considered within the model. Third, experts also pointed out the need to consider the production system, its layout, as well as its flexibility within this method. This is evident, since the production system heavily impacts how production is performed in the first place, but also how a manufacturer may react upon a machine failure. A flexible production system, for example, may have the capability to buffer downtime upon a machine failure more effectively than a highly automated, optimized, and sensitive production line. Hence, the method must allow to consider different production systems. On the other hand, the experts also stressed the importance of the method to be applicable to more than a single production system in order to achieve broad applicability as well as results comparability between different manufacturers and production systems. Hence, we conclude a need for a method that can be instantiated for a large amount of specific production systems. Finally, the experts highlighted that it is important to consider the order list at the time of machine failure, since contractual penalties usually out-weight any other cost. In other words, unplanned downtime due to a machine failure during peak production and full order list leads to different delivery-dependent costs compared to a failure during a production low. This should be reflected within the model.

Table 1. Design requirements

DR 1	The artifact should be abstract enough to be applicable to more than a single production system
DR 2	The artifact should consider current orders and their due dates as well as contractual penalties
DR 3	The artifact should allow the consideration of different shipping alternatives, like *standard* and *express* shipping
DR 4	The artifact should find a balanced trade-off in its level of detail with regard to the effort put in information retrieval

5   Design Cycle: A Meta-Method to Determine the Costs of Downtime Curve

In this section, we highlight our core findings of the different design cycle phases as shown in Fig. 1.

5.1   Awareness of Problem

The awareness of problem has been shown in the previous section. In detail, the objective of this work is to propose an approach that allows industrial manufacturers to determine their cost of downtime curve associated to a specific downtime event. Whilst multiple applications of such cost of downtime curves are possible, this work focuses on its applicability in the context of SOSD. Our literature review has shown that this problem has not been addressed yet. Furthermore, our interview study supports the need for a method to determine the costs of downtime associated to a specific event from an industry perspective. Following common practice in DSR, we formulate additional design requirements based on the findings from the relevance and rigor cycle. The formulated design requirements are shown in Table 1 and will serve as guidelines to evaluate the designed artifact later.

5.2   Suggestion and Development

Given the complexity and diversity of production systems and in order to maintain broad applicability, we propose a meta-method instead of a precise model to determine the costs of downtime curve. Once instantiated for a specific production system, it can easily be applied to determine the cost of downtime curve. In its core, the meta-method relies on point estimates of the costs associated with a specific duration of downtime. Through multiple point estimates, the cost of downtime curve can be fitted, as exemplarily depicted in Fig. 2: For three different durations of downtime (d_1, d_2, and d_3) and their associated cost estimates (c_1, c_2, and c_3), we approximate the overall cost of downtime curve through curve fitting. In this example, the curve is fitted through linear interpolation

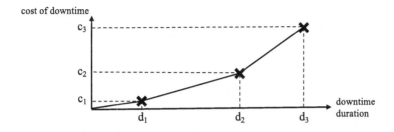

Fig. 2. Exemplary curve fitting through linear interpolation

between the given point estimators. The point estimates themselves are generated through a cost of downtime estimator, for which, in turns, we also propose a standardized approach.

Given this basic understanding, it is evident that the approach can be separated into two steps: First, cost of downtime estimates are derived through repeated application of a costs of downtime estimator for different downtime durations. Second, using those cost of downtime estimates, a cost of downtime curve is fitted. In the following, we briefly explain the cost of downtime estimator and the fitting of the cost of downtime curve.

Cost of Downtime Estimator. The cost of downtime estimator computes the costs associated with the downtime of a specific production machine of duration d. To determine these costs, we propose a simple cost-comparison of two scenarios: First, the initially planned scenario, i.e., without a machine failure, and, second, the scenario with a machine failure resulting in an assumed downtime of duration d. In the following, we refer to the solutions (i.e., production plan and associated costs) of the two scenarios as the first- and second-best solution, respectively. Evidently, the first-best solution is given at the time of machine failure, since it depicts the production plan followed at the time of failure. Let the first-best production plan be derived through a well-defined approach, as, for example, a production or integrated production and distribution planning optimization problem, as introduced in Subsect. 3.2. Figure 3 exemplary depicts a first-best production plan at its top: Given this set of production orders and limited production capacity, its optimal according to the manufacturer's decision making to produce orders 1, 3, 2, 7, 4, and 6 over the planning horizon. Consequently, production order 5 is not planned to be produced during the planning horizon. Upon a machine failure with assumed downtime of duration d, this first-best production plan is no longer feasible. Given the assumed downtime of duration d, we are able to compute a new optimal production plan—i.e., the production plan that is optimal according to the manufacturers priorities given the reduced production capacity. We refer to this new production plan as second-best production plan. Such a second-best production plan is also exemplary depicted in Fig. 3 at the bottom: Compared to the first-best solution, it is now—given the reduced production capacity due to downtime d—optimal for

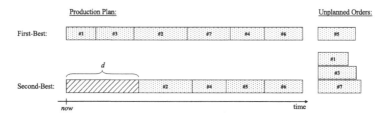

Fig. 3. Exemplary first- and second-best production plans due to a machine failure of duration d

the manufacturer to produce orders 2, 4, 5, and 6. Production orders 1, 3, and 7 are no further planned to be produced during the planning horizon. Given the first- and second-best production plan, we can easily calculate the manufacturers total costs for each plan. Evidently, the cost difference between them depicts the costs of downtime associated to the assumed downtime duration d.

Naturally, the question on how to compute the second-best production plan arises. This, however, is easily done by relying on the already known and well-defined production planning approach used to derive the initial, first-best production plan. As explained in Subsect. 3.2 production planning—in its core—assigns limited production capacity to production orders under a variety of constraints. From an operational perspective, the machine failure that leads to an assumed downtime of duration d does nothing else but to reduce production capacity. Hence, it is evident that the production planning optimization problem used to derive the initial first-best production plan can be re-executed with a reduced production capacity. Deriving the point estimator seems to be easy at first, however, there are two pitfalls that need to be carefully considered: First, the downtime duration d arising at the failed machine is an absolute value. Production, as introduced in Sect. 3.2, however, is not planned at a machine, but instead on a production system level with more than one production machine. Hence, depending on the optimization model used, the absolute downtime d at the failed machine needs to be transformed to reflect how the production system is impacted. Second, it needs to be ensured that the production plan does not intend to resume production on the failed machine prior to the end of its downtime. We argue that both challenges can easily be addressed.

Having discussed the core idea behind the estimator and its pitfalls, we briefly highlight its assumptions: First, the manufacturer determined his initial first-best production plan through a structured approach, as, for example, a production planning optimization problem. Second, it assumes that all information are known and accessible to the system. In practice, we argue that both assumptions are met. Today, manufacturers already rely on complex optimization systems to optimize their production schedules and given the wide adoption of information systems in manufacturing, we also argue that the second assumption is met.

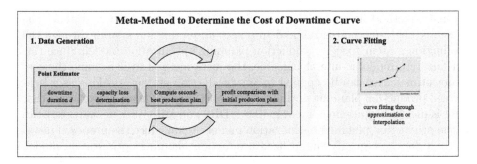

Fig. 4. Meta-model

Curve Fitting. The second step of the meta-method to determine the cost of downtime curve relies on curve fitting. Hence, based on a list of downtime duration and cost of downtime estimates, the cost of downtime curve is computed. Since curve fitting is a well-known technique in mathematics, we do not go into further detail. At this point, however, we do want to highlight two points concerning curve fitting. In detail, these are the number of data points used to and how the curve is generated. First, the quality of the cost of downtime curve increases with increasing number of point estimates. Hence, from this perspective, it seems reasonable to compute a high number of point estimators. Those, however, are expensive to compute, since the production planning optimization problem is known to be NP-hard [34]. Hence, applicants need to trade-off the benefit of additional data points compared to the increased computation power. Second, there are multiple approaches to interpolate the cost of downtime curve. For example, one may use a linear interpolation between the points or an overall interpolation relying on spines. Furthermore, one may also consider to rely on approximation instead of interpolation. The difference between those two approaches lies in the predicted value at the point estimators: Whilst interpolation forces the generated curve to exactly match the value of the point estimators, approximation allows its differentiation. Since the point estimators are only predicted values themselves, we consider approximation to be sufficient.

Final Meta-method. The above described steps are summarized in form of a meta-model as depicted in Fig. 4. Upon a machine failure, data to approximate the cost of downtime curve is generated through point estimates. The point estimates are derived by repeated execution of a point estimator that modifies and relies on well-established methods of production planning.

5.3 Evaluation

In this subsection, we evaluate our artifact with regard to the design requirements introduced in Table 1. Our evaluation is based on logic argumentation, a valid evaluation technique in early DSR steps [29]. DR1 demands that the artifact, i.e., the meta-method, is applicable to more than a single production system. The presented meta-method is tailored to a specific production system through the production system specific production planning optimization model. Since this optimization model is interchangeable, the meta-method to determine the cost of downtime curve can be applied to any production system—as long as a well-defined production planning optimization model exists. Therefore, we argue that DR1 is met by the designed meta-method. DR2 and DR3 refer to characteristics of the production planning optimization model. Again, since the proposed meta-method does not specify the utilized production planning optimization model, these features (e.g., penalties for delayed delivery, express shipping) can be considered. If express shipping and penalties due to delayed delivery are important decision-relevant factors, those should already be considered in the production planning optimization models to derive the first-best solution and, consequently, are also considered when computing the second-best solution. Therefore, even though not required, it is possible to consider express shipping and penalty costs within the cost of downtime curve. Hence, we argue that DR2 and DR3 are met. Finally, we also argue that DR4 is also met. Today, manufacturers already rely on automated production planning; i.e., the decision-relevant data is already available to IT systems. Since we do not need additional data compared to computing the original, first-best production plan, all required data is also available to compute the different second-best production plans according to the different downtime durations.

6   Discussion

In this section, we aim at discussing the developed meta-method. Given its current state, the developed meta-method shows common limitations of conceptual work. For example, the meta-method so far seems reasonable from a theoretical perspective, however, has not yet been applied in practice. Furthermore, the method treats several components as *black boxes*. Given this abstraction, the meta-method increases its applicability at the cost of specification.

Given this lack of further specification, we now aim at highlighting how the components of the meta-method need to be instantiated. Overall, we see two levels of instances, namely on a production system and on a downtime event level. The first level, the production system level, refers to the modelling logic and refers to how the production plan is derived. Given the assumption of a well-defined production planning approach in place, this approach must be adopted to account for a downtime of duration d. Furthermore, if desired, additional characteristics may also be considered, as for example, the integration of distribution planning to account for *standard* and *express* shipping. Once adopted, this approach can be re-used for any downtime event as long as the production system does not change. The second level, the downtime event level, refers to the instantiation of the approach according to the current situation at the time the downtime occurs. From an algorithmic perspective, the production system level described the underlying logic and the downtime event level its execution with given parameters. Whilst the first requires manual modelling, the latter can be automated through accessing existing information systems in place (assuming their existence).

Being a meta-method and given its operational perspective, the proposed approach heavily relies on the defined *black boxes*. Through this dependence, the overall accuracy of the meta-method evidently depends on the accuracy of the *black boxes*. Some costs, for example, may not be decision relevant for the first-best production plan (i.e., the well-defined base model) but indeed for the second-best production plan. An example of such costs may be labor costs: Upon a machine failure, production may collapse. In this case, labor needs still to be paid. Hence, cost components may need to be further specified for the given application.

7 Conclusion

This work proposes a meta-method to derive a cost of downtime curve—i.e., a curve indicating the costs of downtime over the downtime duration—upon an unexpected machine failure. The meta-method is derived by following the well-accepted guidelines of *Design Science Research (DSR)*. As part of the project, we conducted an exploratory interview study to confirm the need for a method to determine the costs of downtime as well as to identify practice-oriented requirements for the model. The derived meta-method itself is characterized by its simple and general approach that ensures its broad applicability in practice.

Given these results, the contribution of this work is two-fold: First, this work contributes to the field of industrial costs of downtime by presenting a standardized approach to determine costs associated with a machine failure based on production planning methods. Since these production planning methods are already applied, the overall effort to apply the proposed approach is low. Second, in a more general sense, this work contributes to the field of service delivery, since it is a necessary requirement for the successful application of system-oriented service delivery.

In its current state, we see three areas of limitations and future work: First, points already highlighted during the discussion in Sect. 6 should be addressed in future work. Second, the presented meta-method is highly conceptual. Consequently, additional work further specifying as well as providing guidelines on how to implement the proposed meta-method are required. Third, the designed meta-method needs to be further evaluated. So far, evaluation was done with regard to the formulated design requirements. Whilst this is an important part of any DSR project, additional evaluation with regard to the usability, applicability, and effectiveness need to be performed. To address this issue, we propose a two-fold evaluation approach consisting of qualitative expert evaluation as well as a quantitative use-case evaluation. Furthermore, additional interview studies may reveal further design requirements. Fourth, we would like to extend the proposed meta-method and create an IT based tool to automatically derive the cost of downtime curve upon a machine failure. In such a design cycle, challenges will be the interface between the new tool and existing IT systems, hence, also addressing DR4. Furthermore, once created, the tool can be used to evaluate the applicability as well as usability from a user perspective. Therefore, future additional design cycles addressing those remarks are already planned.

References

1. Abedinnia, H., Glock, C.H., Grosse, E.H., Schneider, M.: Machine scheduling problems in production: a tertiary study. Comput. Ind. Eng. **111**, 403–416 (2017)
2. Chang, Q., Biller, S., Xiao, G.: Transient analysis of downtimes and bottleneck dynamics in serial manufacturing systems. J. Manuf. Sci. Eng. **132**(5), 051015 (2010)
3. Chen, Z.L.: Integrated production and outbound distribution scheduling: review and extensions. Oper. Res. **58**(1), 130–148 (2009)
4. Chopra, S., Meindl, P.: Supply Chain Management: Strategy, Planning and Operation, vol. 56, 3rd edn. Prentice-Hall, Upper Saddle River (2009)
5. Crumrine, D., Post, D.: Channel chat: when true cost of downtime is unknown. Bad Decis. Ensue **53**(1), 55–60 (2006)
6. Edwards, D.J., Holt, G.D., Harris, F.C.: Predicting downtime costs of tracked hydraulic excavators operating in the UK opencast mining industry. Constr. Manag. Econ. **20**(7), 581–591 (2002)
7. Fahimnia, B., Farahani, R.Z., Marian, R., Luong, L.: A review and critique on integrated production-distribution planning models and techniques. J. Manuf. Syst. **32**(1), 1–19 (2013)
8. Field, J.M., et al.: Service operations: what's next? J. Serv. Manag. **29**(1), 55–97 (2018)
9. Fox, J.P., Brammall, J.R.: Determination of the financial impact of machine downtime on the Australia post large letters sorting process. In: 9th Global Congress on Manufacturing and Management (GCMM 2008), pp. 1–7. Gold Coast, Australia (2008)
10. Geraerds, W.: The cost of downtime for maintenance: preliminary considerations. University of Technology, Department of Industrial Engineering& Management Science, Eindhoven (1983)
11. Gitzel, R., Schmitz, B., Fromm, H., Isaksson, A., Setzer, T.: Industrial services as a research discipline. Enterp. Model. Inf. Syst. Arch. **11**(4), 1–22 (2016)
12. Gregor, S., Hevner, A.R.: Positioning and presenting design science research for maximum impact. MIS Q. **37**(2), 337–355 (2013)
13. Hevner, A., Chatterjee, S.: Design science research in information systems. In: Hevner, A., Chatterjee, S. (eds.) Design Research in Information Systems: Theory and Practice. ISIS, vol. 22, pp. 9–22. Springer, Boston (2010). https://doi.org/10.1007/978-1-4419-5653-8_2
14. Ioannou, G., Sullivan, W.: Use of activity-based costing and economic value analysis for the justification of capital investments in automated material handling systems. Int. J. Prod. Res. **37**(9), 2109–2134 (1999)
15. Kieninger, A., Berghoff, F., Fromm, H., Satzger, G.: Simulation-based quantification of business impacts caused by service incidents. In: Falcão e Cunha, J., Snene, M., Nóvoa, H. (eds.) IESS 2013. LNBIP, vol. 143, pp. 170–185. Springer, Heidelberg (2013). https://doi.org/10.1007/978-3-642-36356-6_13

16. Kieninger, A., Straeten, D., Kimbrough, S., Schmitz, B., Satzger, G.: Leveraging service incident analytics to determine cost-optimal service offers. In: Wirtschaftsinformatik, pp. 1015–1029 (2013)
17. Koltai, T., Lozano, S., Guerrero, F., Onieva, L.: A flexible costing system for flexible manufacturing systems using activity based costing. Int. J. Prod. Res. **38**(7), 1615–1630 (2000)
18. Kuechler, B., Vaishnavi, V.: On theory development in design science research: anatomy of a research project. Eur. J. Inf. Syst. **17**(5), 489–504 (2008)
19. Kuechler, W., Vaishnavi, V.: A framework for theory development in design science research: multiple perspectives. J. Assoc. Inf. Syst. **13**(6), 395–423 (2012)
20. Lincoln, A.R.: Development of a dynamic costing model for assessing downtime and unused capacity costs in manufacturing. Ph.D. thesis, Georgia Institute of Technology (2013)
21. Liu, J., Chang, Q., Xiao, G., Biller, S.: The costs of downtime incidents in serial multistage manufacturing systems. J. Manuf. Sci. Eng. **134**(2), 021016 (2012)
22. Maravelias, C.T., Sung, C.: Integration of production planning and scheduling: overview, challenges and opportunities. Comput. Chem. Eng. **33**, 1919–1930 (2009)
23. March, S.T., Smith, G.F.: Design and natural science research on information technology. Decis. Support Syst. **15**(4), 251–266 (1995)
24. Naiknaware, D.M., Pimplikar, D.S.S.: Equipment costs associated with downtime and lack of availability. Int. J. Eng. Res. Appl. **3**(4), 327–332 (2013)
25. Neely, A.: Exploring the financial consequences of the servitization of manufacturing. Oper. Manag. Res. **1**(2), 103–118 (2008)
26. Ostrom, A.L., et al.: Moving forward and making a difference: research priorities for the science of service. J. Serv. Res. **13**(1), 4–36 (2010)
27. Ostrom, A.L., Parasuraman, A., Bowen, D.E., Patrício, L., Voss, C.A.: Service research priorities in a rapidly changing context. J. Serv. Res. **18**(2), 127–159 (2015)
28. Park, B.J., Choi, H.R., Kang, M.H.: Integration of production and distribution planning using a genetic algorithm in supply chain management. In: Melin, P., Castillo, O., Ramírez, E.G., Kacprzyk, J., Pedrycz, W. (eds.) Analysis and Design of Intelligent Systems using Soft Computing Techniques. AINSC, vol. 41, pp. 416–426. Springer, Heidelberg (2007). https://doi.org/10.1007/978-3-540-72432-2_42
29. Peffers, K., Rothenberger, M., Tuunanen, T., Vaezi, R.: Design science research evaluation. In: Peffers, K., Rothenberger, M., Kuechler, B. (eds.) DESRIST 2012. LNCS, vol. 7286, pp. 398–410. Springer, Heidelberg (2012). https://doi.org/10.1007/978-3-642-29863-9_29
30. Ritchie, J., Lewis, J.: Qualitative Research Practice, 1st edn. SAGE, London (2003)
31. Saldana, J.: The Coding Manual for Qualitative Researchers, 2nd edn. SAGE, London (2013)
32. Salonen, A., Tabikh, M.: Downtime costing—attitudes in swedish manufacturing industry. In: Koskinen, K.T., et al. (eds.) Proceedings of the 10th World Congress on Engineering Asset Management (WCEAM 2015). LNME, pp. 539–544. Springer, Cham (2016). https://doi.org/10.1007/978-3-319-27064-7_53
33. Saunders, M., Lewis, P., Thornhill, A.: Research Methods for Business Students, 5th edn. Pearson Education, Harlow (2009)
34. Solomon, M.M.: Algorithms for the vehicle routing and scheduling problems with time window constraints. Oper. Res. **35**(2), 254–265 (1987)
35. Wolff, C., Kühl, N., Satzger, G.: System-oriented service delivery: the application of service system engineering to service delivery. In: 26th European Conference on Information Systems (ECIS), Portsmouth, United Kingdom (2018)

36. Wolff, C., Reuter-Oppermann, M., Kühl, N.: On the impact of the customer base on the added value through system-oriented service delivery in industrial maintenance. In: 52nd Hawaii International Conference on System Sciences, Grand Wailea, Hawaii, USA, pp. 1896–1905 (2019)

37. Wolff, C., Schmitz, B.: Determining cost-optimal availability for production equipment using service level engineering. In: 19th IEEE Conference on Business Informatics (CBI), Thessaloniki, Greece, pp. 176–185 (2017)

38. Wolff, C., Voessing, M.: A framework for the simulation-based estimation of downtime costs. In: Satzger, G., Patrício, L., Zaki, M., Kühl, N., Hottum, P. (eds.) IESS 2018. LNBIP, vol. 331, pp. 247–260. Springer, Cham (2018). https://doi.org/10.1007/978-3-030-00713-3_19

39. Özbayrak, M., Akgün, M., Türker, A.: Activity-based cost estimation in a push/pull advanced manufacturing system. Int. J. Prod. Econ. **87**(1), 49–65 (2004)

Half-Empty or Half-Full? A Hybrid Approach to Predict Recycling Behavior of Consumers to Increase Reverse Vending Machine Uptime

Jannis Walk[1]([✉]), Robin Hirt[1], Niklas Kühl[1], and Erik R. Hersløv[2]

[1] Karlsruhe Service Research Institute, Karlsruhe Institute of Technology,
Kaiserstraße 12, 76131 Karlsruhe, Germany
{walk, robin.hirt, kuehl}@kit.edu
[2] TOMRA Systems ASA, Drengsrudhagen 2, 1385 Asker, Norway
erik.reinhardt.herslov@tomra.com

Abstract. Reverse Vending Machines (RVMs) are a proven instrument for facilitating closed-loop plastic packaging recycling. A good customer experience at the RVM is crucial for a further proliferation of this technology. Bin full events are the major reason for Reverse Vending Machine (RVM) downtime at the world leader in the RVM market. The paper at hand develops and evaluates an approach based on machine learning and statistical approximation to foresee bin full events and, thus increase uptime of RVMs. Our approach relies on forecasting the hourly time series of returned beverage containers at a given RVM. We contribute by developing and evaluating an approach for hourly forecasts in a retail setting – this combination of application domain and forecast granularity is novel. A trace-driven simulation confirms that the forecasting-based approach leads to less downtime and costs than naïve emptying strategies.

Keywords: Machine learning · Time series forecasting · Retail forecasting · Plastic packaging recycling

1 Introduction

According to the Ellen MacArthur Foundation [1], 78 million tons of plastic packaging were produced in 2013. Only 2% of this material went to closed-loop recycling, i.e. it was recycled into same or similar-quality applications. It is projected that by 2050 the oceans will contain more plastic than fish, by weight. This is a serious threat regarding food safety, nature and the world economy.

Reverse Vending Machines (RVMs) are a proven instrument for facilitating a circular economy. At our cooperating firm, TOMRA Systems ASA it is internally estimated that their RVMs facilitate 27% of the global closed-loop plastic packaging recycling. A recent study from the state of New South Wales in Australia confirms the immediate effect of deposit schemes using RVMs [2]. The deposit system was introduced in December 2017 and by May 2018 drink container litter, the largest proportion of litter volume in New South Wales, has been reduced by one third.

To facilitate a good customer experience at the RVM its uptime is crucial to avoid waiting times [3]. Additionally, an unexpected downtime of an RVM might bind an

© Springer Nature Switzerland AG 2020
H. Nóvoa et al. (Eds.): IESS 2020, LNBIP 377, pp. 107–120, 2020.
https://doi.org/10.1007/978-3-030-38724-2_8

employee's capacity in undetermined and inadequate situations. The research at hand is instantiated in cooperation with TOMRA Systems ASA, the world leader on the RVM market with 75% estimated market share [4]. According to an expert interview and based on internal data, full bins are currently the number one reason for downtimes of RVMs—thus, preventing customers from entering beverage containers, which can lead to a negative connotation with RVMs in particular and recycling in general. The RVM can only continue to work after the "bin full" event is handled by an employee, who needs to stop his current activity and empty the bin.

Consequently, avoiding bin full events is the biggest lever for increasing the uptime of RVMs and, thus, improving the customer experience. The goal of the research at hand is to develop and evaluate a data-driven method for reducing bin full events, thus, improving the customer experience and optimizing an employee's workflow by warning them about bin full events beforehand. Modern RVMs are communicating the number and time of returned items. In this work, historic real-world data is first used to understand the nature of consumer's recycling behavior and the occurrence of bin full events. Then, the data serves as a basis for developing and evaluating a predictive model to avoid those situations by informing an employee beforehand. In order to solve this problem, we follow a Design Science Research approach to guide the development of our artifact [5, 6]. As we aim to solve a mature problem (the prediction of bin full events) with new solutions (analyzing RVM sensor data and deriving predictive models), thus contributing with an "improvement" [7] to the body of knowledge. To guide our artifact design, we rely on justificatory knowledge [8] from the fields of time series forecasting [9] and machine learning [10–13]. To evaluate the developed artifact from three perspectives, we perform a technical experiment [14]: First, we determine the technical performance of the predictive models and compare them to benchmarks. Second, we determine the business performance of the approach to show its utility. Third, we assess the applicability of the developed artifact to data sets from other bins.

The remainder of the paper is structured as follows. First, related work regarding the forecasting task is presented in Sect. 2. In Sect. 3, the use case, the data itself and the data exploration are described, and the resulting tentative design is presented. Subsequently, the evaluation and results are described in Sect. 4. Finally, we draw conclusions and directions for further research in Sect. 5.

2   Related Work

The main goal of this study is to solve the problem of unexpected bin full events based on historic data that describes the number and time of returned items. Thus, this is a time series forecast problem. Usually, several bin full events per day occur on a given RVM. Hence forecasts with an hourly forecast granularity are required. Related work can be structured along the application domain, and the forecast granularity (e.g. hourly, daily or weekly periods). An excerpt of related work is shown in Table 1. Due to limited relevance of larger forecast granularities only papers addressing forecasts on hourly or daily level are included in the table.

The work at hand is very closely related to retail forecasting since RVMs are mostly placed in supermarkets. Thus, customers usually combine shopping and returning empty beverage containers. However, in the domain of retail forecasting no published work with an hourly forecasting is found. Aburto and Weber [10] and Hasin et al. [15] forecast the daily demand of various products in a supermarket. Taylor [16] forecasts the daily demand for one product. Thiesing and Vornberger [17] forecast the demand of 20 products in a supermarket on a weekly basis. Also in the financial forecasting domain no hourly forecasts are found, e.g. Kim [18] forecasts the direction of change in the daily Korean composite stock price index. Other papers like Wang et al. [19] forecast trends of indices on a monthly basis and are therefore not included in the table. Hourly forecasts are used for example in the meteorology domain: Sfetsos [20] forecasts the mean hourly wind speed series on the island of Crete. Sfetsos and Coonick [21] forecast the hourly solar radiation. However, these time series are not generated by human behavior like in the case at hand. In other application domains hourly forecasts for time series generated by human behavior exist, however our forecasting problem poses special challenges explained in the following. In the domain of bike sharing demand forecasting hourly forecasts for time series reflecting human behavior are used (compare Li et al. [22]). However, the domain is different, for example, weather has an immediate effect on bike sharing demand. This is not the case for recycling behavior. In electricity load and price forecasting hourly forecasts are already used for a long time (the interested reader finds a review of electricity price forecasting in [11]). However, the domain is different, and forecasts are in a less fine granularity regarding the spatial dimension: Fan and Chen [23] forecast the hourly electricity demand in New York City. Crespo Cuaresma et al. [24] and Kristiansen [25] forecast the hourly electricity prices for whole energy exchanges spanning several European countries. For the problem at hand we consider only single bins of single RVMs and thus the behavior of drastically fewer people. Consequently, even though both forecasts have an hourly granularity, our forecasting problem has a finer granularity.

Hitherto, hourly forecasts have not been examined in the application domain of retail or a comparable application domain. Hence, we contribute by developing and evaluating an approach for hourly forecasts in a retail setting where the time series are fine granular in the sense that they reflect the behavior of relatively few humans.

Table 1. Excerpt of related work regarding the forecasting task.

Source	Application domain					Forecast granularity	
	Retail	Finance	Meteorology	Bike sharing	Electricity	Hourly	Daily
[10] [15, 16]	X						X
[18]		X					X
[20, 21]			X			X	
[22]				X		X	
[23–25]					X	X	
This work	X					X	

3   Predicting Bin Full Events by Forecasting Consumer Recycling Behavior

In this study we want to decrease the downtime of RVMs by predicting bin full events based on historic data. In the following, we first describe the use case, the data we used as a basis for our forecasting method and the data exploration. Then we present our tentative design that levers a hybrid prediction model combining machine learning and statistical approximation and is using endogenous and exogenous data.

3.1   Use Case, Data Description and Exploration

In practice, two data sources reflecting the fill level of a bin are available: The measurement of a sensor positioned above a bin and the time and type of all returned items. 90% fill level is the last signal from the bin level sensor before the bin full signal. Thus, relying solely on the bin level sensor would be too inefficient: The employee would need to empty the bin when there is still capacity left. Furthermore, the sensor data would be too coarse-grained as basis for a prediction model. Consequently, our envisioned system uses the bin level sensor only to know, when a fill level is at 90%. This reduces uncertainty since "the further ahead we forecast, the more uncertain we are" [9]. This 90% fill level signal will be used to trigger a forecast about the remaining 10% fill level which is based on detailed item return data. This is based on the fact that every fill level change is associated with a certain number of items going into that bin. Thus, instead of predicting the fill level directly, we predict the number of returned items. This prediction output can be mapped back to a corresponding fill level.

In addition to the data about returned items, we consider exogenous data sources, such as the weather or public holidays, as potential candidates for input features in the prediction model. Having stated which data sources are used now the data set and the results of the data exploration are presented.

The data used to build the artifact stems from one bin of one RVM in a supermarket in Norway and covers the period from the 12th of June 2014 to the 29th of May 2017. To make the data interpretable it needs to be aggregated to fixed time intervals. Thus, all item returns within these time intervals (monthly, daily and hourly) are added up. The monthly number of returned items shows an upward trend. Additionally, a monthly seasonality is present: there are more item returns in summer than in winter, which seems reasonable due to the higher temperatures in summer. These lead to a higher consumption and return of beverage containers. Also, it is observed that there are less item returns in July than in June and August, which could be explained by July being the main vacation month in Norway.

When the data is aggregated to a daily interval it can be observed that the 25% quantile of returned items on Saturdays is higher than the median of all other weekdays. The other weekdays do not show significant differences. Also, the distribution of the number of returned items per day has a long tail, i.e. on few days, an abnormal high number of returned items can be observed. In a second stage of data exploration further possible predictors for a prediction model for the daily time series are identified. The auto-correlation plot (Fig. 1) shows that the number of returned items today has the strongest correlation with the number of items returned yesterday and on the same weekday 1 to 5 weeks ago (lag of 6, 12, 18, 24 and 30 business days since the supermarket considered is closed on Sundays). These correlations are larger than 0.3.

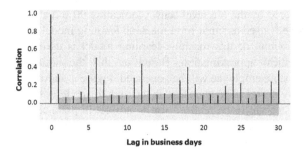

Fig. 1. Auto-correlation plot of daily item returns with maximum 30 business days lag

Previous research indicates that shopping behavior is influenced by the weather [26, 27]. Since recycling behavior is assumed to be very similar to shopping behavior weather variables are included in our predictive model. Another exogenous influence are holidays. It seems that this variable is also relevant for the problem at hand: The 25% quantile of days after holidays is higher than the 75% quantile of all other days. However, there are only 34 holidays in the data set. Thus, the sample size is relatively small and consequently we cannot be very confident about this statement. For days after holidays no significant difference can be found.

The hourly time series is more variable than the daily item returns time series (coefficient of variation of 125.90% vs. 43.55%). Also, it can be observed that in almost 17% of the hours no items are returned. The hourly item returns can be interpreted as observed hourly demand for returning empty beverage containers at a given store. As such it can be characterized as lumpy demand according to e.g., Gutierrez et al. [28], since it is characterized by many intervals with zero demand and periods with actual demand occurrences. It is a known problem to produce reliable forecasts for time series which are characterized by lumpy demand [29, 30]. Besides the autocorrelations for the hourly time series are all below 0.3. Overall, the higher coefficient of variation, the lumpiness and the lower autocorrelations suggest that the hourly time series is more difficult to forecast than the daily forecast. Consequently, the predictive modeling is based on a forecast of the number of returned items per day.

Based on the data exploration a data set is constructed for the predictive modeling. The number of returned items per day is the target variable of the prediction model. The first group of predictors are the lagged variables: the number of returned items 1, 6, 12, 18, 24 and 30 days ago are used. For the seasonalities date-related variables are used: the weekday as numeric variable, the month and the year. Besides two types of exogenous variables are used: boolean variables indicating a day before or after a public holiday and the precipitation intensity and the apparent maximum temperature as weather variables.

3.2   Tentative Design: A Hybrid Prediction Model Based on Endogenous and Exogeneous Data Sources

The tentative design is a hybrid prediction model which uses machine learning and statistical approximation and is based on different data sources. The prediction of bin full events is triggered by the bin level sensor indicating 90% bin level. Then the data set described above is used as input for a machine learning model to forecast the daily item returns. The output of this machine learning model is used to compute hourly forecasts by statistical approximation. Based on this the notification sent to the smartphone of the store employee will be generated. Figure 2 shows an overview of the envisioned tentative design. The two steps in the middle are described in the following. The first and last step are to be realized for an artifact in practice. This realization is viable: the sensors for the first step already exist. Also, there is already a smartphone app which can be used for the notification of the store employee. The envisioned productive system is illustrated in the lower part of Fig. 2.

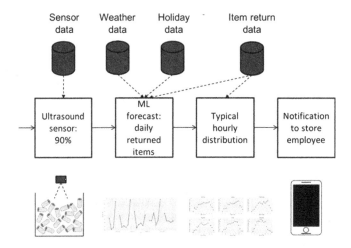

Fig. 2. Tentative design: a hybrid prediction model to predict bin full events

In the following it is described how the forecasting model was selected. The recently most widely used techniques for time series forecasting can be categorized into time series models, like Autoregression Moving Average (compare Brockwell and Davis [31]) and machine learning models. Time Series models are specifically designed to forecast time series and usually use only information stemming directly from the time series. Machine learning models such as Neural Networks and Support Vector Machines are applicable for a wider set of tasks like classification and clustering. For time series forecasting regression algorithms are applied.

In this work, we are analyzing long series with high-frequency data, where machine learning models work better according to Bergmeir and Benítez [13]. Also, research implies that machine learning models can deal better with the complexity and non-linearity of electricity price time series than time series models [11]. According to this paper, Neural Networks, Support Vector Regression and Multiple Linear Regression have shown to produce reliable forecasts for electricity prices. Research in the closely related application domain of retail forecasting also suggests that machine learning models (e.g. Neural Networks) outperform classical time series models [6]. Consequently, machine learning models, namely Support Vector Regression, Artificial Neural Networks and Multiple Linear Regression, are used for the forecasting task at hand. Additionally, a Gradient Boosting Regressor (compare Friedman [12]) is considered due to its successful application in classical research as well as data science competitions [32].

To predict bin full events, hourly forecasts are necessary. We implement and evaluate several approaches and present the most promising one. First, several approaches for direct forecasts of the number of returned items per hour are implemented and evaluated. This is a wide-spread approach in application domains like electricity forecasting [23–25]. Second, we implement and evaluate a hybrid approach based on machine learning and statistical approximation (averaging). This approach is similar to the one described by Gross and Sohl [33]. Their disaggregation based on statistical approximation is from product families to products. In the case at hand the disaggregation is in a temporal dimension: from daily forecast to hourly forecast. The second approach outperforms the direct hourly forecasts and is described in more detail in the following. As depicted in Fig. 3, typical distributions for each weekday are used to convert the daily forecast into an hourly one. The upper plot shows the daily forecast. The middle plot shows the distribution which is used for mapping. It is obtained by calculating the mean of the percentage of day values for each hour of all preceding same weekdays. The daily forecast is multiplied with these percentage values to obtain the forecast for the respective hour. This resulting forecast is shown in the lower plot.

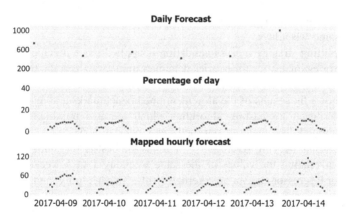

Fig. 3. Illustration of mapping from daily to hourly forecast

4 Evaluation and Results

We evaluate the presented artifact in three steps: First, we perform a technical experiment to determine the performance of the prediction model for daily and hourly returned items and compare it to forecast benchmarks. Second, we evaluate the performance of the proposed approach from a business perspective, considering the cost for RVM downtime and the cost for perished bin capacity. Third, we briefly evaluate the performance of the prediction model on other RVMs at different locations to test the degree of generalization of the approach.

4.1   Performance of the Prediction Model

As first evaluation measure we use the Mean Absolute Error (MAE) that is calculated by

$$\text{MAE} = \frac{\sum_{i=1}^{n} |y_i - \widehat{y}_i|}{n} \tag{1}$$

as it is easily understandable. Because it is scale-dependent, it cannot be used to compare performances across data sets. One of the most widely used scale-independent measures is the Mean Absolute Percentage Error. However, it cannot be used for the data at hand since it is not suitable if $y_i = 0$ or close to zero. Then it yields infinite or very high numbers. For data sets with this property the MAE can be divided by the mean of the actual values to obtain a scale-dependent error measure [34]. These two evaluations measures are used in the following.

In time series forecasting it is good practice to report the performance of simple forecasting methods as benchmarks in addition to the evaluation measures [9]. The following three benchmarks are reasonably applicable for the data set used in the following. A daily time series with weekly seasonality is used for illustration purposes. The naïve forecast uses the value of yesterday as forecast for today: $\widehat{y}_i = y_{i-1}$. The seasonal naïve forecast is a similar approach. Its forecast for today is the value of the same day last week: $\widehat{y}_i = y_{i-m}$, with m = seasonality period. Another applied benchmark is the Seasonal Moving Average, it computes the forecast by averaging the value of the last x same weekdays.

As data splitting strategy a cross-validation is applied since it achieves more precise error estimates for time series data than time series specific procedures [13, 35].

Table 2 shows the results of the aforementioned benchmarks and machine learning algorithms. For each forecasting algorithm which requires training 10-fold cross-validation is applied. The mean of returned items per day is 512.11.

For the Seasonal Moving Average $x = 5$ yields the best results, thus the forecast is computed by averaging the values of the same weekday 1 to 5 weeks ago. For the Support Vector Regressor all available kernels ('rbf', 'linear', 'poly' and 'sigmoid') are tested. The linear kernel performs best and is used to obtain the results in Table 2. Due to limited computing capabilities on the available hardware no grid-search for parameter tuning is conducted and therefore the standard parameters from scikit-learn [36] are used. Parameter tuning can and most probably will further enhance the performance of the machine learning algorithms.

As one can observe in Table 2 all machine learning models yield better results than the simple benchmarks. The differences between the machine learning models are relatively small. When inspecting the machine learning models, the weekday and the lagged variables 1 to 5 weeks ago are always among the most important predictors.

This explains the similar performance of the Seasonal Moving Average and the machine learning models.

Table 2. Results of different benchmarks and machine learning forecasts for daily item returns.

Method	MAE	MAE/Mean
Naïve forecast	194.51	37.98%
Seasonal naïve forecast	162.22	31.68%
Seasonal Moving Average	136.49	26.65%
Multi-layer Perceptron Regressor	134.85	26.33%
Support Vector Regressor	129.66	25.32%
Multiple Linear Regression	129.00	25.19%
Gradient Boosting Regressor	125.08	24.42%

In case the Seasonal Moving Average performs almost as good as the best performing model for more data sets it is a promising candidate for the productive use of such a system. It does not require any learning and thus facilitates the development of a productive system considerably.

The results of the hourly forecast are shown in Table 3 for a selection of algorithms. The mapping approach is the one explained in Sect. 3.2. The mean of returned items per hour is 32.24.

Table 3. Results of different forecasting methods for hourly item returns.

Method	MAE	MAE/Mean
Naïve forecast mapped to hourly forecast	25.74	79.59%
Seasonal Moving Average mapped to hourly forecast	24.22	74.89%
Gradient Boosting Regressor mapped to hourly forecast	23.82	73.65%

The MAE/Mean evaluation measures for hourly item returns are approximately three times worse than for the daily item returns. In the data exploration in Sect. 3.1 the lumpiness of the hourly item returns time series is already mentioned as a factor which negatively impacts its forecastability. In addition, there is the use-case-specific challenge of bulk arrivals. A bulk arrival occurs when one customer returns a high number of empty beverage containers at once. In the data set considered the most extreme bulk arrival contains 285 items. On average just 32.24 items are returned per hour. Hence, these events which cannot be foreseen with the data at hand yield extreme spikes in the hourly time series.

The similarity of these performance values shows that the forecasting accuracy for the daily item returns has little influence on the forecasting accuracy on hourly basis. The inaccuracy introduced by mapping via past hourly distributions is the major inaccuracy. The results of the naïve forecast are reported to highlight this. The relatively large performance differences on a daily level between the Gradient Boosting Regressor and the naïve forecast do not translate into big performance differences on an hourly level.

4.2 Artifact Evaluation from a Business Perspective

The evaluation measures reported in the section above cannot be interpreted regarding the utility of the approach. Thus, to evaluate the artifact from a business perspective, key performance indicators reflecting the different costs of emptying strategies are selected with domain experts: the percentage of avoided bin full events and the average number of hours the notification is too early (compare Fig. 4).

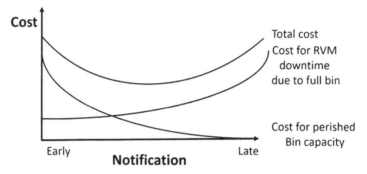

Fig. 4. Trade-off between different costs for bin full notifications

A bin full event is avoided, if the employee is notified about an upcoming bin full event in advance. A simple strategy to avoid a high percentage of bin full events is to use very early notifications. Therefore, it is also evaluated how many hours before the actual bin full event the notifications are on average. This KPI can then be used to assess how much perished bin capacity and additional labor cost there is to expect for a certain combination of policy and forecast. To compare different emptying strategies regarding the aforementioned KPIs a simulation is implemented. The simulation is trace-driven, i.e. real data is used as input, compare Sargent [37].

As for the forecasting, first simple benchmarks are constructed. The benchmarks are different hour-based policies, they do not rely on any forecast. For example, the time in the notification is always two hours after the 90% bin level signal (compare the upper rows of Table 4).

Table 4. Simulation: results of hour-based and forecast-based policies.

Hours between 90% bin level & Notification	Forecasting method	% avoided bin full events	Average hours too early
2	–	79.90	3.37
0	–	100	4.28
–	Naïve forecast	77.95	2.38
–	Seasonal moving average	80.33	2.24
–	Gradient Boosting Regressor	81.17	2.16

Then, these benchmarks are compared to policies in which the time in the notification is determined based on different forecasts. For that, a data set with the actual and forecasted hourly item returns is used (since cross-validation is applied forecasts exist for each business hour of the observed period). To obtain the time in the notification, the forecasted item returns time series is accumulated until it reaches 100 items. On average the considered bin takes 131 items between 90 and 100% bin level, the 100 items are arbitrarily chosen to have a buffer. Compare the bottom rows of Table 4 for the corresponding results. For each emptying policy there were minimum 2354 simulated bin full events.

The most important observation is that the simulation shows that forecast-based policies yield better KPIs than hour-based policies. The two-hour policy and the forecast-based policies with a threshold of 100 can be compared very well. Using the Gradient Boosting Regressor and the Seasonal Moving Average leads to an improvement of both KPIs: the percentage of avoided bin full events is slightly higher while the notifications are on average more than one hour less too early. Consequently, both the cost for downtime and the cost for perished bin capacity are lower. Thus, the total cost curve is lower for forecast-based policies than for hour-based policies.

Also, it can be observed that there are only little differences between the different forecasting methods. In case of putting the artifact into practice, it needs to be assessed if the relatively small enhancement of the Gradient Boosting Regressor over the Seasonal Moving Average justifies the necessary additional effort and the potential cost for weather data from external providers.

One rather strong assumption is that the considered bin takes always exactly 131 items after a 90% bin level signal until the bin is full. This is only an average value, there will be deviations in real life. Additionally, a perfectly functioning bin level sensor is assumed in the sense that the indicated 90% bin level is accurate. It is expected that the bin level sensor works reliable, but it is not capable of detecting the current bin level precisely. As a consequence of these limitations, the results of the simulation are not to be understood as expected real life behavior. But all limitations apply both for the hour and forecast-based policies. Consequently, they provide a valid guideline regarding the difference between these two approaches. They clearly demonstrate the benefit of forecasting the number of returned items per hour instead of relying on hour-based policies.

4.3 Applicability to Other RVMs

So far, only one bin of one RVM is considered. To assess the generalizability the same forecasting model is trained and evaluated on the data of four more bins. The according supermarkets are chosen to have different location characteristics (city center & rural area) and to be in different countries. For the sake of brevity only the results of the hourly forecast are mentioned: the respective best forecasting approaches yield MAE/Mean values between 50,75 and 63,03%. The MAE/Mean for the time series considered so far is 73,62%. Thus, the approach works better for all additionally considered time series, which confirms the generalizability of the approach.

5 Conclusion and Outlook

Bin full events are the major reason for downtime of RVMs and yield waiting times for supermarket customers as well as negative customer experience—and might bind an employee's capacity in undetermined and inadequate situations. To address this problem, the paper at hand presents and evaluates a hybrid approach for predicting events of full bins by applying a Design Science Research approach. The presented hybrid approach combines machine learning and statistical approximation and uses endogenous and exogenous data sources. The obtained forecasts are not very accurate due to lumpiness and bulk arrivals negatively impacting the forecastability of the hourly item returns. However, a trace-driven simulation shows that bin emptying strategies based on the hybrid approach yield a higher utility compared to naïve preventive bin emptying strategies. Overall, we contribute by developing and evaluating an approach for hourly forecasts in a retail setting where the time series granularity reflects the behavior of relatively few humans. Such studies on forecastability are encouraged contributions both in Information Systems research in general [38] and in the field of reverse logistics [39] for product returns. Also, the paper at hand shows how machine learning and statistical approximation can be leveraged to facilitate a good customer experience and increase process efficiency. Ultimately, this enables a further proliferation of RVMs which help to increase the rate of plastic packaging being recycled in a closed-loop.

Besides the contribution, this work also has limitations. So far, the approach has only been tested for item return time series of five different bins. Evaluations on data sets from more bins are to be performed to confirm the approach. Also, the simulation assumes that the considered bin always takes exactly 131 items between 90 and 100% bin level and the sensor indicating the current bin level works faultlessly.

However, this work lays ground for future research in the field of hourly forecasts for lumpy time series. It shows that the hybrid approach combining machine learning and statistical approximation yields better results than direct hourly forecasts. This can be tested for time series with similar characteristics like supermarket sales on stock keeping unit level. Another possibility for future research is the realization of the tentative design by integrating the live bin level sensor data and the smartphone app for the notification of the supermarket employee.

Our study informs managers of supermarkets and RVM manufacturers that predicting bin full events is possible and relying on a forecasting-based approach to do so seems reasonable. This facilitates proactive bin emptying strategies instead of the currently used reactive ones. Ultimately, when implemented, the resulting artifact can facilitate customer experience with reduced waiting times—and contribute to the general experience of recycling as a positive one. Also, it can facilitate a more sustainable operation of further parts of the value chain: in some markets like Australia RVMs are placed outside supermarkets. For such RVMs the forecast can be used to optimize the schedule for the trucks picking up the empty beverage containers. Thus, the overall reverse vending logistics chain can be optimized regarding its sustainability.

Due to the limitations of the simulation, a real-world evaluation is necessary to confirm the utility of the approach. In this evaluation it should also be assessed how precisely the proposed system affects employee and customer satisfaction. For this real-world evaluation the usage of the Seasonal Moving Average for the forecast of the daily time series seems most reasonable. It considerably facilitates the development of the proposed system since no model training is necessary while the MAE/Mean measure for the hourly time series decreases only 1.24% points. Later, if the system is deployed to more RVMs a more sophisticated, machine learning based method can be beneficial due to economies of scale.

References

1. Ellen MacArthur Foundation: The New Plastics Economy - Rethinking the future of plastics (2016)
2. Waste Management Review: NSW litter reduced by a third with help from Return and Earn. http://wastemanagementreview.com.au/nsw-litter-reduce-third/
3. Taylor, S.: Waiting for service: the relationship between delays and evaluations of service. J. Mark. **58**, 56 (1994)
4. TOMRA System ASA: Key Facts. https://www.tomra.com/en/news-and-media/key-facts
5. Hevner, A.R., March, S.T., Park, J., Ram, S.: Design science in information systems research. MIS Q. **28**, 75–105 (2004)
6. Peffers, K., Tuunanen, T., Rothenberger, M., Chatterjee, S.: A design science research methodology for information systems research. J. Manag. Inf. Syst. **24**, 45–77 (2007)
7. March, S.T., Smith, G.F.: Design and natural science research on information technology. Decis. Support Syst. **15**, 251–266 (1995)
8. Gregor, S., Jones, D.: The anatomy of a design theory. J. Assoc. Inf. Syst. **8**, 312–335 (2007)
9. Hyndman, R.J., Athanasopoulos, G.: Forecasting: Principles and Practice, 2nd edn. OTexts, Melbourne (2018). OTexts.com/fpp2
10. Aburto, L., Weber, R.: Improved supply chain management based on hybrid demand forecasts. Appl. Soft Comput. J. **7**, 136–144 (2007)
11. Weron, R.: Electricity price forecasting: a review of the state-of-the-art with a look into the future. Int. J. Forecast. **30**, 1030–1081 (2014)
12. Friedman, J.H.: Greedy function approximation: a gradient boosting machine. Ann. Stat. **29**, 1189–1232 (2001)
13. Bergmeir, C., Benítez, J.M.: On the use of cross-validation for time series predictor evaluation. Inf. Sci. (NY) **191**, 192–213 (2012)
14. Peffers, K., Rothenberger, M., Tuunanen, T., Vaezi, R.: Design science research evaluation. In: Proceedings of the 7th International Conference on Design Science Research in Information Systems: Advances in Theory and Practice, pp. 398–410 (2012)
15. Hasin, M.A.A., Ghosh, S., Shareef, M.A.: An ANN approach to demand forecasting in retail trade in Bangladesh. Int. J. Trade Econ. Finance **2**, 154–160 (2011)
16. Taylor, J.W.: Forecasting daily supermarket sales using exponentially weighted quantile regression. Eur. J. Oper. Res. **178**, 154–167 (2007)
17. Thiesing, F.M., Vornberger, O.: Sales forecasting using neural networks. In: International Conference on Neural Networks, pp. 2125–2128 (1997)
18. Kim, K.J.: Financial time series forecasting using support vector machines. Neurocomputing **55**, 307–319 (2003)
19. Wang, J.J., Wang, J.Z., Zhang, Z.G., Guo, S.P.: Stock index forecasting based on a hybrid model. Omega **40**, 758–766 (2012)

20. Sfetsos, A.: A comparison of various forecasting techniques applied to mean hourly wind speed series. Renew. Energy **21**, 23–35 (2000)
21. Sfetsos, A., Coonick, A.H.: Univariate and multivariate forecasting of hourly radiation with artificial intelligence techniques. Sol. Energy **68**, 169–178 (2000)
22. Li, Y., Zheng, Y., Zhang, H., Chen, L.: Traffic prediction in a bike-sharing system. In: Proceedings of the 23rd SIGSPATIAL International Conference on Advanced Geographic Information Systems - GIS 2015, pp. 1–10 (2015)
23. Fan, S., Chen, L.: Short-term load forecasting based on an adaptive hybrid method. Power Syst. IEEE Trans. **21**, 392–401 (2006)
24. Crespo Cuaresma, J., Hlouskova, J., Kossmeier, S., Obersteiner, M.: Forecasting electricity spot-prices using linear univariate time-series models. Appl. Energy **77**, 87–106 (2004)
25. Kristiansen, T.: Forecasting Nord Pool day-ahead prices with an autoregressive model. Energy Policy **49**, 328–332 (2012)
26. Murray, K.B., Di Muro, F., Finn, A., Popkowski Leszczyc, P.: The effect of weather on consumer spending. J. Retail. Consum. Serv. **17**, 512–520 (2010)
27. Parsons, A.G.: The association between daily weather and daily shopping patterns. Australas. Mark. J. **9**, 78–84 (2001)
28. Gutierrez, R.S., Solis, A.O., Mukhopadhyay, S.: Lumpy demand forecasting using neural networks. Int. J. Prod. Econ. **111**, 409–420 (2008)
29. Chopra, S., Meindl, P.: Supply Chain Management: Strategy, Planning, and Operation. Pearson Education, Inc., London (2007)
30. Ghobbar, A.A., Friend, C.H.: Evaluation of forecasting methods for intermittent parts demand in the field of aviation: a predictive model. Comput. Oper. Res. **30**, 2097–2114 (2003)
31. Brockwell, P.J., Davis, R.A.: Introduction to Time Series and Forecasting (2016)
32. Kaggle Inc.: Kaggle. https://www.kaggle.com
33. Gross, C.W., Sohl, J.E.: Disaggregation methods to expedite product line forecasting. J. Forecast. **9**, 233–254 (1990)
34. Kolassa, S.: Evaluating predictive count data distributions in retail sales forecasting. Int. J. Forecast. **32**, 788–803 (2016)
35. Bergmeir, C., Hyndman, R.J., Koo, B.: A note on the validity of cross-validation for evaluating time series prediction. Monash University, Working Papers 10/15 (2015)
36. Pedregosa, F., et al.: Scikit-learn: machine learning in python. J. Mach. Learn. Res. **12**, 2825–2830 (2012)
37. Sargent, R.G.: Verification and validation of simulation models. J. Simul. **7**, 12–24 (2013)
38. Shmueli, G., Koppius, O.R.: Predictive analytics in information systems research. MIS Q. **35**, 553–572 (2011)
39. Govindan, K., Soleimani, H., Kannan, D.: Reverse logistics and closed-loop supply chain: a comprehensive review to explore the future. Eur. J. Oper. Res. **240**, 603–626 (2015)

Predicting Market Basket Additions as a Way to Enhance Customer Service Levels

Vera L. Migueis[(✉)] and Ricardo Teixeira

INESC TEC and Faculty of Engineering, University of Porto,
Rua Dr. Roberto Frias, 4200-465 Porto, Portugal
{vera.migueis,ei10045}@fe.up.pt

Abstract. It is imperative that online companies have a complete in-depth understanding of online behavior in order to provide a better service to their customers. This paper proposes a model for real-time basket addition in the e-grocery sector that includes predictors inferred from anonymous clickstream data, such as a Markov page view sequence discrimination value. This model aims at anticipating the addition and the non-addition of items to customers' market basket, in order to enable marketers to act conveniently, for example recommending more appropriate items. Two classification techniques are used in the empirical study: logistic regression and random forests. A real sample of anonymous clickstream data taken from the servers of a European e-retailing company is explored. The empirical results reveal the high predictive power of the model proposed, based on the explanatory variables introduced, as well as the supremacy of random forests over logistic regression.

Keywords: Clickstream data · Web usage mining · Data mining and customer service

1 Introduction

Due to the increase in significance and market share, e-commerce companies have to adopt new strategies that fit the needs of online customers [1]. This type of customers has different behavior from the physical ones [2]. Web users have a much easier job in comparing different companies through simple online queries, making it hard to retain customers the same way it is done in traditional stores [3]. This fact, combined with the high competition levels and low costs of switching product or service provider, accentuate the need to perform a thorough analysis of customer behavior, based mostly on anonymous data, in order to avoid customer attrition [4]. This may be achieved by providing engaging navigation experiences.

To increase profit, online retailers should not only increase traffic but should also increase conversion rates, i.e. the proportion of orders to website visitors. However, in contrast with the pattern of a substantial increase in traffic, some

© Springer Nature Switzerland AG 2020
H. Nóvoa et al. (Eds.): IESS 2020, LNBIP 377, pp. 121–134, 2020.
https://doi.org/10.1007/978-3-030-38724-2_9

websites experience stagnant conversion rates or the conversion rates are not increasing significantly [5]. For example, in 2015 the average conversion rate for shoppers at the top 500 largest online merchants in North America was 3.32% [6].

Online retailers use several methods to convert traffic into sales. One of the most important aspects is to ensure that the right product is available when requested, at the right place and at the right price. Furthermore, online retailers are aware of the importance of providing a good shopping experience that leads customers to purchase and repurchase in the future and consequently are investing in enhancing their e-commerce platforms [7]. Some popular features of customer service include greeting customers upon store entry, providing product knowledge, recommending alternatives, helping customers find the product they are looking for and facilitating the checkout process [8].

With the computer power flourishing in the early nineties online companies started to uncover the power of user behavior mining, which can be crucial to promote users' conversion. A new kind of direct marketing based on customer knowledge resulting from the sheer volume of data stored from the clients' interaction with the company is gaining momentum [9]. E-commerce managers and marketers are starting designing plans to improve electronic Customer Relationship Management (e-CRM) through data analytics.

As the Internet works on the basis of interchangeable data, there are new data sources that companies ought to exploit. This data enables e-commerce managers to overview the business in ways that were not previously possible. Through an online store, it is possible to track much more data that is a direct result of how the customer interacts with the company. The so-called clickstream data, i.e. the collection of web logs that compose the session of a specific customer on the company website, is key in understanding customer behavior and it is also the main source of information for companies to adapt their service according to their audience [10].

In this context, this study proposes a new methodology to support e-CRM, with a focus on the e-grocery sector. This study applies data mining techniques to extract knowledge from databases containing clickstream data. Unlike the traditional recommender systems, the models developed in this study aim to use clickstream navigation data to provide future insights that might help decision makers to increase service level. In contrast to the existing literature regarding e-commerce and in particular e-grocery retail market, we contribute to the scientific community by proposing a model to predict the likelihood of basket addition on a specific page, considering user's session features. Two data mining techniques are used as classifiers for the proposed model: random forests and logistic regression. The model enables the company to have real time information over what and when their customers are or are not going to add something to their baskets, and act accordingly. For example, targeted pop-up windows could be triggered to recommend related products.

The reminder of this paper is as follows. The next section introduces a brief literature review on the applications of web usage mining. Section 3 presents the methodology used in this study, including the preprocessing approach proposed, the explanatory variables derived, the data mining classification techniques used and the evaluation criteria. Section 4 presents the application, i.e. the data used and the results. The paper closes with conclusions and some ideas for future research.

2   Web Usage Mining Fundamentals and Related Work

Web mining is the application of data mining methodologies, techniques and models to extract knowledge from web data (e.g. [11,12]) so that web personalization can be achieved. Web mining is a broad field of study as it can be decomposed into sub-topics according to the different types of web data available, i.e. Web usage data, also known as clickstream data, web content data, and web structure data [13]. This paper falls within the scope of web usage mining.

Web usage mining refers to the automatic discovery and analysis of patterns in clickstream data and associated data collected or generated as a result of user interaction with web resources, typically, a web server [14]. The goal is to capture, model and analyze the behavioral patterns and the interaction between users and a Website. The primary data sources in web usage mining are the server log files, which include web server access logs and application server logs. Each click made by the user while navigating the Internet, corresponds to an HTTP request to the website's server, and it produces a new entry in the server entry log.

Although still in its infancy, during the past decade, academic literature has been contributing to the study of customer behavior online, particularly with some publications in the field of Web usage mining. The low conversion percentages observed has justified the identification and analysis of customer profiles [15]. Moe and Fader [16] elaborated on this and developed a typology of website visits, using navigational patterns, which identified four types of browsing strategies: directed buying, search/deliberation, hedonic search and knowledge building. They concluded that more frequent visitors have a greater propensity to buy. In line with the previous study, Montgomery et al. [17] tried to dichotomize customer behavior as either browsing or deliberation. The most relevant conclusion of their research is that visitors might alternate between different behavioral modes during a single session. Sismeiro and Randolph [18] went further and decompose online ordering of the website into sequential user tasks and examine purchase behaviors. Results indicate that visitors' browsing experiences and navigational behavior influence task completion. Examples of other studies that explore relevant attributes for predicting purchase behavior include [19,20]. Ding et al. [21] go further and propose a model that considers each individual user's browsing behavior, tests the effectiveness of different marketing and web stimuli and performs optimal webpage transformation. The results obtained reveal that the designed interventions increase purchase conversions.

The above-mentioned studies are mainly focused in understanding customer behavior and extracting relevant features upon which patterns can be drafted, through exploratory models. However, there is another branch of literature that apply and derive different data mining algorithms to explore and predict users behavior. For example, Chang et al. [22] proposes a model to predict the interest in the most profitable products. The key characteristics of loyal customers' personal information are collected through a clustering analysis and are used to locate other potential customers. Afterwards, association rules analysis derive knowledge of loyal customers' purchasing behavior, which is used to detect potential customers' likelihood to buy a star product. Wang [23] introduces a clustering algorithm to support data stream analysis, which turns to produce better results from large, high-dimensional data streams than some other data stream algorithms. Classification algorithms have also been used in web mining usage context. For example, Cho et al. [24] use a classification model to target their recommendations to those customers who are likely to buy the recommended products. Aguiar et al. [25] proposes the development of a model that can learn the individual preferences and characteristics of each user, and utilize this information to predict how engaged they will be to a particular video stream. In the same line, Pachidi et al. [26] develop a model to identify those customers who, having used a trial version of a software, will buy a license.

Despite the growing volume of literature on web usage mining, there seems to be room for further research particularly in what concerns the use of data mining techniques to study navigational patterns. To the best of our knowledge, the addition of new items to the market basket throughout an anonymous user session in real-time has never been addressed in the literature. In fact, in this context, the most relevant work is elaborated by Ruud et al. [19]. This used anonymous clickstream data to predict customer conversion over time. Furthermore, this study uses as case study a retailer selling lingerie, nightwear and bathing fashion, whose nature differs significantly from an e-grocery store.

3   Methodology

The methodology followed in this paper seeks to explore the use of anonymous clickstream data to predict a basket addition in the next request. Each observation used in the prediction model corresponds to the instant after a HTTP request (or click) to the website's server, and the explanatory variables enable to characterize the user's session in that instant of time. The target variable is binary and refers to the addition or not of a good to the user's basket in the next request.

The methodology proposed encompasses a preprocessing stage to create a workable dataset that feed the next stages, i.e. predictors inference and model construction. We identify a set of predictors that refer to the characterization of the users' session. Two data mining classification techniques are used and compared in the empirical study: logistic regression and random forests.

In the remainder of this section we present the preprocessing process proposed, the explanatory variables proposed, the classification techniques used and the evaluation criteria used to compare their performance.

3.1   Preprocessing

Preprocessing the data is of utmost importance in order to build a solid clickstream prediction model, especially because the raw web logs, the main source of data, are not usable in web usage mining due to their unstructured nature. Dtaa mining algorithms require the data to be structured.

Figure 1 depicts a pipeline of the preprocessing of the web logs.

Fig. 1. Preprocessing of the web logs.

Data Cleaning involves the identification and elimination of all these unimportant elements. Despidering consists of removing entries generated by requests from crawlers (i.e. web robots that browse the web in a methodical, automated manner) or web robots. User Identification aims at identifying repeated visits. Sessionization implies a segmentation of user activity records from each identified user into sessions, each representing a complete visit to the website. Path Completion addresses the problem of client-side caching, which occurs when a user returns to a previously requested or downloaded page. Page view Identification is the process of aggregating meaningful page references, i.e. pages linked to some action performed on a website.

Apart from the standard preprocessing pipeline, there are other relevant transformations. These transformations are linked to some domain restrictions and are fundamental in order to keep dataset coherence for further analysis. Data integration is also a relevant stage which involves the combination of different types of data onto one entity with diversified and relevant attributes.

All the steps shown in Fig. 1 can be automatically executed, enabling to obtain a tool that can operate in real time.

3.2   Explanatory Variables

Having preprocessed the clickstream data, we identify a set of relevant predictors. This process is linked to an exploratory analysis to be developed. More specifically, we propose to use a filter approach that enables to select the variables that are significantly related to the dependent variable.

This section will only focus on the final set of features that were used in the prediction model (see Table 1). The rationale behind the use of these features is detailed below. All these variables are continuous.

Table 1. Variables used in the prediction model.

Feature	Label
PBA	Previous basket additions
RBA	Requests between basket additions
subCat	Requests related to Sub Category Menu
searchRes	Requests related to Search Results
prodPage	Requests related to Product Page
shipPay	Requests related to Payment
markovLik	Markov likelihood value
markovDis	Markov discrimination value

Previous Basket Additions - PBA. It is our understanding that a key factor driving basket additions is the number of items a customer has already added. Furthermore, we believe that the users' purchase items on e-retailers follow similar patterns, namely, in terms of the distribution of the number of products they buy. Therefore, we explore the hypothesis that the number of previous basket additions influences the probability of further additions.

Requests Between Basket Additions - RBA. We also consider the number of requests between basket additions a relevant information. From our explanatory analysis we have concluded that users tend to follow a purchase cadence, i.e. on average, they perform the same number of requests before adding an item to the cart. It seems that each customer tends to follow the same search method to find products. Furthermore, e-grocery websites mimic the organization of physical stores, that is, different sections containing products from similar categories. As users try to optimize the time they spend at the store, they will keep adding products as they travel from section to section. On the website, we believe that users also try to minimize their shopping time by engaging a certain buying pattern. In order to inform the prediction models how every request stands regarding a new basket addition, we consider the ratio between the number of requests since the last basket addition and the current session's average number of requests between basket additions. So, for instance, if the last item was added to the cart three requests ago, and the current average number of requests between basket additions is 6.70 requests, then, the value passed to the predictors would be 0.45, which means that only about 45% of the average number of requests were performed since the last addition.

Requests Related to Sub Category Menu, Search Results, Product Pages and Shipping and Payment - SubCat, SearchRes, ProdPage and ShipPay. During a session, users tend to execute two distinct types of tasks: search/purchase and shipping/payment. Naturally, basket additions occur during the search/purchase phase, thus the need to distinguish both stages.

In order to do so, we pinpoint some specific page views (unitary elements of clickstream navigation) that may explain how users' behavior alters throughout the session. Furthermore, we use a sliding window strategy to evaluate the ten most recent sessions' page view requests. During this request window, we look for specific page view requests that pinpoint specific user behaviors:

– Sub Category Menu: Navigating through product categories' menus is associated with users that are still in the search/purchase stage.
– Search Results: Users who submit a search term request are also linked to the search/purchase behavioural pattern.
– Product Page: Requesting the page with detailed product information displays an active interest in a certain product, thus the presence of this pageview within the 10 request sliding window displays clear search/purchase intentions.
– Shipping & Payment: All requests related to shipping/payment refer to post-addition events. These requests relate to choosing delivery dates, reviewing basket items, handling payment, etc.

For each of these four navigational stages, we propose to use an indicator that represents the percentage of requests, within the sliding window of ten requests, that match the given page view.

Markov Basket Addition Likelihood - MarkovLik. Thus far, every feature introduced has been directly related to session metrics or specific actions. However, we can also retrieve information from the sequence of clickstreams. Therefore, we propose a measure of basket addition likelihood by using first order Markov chains.

Markov chains, introduced by Markov [27], are mathematical systems that model the transitions between stages through transition probabilities. First order Markov chains, also known as memoryless Markov chains, are systems where the probability of the next state depends only on the current state and disregards everything that happened before. Clickstream data is naturally adapted to be modeled by Markov chains, as each session is an ordered sequence of requests. Moreover, assuming the non-randomness of these requests, Markov modeling yields detailed information regarding the probability of requesting a page view j after page view i.

We use clickstream data to build a first order Markov chain that models the transitions among different page view requests. Since we are concerned with basket addition likelihood, markovLik feature uses the values of the transition matrix that specifies the probability of the next request being a basket addition depending on the current page view. This is a static feature as the probability of requesting a basket addition, regardless of the current state, is pre-computed into the transition matrix based on historical data.

Page View Sequence Likelihood - MarkovDis. In addition to the last predictor, we try to extract further knowledge from the sequence of page views already requested by the user. We believe that, using a sliding window of 10 requests, it is possible to distinguish different phases of the buying process, namely between the search/purchase and shipping/payment stages. We split buyers' sessions so that, the requests made until the last basket addition (befAddition) and the requests made after the last basket addition (postAddition) are separated. Then, we model Markov processes for both sets of requests, in order to obtain distinct probabilistic models that describe each set. A measure of session similarity is then estimated by using Markov for discrimination, introduced by Durbin et al. [28] and used, for instance, by Migueis et al. [29] for identifying churners in offline stores.

We build, for each of the two groups of requests (befAddition and postAddition), a different transition matrix that reflects its specific page view sequences. Following Durbin et al. [28], we use these transition matrices to compute the log-odds ratio between the odds of observing a sequence x given it originates a befAddition path and the odds of observing sequence x given it belongs to the postAddition path:

In order to compute the log-odds ratio for a sequence of page views, within the sliding window of 10 requests, we use the sum operation to join the log-odds of every transaction, as depicted in Eq. 1. The outcome of S(x) allows the affinity of a visitor to be measured with respect to pre-addition and pos-addition stages of the navigation. A positive ratio indicates that the visitor is likely to add something to the basket while a negative ratio means the opposite.

$$S(R_1 \rightarrow R_2 \rightarrow (...) \rightarrow R_n) = \sum_{k=1}^{n} log \frac{P(R_k|\text{befAddition})}{P(R_k|\text{postAddition})} \qquad (1)$$

3.3   Classification Techniques

In this study we use both random forests and logistic regression as classification techniques. We have adopted these techniques as they usually represent a good compromise between prediction performance and simplicity, namely in terms of parametrization [30]. These techniques have also been benchmarked in several contexts (e.g. [31]).

3.4   Evaluation Criteria

In order to evaluate the performance of the techniques used to handle this problem, we use a validation procedure that aims to avoid overfitting. We sample 10 unique samples of the dataset and then, for each sample, we feeds our models with 80% of the data during the training phase and test with the remaining 20%.

Considering the nature of the problem tackled, the data involved is usually heavily imbalanced. In fact, considering the conversion rates already discussed only a small proportion of user session's requests correspond to basket additions. For this reason, we balanced the dataset, using a undersampling approach, in order to have about the same percentage of requests that lead to a basket addition and requests that do not.

In order to measure the performance of the proposed prediction models, we use the test set and compute the well known Receiver Operating Characteristic curve (ROC) and analyze the Area Under Curve (AUC) [32]. Moreover, we use the precision/recall curve in order to understand how this trade-off varies. We also evaluate each model with a measure that specifies the accuracy for each class separately. Finally, we also use a simple yet informative accuracy curve as an evaluation metric.

4   Case Study

This paper uses a case study to illustrate the predictive ability of the model proposed. This case study is based on clickstream data from a major European e-grocery. This clickstream data was retrieved from their commercial website that sells a vast collection of grocery products, as well as other goods such as clothes and electronic equipment. We will refer to this data as the CLM dataset throughout this paper.

4.1   Data

The CLM dataset is a month's user-session data, i.e. logs, from the company's servers. This data is both disaggregated and anonymous, in other words, we do not have access to registered user data such as demographic information and other session dependent attributes. The fact that we are working with anonymous data is important because it enables the model proposed to be applied in other grocery retailing settings without having a market over fitting. On the other hand, if further user information was available, new features could be designed and the model could perform better.

Originally, the CLM dataset contained all requests sent to the server, i.e. 212,675,331 rows of data corresponding to different session stages. Data cleaning and despidering processes resulted in 33,105,835 entries. As mentioned previously, we assumed that each session was performed by a new user. The sessionization was based on a local timeout technique based on a time threshold of ten minutes, that resulted in a total of 1,257,249 unique user sessions. Considering the page view identification method proposed (see Sect. 3.1 for further details), the total 33,105,835 entries were grouped into 876 unique page requests. These entries were then grouped according to their inherent action. This process involved a categorization of 99.32% of the requests. The remaining 0.68% were classified as undetermined/other requests and are mainly composed of unique urls to really specific pages within the site, rather than a broad user action. Table 2 shows the 18 different types of requests identified. The additional preprocessing stage introduced in Sect. 3.1, reduced the CLM dataset to a total of 25,691,403 rows that correspond to 422,618 sessions or users.

Table 2. Types of requests identified.

Action	Meaning
Homepage	Requests for the website's homepage
Login	Request to authenticate a registered user
Logout	Request to terminate an authenticated session
Sub Category Menu	Request to display a descendant sub menu, e.g. clicked on Bakery to access a sub menu with different descendants such as Bread or Cakes & Pies
Category Page	Request for the display of all products under this category
Filter	Request to apply a filter to the present list of products
Search Results	Request for a text search (through the proper input box)
Autocomplete	Request for a list of terms that complete the current search term
Product Page	Request for the specific page of a particular product
Add to Basket	Request to add a specific product to the user's basket
Remove from Basket	Request to remove a specific product to the user's basket
Checkout	Request to checkout current basket and proceed to payment
Shipping & Payment	Collection of possible requests regarding shipping, scheduling and payment
Submit	Request for final submit of current purchase
Shopping List	Collection of possible requests regarding the creation and edition of personalized user shopping lists
User Profile	Collection of possible requests regarding different modifications to the user profile
Flyers	Request for a flyer with company promotions and discounts
Other	All requests that did not fit in the previous categories

4.2 Results and Managerial Implications

As mentioned in Sect. 3.4, we used a validation strategy based on ten unique samples. Due to memory constraints, each sample was composed by 50 000 entries, i.e. session stages, out of the whole data. For each sample, we fed the classification techniques with 80% of the data during the training phase and test with the remaining 20%.

Figure 2 depicts the performance measures of both models, logistic regression and random forests, after these were trained with the balanced dataset and tested with the reserved test set. From the analysis of Fig. 2, we can conclude that anonymous purchase likelihood prediction in this context is promising. The AUC values are high both when logistic regression and random forests are used. The results obtained also show that random forests outperforms logistic regression in all performance metrics. Moreover, the AUC value for random forests is close to 2.6% higher than the logistic regression's AUC. The accuracy rate for the class that corresponds to a basket addition in the next request (positive values), is relatively high for logistic regression and random forests. For example, a cut-off value of 0.5, yields an accuracy above 80%, for both learning models. Furthermore, for both classification techniques, the accuracy related to negative cases is also high.

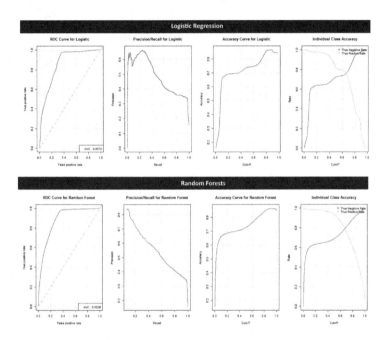

Fig. 2. Performance metrics for predicting basket additions with a balanced dataset.

From the CRM point of view, these results are very encouraging as new methods of targeting customers could be derived from the methodology proposed. In fact, we believe that the model developed within this study, from a business standpoint, is even more interesting than the models predicting customer conversion. Being able to anticipate basket additions opens new possibilities for marketing maneuvers. For example, if the model predicted a basket addition, cross-promotions can be highlighted in order to provide a better service. Furthermore, in the case of prediction of non-addition, marketers may for example suggest similar products to the one viewed and whose affinity with the user is higher.

5   Conclusions and Future Research

This study proposes a model based on data mining classification techniques, namely random forests and logistic regression, to predict whether users will add an item to their basket, given their past navigational patterns. The methodology proposed encompasses the analysis and processing of anonymous ,clickstream data gathered from an e-grocery store used as case study. The model introduced constitutes a new tool for e-commerce web analysts, as it enables to follow users' sessions and continuously evaluate purchasing intentions.

The obtained results reveal that prediction models are effective in this domain. The results show that anonymous clickstream data from online grocery retailers is enough to develop a solid model for purchase engagement prediction. The process of feature engineering revealed to be fruitful as, for instance, the model supported by random forests was able to reach performance levels of about 89% in terms of AUC.

From a business standpoint, we believe the model proposed is extremely relevant as it helps to infer users' behavior and may result in an enhancement of users' experience. Using this tool, the web owner is able to anticipate product basket additions and act according to the users' intentions. For example, in order to assist those users whose propensity to buy is low, the web owner could recommend related products whose propensity for purchase is higher.

Regarding future work, from a technical perspective, there are alternative approaches including neural networks, deep learning, categorical principal analysis and further decision tree-based methods which could be explored in order to find better predicting customer engagement performances. Furthermore, we believe that future studies may explore new methodologies for preprocessing, feature engineering and modeling. We also believe that the incorporation of other variables, e.g. demographic variables, and the access to a dataset containing all regular clickstream variables would enable to include more information in the prediction models, enhancing the resulting performance. Moreover, we believe that any information regarding the categories and/or products that users search for could also increase the performance rates as, for instance, the motivation for buying products from milk section is totally different from the motivation to buy products from the light-bulbs section.

References

1. Juliá-Igual, J.F., Cervelló-Royo, R., Berné-Lafuente, I.: Market value analysis of a Chinese e-commerce holding group: a multicriteria approach. Serv. Bus. **11**(3), 475–490 (2017)
2. Cranshaw, J., Toch, E., Hong, J., Kittur, A., Sadeh, N.: Bridging the gap between physical location and online social networks. In: Proceedings of the 12th ACM International Conference on Ubiquitous Computing, Series Bridging the Gap Between Physical Location and Online Social Networks, pp. 119–128, Copenhagen (2010)
3. Fernández-Sabiote, E., Román, S.: The multichannel customer's service experience: building satisfaction and trust. Serv. Bus. **10**(2), 423–445 (2016)
4. Reichheld, F.F., Schefter, P.: E-loyalty. Harv. Bus. Rev. **78**(4), 105–113 (2000)
5. Gudigantala, N., Bicen, P., Eom, M.: An examination of antecedents of conversion rates of e-commerce retailers. Manag. Res. Rev. **39**(1), 82–114 (2016)

6. Internet Retailer: Top 500. Technical report, Vertical Web Media, Chicago (2016). https://www.internetretailer.com/top500/#!/

7. Agudo-Peregrina, Á.F., Pascual-Miguel, F.J., Chaparro-Peláez, J.: It's never the same: the role of homogeneity in online services. Serv. Bus. **8**(3), 453–464 (2014)

8. Perdikaki, O., Kesavan, S., Swaminathan, J.M.: Effect of traffic on sales and conversion rates of retail stores. Manufact. Serv. Oper. Manag. **14**(1), 145–162 (2012)

9. McCarty, J.A., Hastak, M.: Segmentation approaches in data-mining: a comparison of RFM, CHAID, and logistic regression. J. Bus. Res. **60**(6), 656–662 (2007)

10. Bucklin, R.E., Sismeiro, C.: Click here for Internet insight: advances in clickstream data analysis in marketing. J. Interact. Mark. **23**(1), 35–48 (2009)

11. Koeltringer, C., Dickinger, A.: Analyzing destination branding and image from online sources: a web content mining approach. J. Bus. Res. **68**(9), 1836–1843 (2015)

12. Wang, K.-Y., Ting, I.-H., Wu, H.-J.: Discovering interest groups for marketing in virtual communities: an integrated approach. J. Bus. Res. **66**(9), 1360–1366 (2013)

13. Cooley, R., Mobasher, B., Srivastava, J.: Data preparation for mining world wide web browsing patterns. Knowl. Inf. Syst. **1**(1), 5–32 (2013)

14. Liu, B.: Web Data Mining: Exploring Hyperlinks, Contents, and Usage Data. Springer, Heidelberg (2007). https://doi.org/10.1007/978-3-642-19460-3

15. Laudon, K.C., Traver, G.C.: E-commerce. Pearson/Addison Wesley (2007)

16. Moe, W.W., Fader, P.S.: Dynamic conversion behavior at e-commerce sites. Manag. Sci. **50**(3), 326–335 (2004)

17. Montgomery, A.L., Li, S., Srinivasan, K., Liechty, J.C.: Modeling online browsing and path analysis using clickstream data. Mark. Sci. **23**(4), 579–595 (2004)

18. Sismeiro, C., Bucklin, R.E.: Modeling purchase behavior at an e-commerce web site: a task-completion approach. J. Mark. Res. **41**(3), 306–323 (2004)

19. Verheijden, R.: Predicting purchasing behavior throughout the clickstream. Master thesis, Technische Universiteit, Eindhoven (2012)

20. Lim, M., Byun, H., Kim, J.: A web usage mining for modeling buying behavior at a web store using network analysis. Indian J. Sci. Technol. **8**(25), 1–7 (2015)

21. Ding, A.W., Li, S., Chatterjee, P.: Learning user real-time intent for optimal dynamic web page transformation. Inf. Syst. Res. **26**(2) 339-359 (2015)

22. Chang, H.-J., Hung, L.-P., Ho, C.-L.: An anticipation model of potential customers' purchasing behavior based on clustering analysis and association rules analysis. Expert Syst. Appl. **32**(3), 753–764 (2007)

23. Wang, H.B.: An Efficient K-Means Algorithm in the Data Stream Model. CSRE Press, Athens (2005)

24. Cho, Y.H., Kim, J.K., Kim, S.H.: A personalized recommender system based on web usage mining and decision tree induction. Expert Syst. Appl. **23**(3), 329–342 (2002)

25. Aguiar, E., Nagrecha, S., Chawla, N.V.: Predicting online video engagement using clickstreams. In: 2015 IEEE International Conference on Data Science and Advanced Analytics (DSAA) 2015, vol. 36678, pp. 1–10 (2015)

26. Pachidi, S., Spruit, M., van de Weerd, I.: Understanding users' behavior with software operation data mining. Comput. Hum. Behav. **30**, 583–594 (2014)

27. Markov, A.A.: Investigation of an important case of dependent trials. Izv. Acad. **1**(6), 61–80 (1907)

28. Durbin, R., Eddy, S.R., Krogh, A., Mitchison, G.: Biological Sequence Analysis: Probabilistic Models of Proteins and Nucleic Acids. Cambridge University Press, Cambridge (1998)

29. Miguéis, V.L., den Poel, D.V., Camanho, A.S., Cunha, J.F.: Predicting partial customer churn using Markov for discrimination for modeling first purchase sequences. Adv. Data Anal. Classif. **6**(4), 337–353 (2012)
30. Trigila, A., Iadanza, C., Esposito, C., Scarascia-Mugnozza, G.: Comparison of logistic regression and random forests techniques for shallow landslide susceptibility assessment in Giampilieri (NE Sicily, Italy). Geomorphology **249**, 119–136 (2015)
31. Thalji, N.M., et al.: Assessment of coronary artery disease risk in 5463 patients undergoing cardiac surgery: when is preoperative coronary angiography necessary? J. Thorac. Cardiovasc. Surg. **146**(5), 1055–1064.e1 (2013)
32. Hanley, J.A., McNeil, B.J.: The meaning and use of the area under a receiver operating characteristic (ROC) curve. Radiology **143**(1), 29–36 (1982)

Emerging Service Technologies

Artificial Intelligence Theory in Service Management

João Reis[1]($\boxtimes$), Paula Espírito Santo[1], and Nuno Melão[2]

[1] Institute of Social and Political Sciences (ISCSP) and CAPP,
University of Lisbon, Lisbon, Portugal
jcgr@campus.ul.pt, paulaes@iscsp.ulisboa.pt
[2] Department of Management and CISED, School of Technology
and Management of Viseu, Polytechnic Institute of Viseu, Viseu, Portugal
nmelao@estgv.ipv.pt

Abstract. Artificial intelligence (AI) is expected to be more promising in the coming years, with, for example, notable gains in productivity, although there may be a significant impact on job reduction, which may jeopardize labor sustainability. Accordingly, there is a need to better understand this phenomenon and to analyze it in the light of a particular theory. However, there is a scarcity of AI theories in the service management literature. In order to obtain a better understanding of the subject, we have conducted a systematic review of the literature to provide a comprehensive analysis of the theories developed regarding AI in service management. The results have showed a wide range of theories, but not all directly related with AI; the latter are smaller in number making it difficult to draw a clear pattern. At current days, researchers are slowly advancing with new AI theories and moving away from those already in use, such as in computer science, ethics, philosophical theories, and so on.

Keywords: Artificial intelligence · Service management · Systematic literature review · Theory

1 Introduction

Artificial intelligence (AI), often associated to theory and developments of computer systems that are able to perform human tasks where intelligence is required, has captured the attention of researchers interested in learning how it could be applied to service management [1]. The extent and nature of explicit theory-driven research has been investigated in a number of social sciences disciplines. Walker *et al.* [2] are a notable example, as they have reviewed decades of publications across three operations management (OM) journals and analysed over 3,000 articles to identify which theories over time had been adopted by authors with a view to understand OM topics. To the best of the authors' knowledge, there is no similar work in the service management literature regarding AI. As result, the intention of this paper is to examine existing theories of AI in the service management literature. This theoretical advancement not yet reached to present days will certainly provide avenues for future empirical research. To that purpose, this paper suggests the following exploratory steps: first, to review

© Springer Nature Switzerland AG 2020
H. Nóvoa et al. (Eds.): IESS 2020, LNBIP 377, pp. 137–149, 2020.
https://doi.org/10.1007/978-3-030-38724-2_10

three decades of publications across service management journals and analyse over 86 articles to identify which theories have been conceptualized and adopted by researchers and practitioners so far; second, to summarize the existing body of AI theories in order to provide guidance in developing further empirical research in the field of service management. An example of AI theory is well-illustrated by Huang and Rust [3], which have developed a theory regarding jobs replacement caused by AI, as the authors relate innovation and the threat it poses to human employment. Specifically, their theory focus on four intelligence requirements for service tasks – mechanical, analytical, intuitive, and empathetic – and lays on the ways firms should decide between humans and machines for accomplishing those tasks.

The paper is structured into five sections: it starts with a general overview of the topic; then, similar research in the field is discussed, followed by a section that presents and explains the methodological procedures; it continues with a presentation of the main results, where the main theories and advancements in the field are discussed; lastly, the authors discuss the implications to theory and practice, as well as some suggestions for future research.

2 Conceptual Background

AI research has its roots in the theory of computability that was developed in 1930 [4]. Since then, the objective of developing non-biological intelligence has been a goal, especially after AI has been formally established as a field of knowledge since 1956 [5]. At first, scientists tried to frame AI as a field of applied sciences. A notable work was conducted by Simon [6], which tried to delimit the boundaries of AI, hence distinguishing it as (at least) a science, rather than engineering.

As this paper aims to analyse AI theories, we found it necessary to distinguish some elementary concepts, notably science, theory, and artificial intelligence.

Therefore, *science* is usually associated to theories or models, but science is also frequently built on mathematical formulas shaped by scientist and refined through experiments [7].

In turn, *theory* has been described as "a coherent description, explanation and representation of observed or experienced phenomenon" [8]. Thus, theories usually come from new ideas or metaphors that lead to the advancement of conceptual models, which help to better explain a topic [2].

Regarding *artificial intelligence*, it stimulated several discussions, specifically, in its relationship with humans. In this context, Turing [9] made significant efforts to distinguish "computing machinery" and "intelligence". Despite Turing's achievements in defining "computing machinery", he did not come up with a consensual definition of "intelligence". The relevance of the aforementioned research becomes even more evident from the moment AI developed computational skills to a degree that could emulate human cognition, such as medical diagnosis [10]. The above argument started from raising several philosophical questions, and challenged the theoretical basis due to the complexity required to distinguish humans and machines. The theory of mind-brain identity, which is used to claim naturalism and materialism, argued that the types of mental states are identical to the types of brain states [11]. According to Clark [11], a cognitive function (e.g. feeling pain) is identified with being in some specific neural state, thus "this leads to an uncomfortable kind of species chauvinism, because there is no reason to suppose that a being lacking neurons but equipped with a silicon brain could not feel pain or have some other ability – a simple view has it that AI vindicates the basic claims of the materialist-naturalist" [11, p.1].

Over time, AI has challenged the human brain in competitive games, raising expectations beyond imagined. AlphaGo, recently developed by Google DeepMind, a program that "combines deep neural networks that are trained by a novel combination of supervised learning (human experts) and reinforcement learning (self-play) played Go at the level of state-of-the-art Monte Carlo tree search programs" [12, p. 484]. In March 2016, the AlphaGo defeated the world Go champion, Sedol Lee, with a 4:1 score, attracting a new wave of global attention to the system's ability in exceeding humans' intelligence [13]. Cumulatively, much progress towards AI has been made using supervised learning systems and on current days it has evolved to unsupervised forms, where AlphaGo became its own teacher [14].

In addition to technological advances, the latest scientific research indicates that fears begin to emerge in several areas [15–17], such as the replacement of human labour by AI and/or robotics [18]. According to Frank et al. [19], while AI and automation can improve and augment productivity of some workers, it can also replace the work performed by others and will likely transform almost all occupations at least to some degree [19]. In that regards, there are numerous studies, with some examples being related to the: (1) healthcare sector – there is already a robotic revolution happening in healthcare wherein robots have made tasks and procedures more efficient and safer [20]; (2) Manufacturing and production – the McKinsey Institute is expecting that occupations incorporating significant amounts of physical work in predictable environments, including production workers, are likely to face significant displacement of their activities by automation [21]. Instead, healthcare providers and professionals including engineers and business specialists are less likely to experience as much displacement [21]; (3) Transportation and logistics occupations – Frey and Osborne [22] are estimating that around 47 percent of US employment is in high risk category among 702 occupations. According to the aforementioned authors, their model predicts that most workers in transportation and logistics occupations, together with the bulk of office and administrative support workers, and labour in production occupations, are at risk [22, p. 268]. More surprisingly, they found that a substantial portion of employment in service occupations, where most US employment growth has occurred in recent decades, is highly susceptible to robotization (i.e. service robots) and the gradual decline in the comparative advantage of human labour [22, p. 269]. As "AI is increasingly reshaping service by performing various tasks, constituting a major source of innovation yet threatening human jobs" [3, p. 1], Huang and Rust [3] proposed a theory of AI job replacement to address this two-edged impact. It is essentially to address an observed phenomenon (i.e. increase in unemployment) that the aforementioned theory came to better explain the topic and align current and future scientific research.

3 Methodology

This paper systematically retrieved a sample of literature from two different peer-reviewed databases, namely: Thomson Reuters' Web of Science (WoS) and Scopus. The first database was selected as it provides a detailed set of meta-data, essential to the analysis (e.g. journals, dates, citations) that is not readily available in other databases [23]. In turn, Scopus has the ability to provide a wider coverage as it indexes a broader range of AI journals [17], when compared with the previous one. The database selection method was motivated by the following *systematic literature review* conceptualization: "systematic, explicit, and reproducibly methods for identifying, evaluating, and synthesizing the existing body of completed and recorded work produced by researchers, scholars and practitioners" [24, p. 3]. Accordingly, the researchers used the "Preferred Reporting Items for Systematic Reviews and Meta-Analysis" (PRISMA), which helps summarizing the existing evidence [25] according to an explicit, rigorous, and transparent step-wise process [26], and is also used by Dekker and Bekker [27], Tursunbayeva *et al.* [28], and Mergel *et al.* [29], among others.

The data search was conducted on May 5th, 2019, and traced 1,873 manuscripts, by using the keywords "Artificial Intelligence" AND "Theory" AND "Service" in topic (Title-Abstracts-Keywords). Table 1 shows the literature review process.

A relevant discrepancy between Scopus (166,133) and WoS (477) was observed, since the search in WoS focused on its core collection, which excluded e.g. Derwent Innovations Index, KCI-Korean Journal Database, among others. We reduced the number of hits by limiting the search to language (English), in order to avoid misinterpretations, but also the source type (Journals) and documents type (Article) to ensure scientific rigor, focusing on peer-reviewed articles. After removing repeated articles, the final search lead to a number of 86 references. Although the reduction from 1873 articles to 86 references seems drastic, this cut is mainly due to the application of filters that implied a higher quality of the systematic review, such as the focus on journal articles instead of including proceedings, among other sources.

4 Findings

AI is a "field whose ultimate goal has often been somewhat ill-defined and subject to dispute" [30, p. 57]. It may be currently defined as an "interdisciplinary science studying the development of the simulation, extension, and expansion of human intelligence applications, methods, theories, and technologies" [31, p. 286]. While AI and cognitive science, at a first glance, are two distinct disciplines with different goals [32], on the classical perspective they are two sides of the same coin, and is now acceptable to discuss human-level AI in machines as a real possibility [33]. Therefore, it is not surprising that we have verified a wide range of theories, not all directly related to AI. In fact, AI theories are much smaller in number, which makes it difficult to draw a clear pattern. At current days, AI is slowly advancing new theories (Table 2) under the subject areas of "business, management & accounting", "decision sciences" and "social sciences" – our own selection (Table 1). Moreover, Scopus automatically provided the results according to minor subject areas[1], as shown in Fig. 1. Not surprisingly, 20.5% of the articles are published in titles that are classified under the computer science topic – which is still the predominant subject area; 19% to business, management & accounting; and 17,9% to decision sciences.

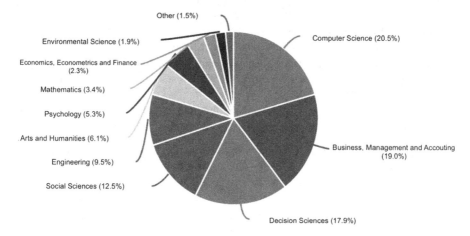

Fig. 1. Document by subject area (Source: Scopus database)

We have also noticed that the 86 references were published in 58 different source titles, and from those source titles just 4 of them are categorized under the "Artificial Intelligence" subject category[2], namely: Knowledge-based Systems – Elsevier (6 published articles); Information Sciences – Elsevier (3 published articles); Artificial Intelligence – Elsevier (2 published articles); Kybernetes – Emerald (1 published article). Unfortunately, from a multidisciplinary perspective, the AI own theoretic field has still no significant expression in the service management literature and, consequently, that has been reflected in the theories that we found. As a result, this situation may be the major gap we found in the literature, although we are convinced that this may be the moment of the inflection stage.

4.1   Newest AI Theories

In line with the above, we identified just a couple of AI theories. The analysed articles often discussed the application of AI-based techniques to solve specific management issues; for instance, in logistics or transports [34], rather than to develop new theories (Table 2). However, we do believe that AI theories and topics are changing over time,

---

[1] Titles on Scopus are classified under four board subject clusters (life sciences, physical sciences, health sciences and social sciences & humanities), which are further divided into 27 major subject areas and 300+ minor subject areas [35].

[2] The subject category "Artificial Intelligence" includes 797 titles (source: Scimago).

gaining terrain in their own theoretic field. Why? Much similar to Walker et al.'s [2] work, we expect that theories may be influenced at the level of the AI community, such as:

- Journals calling for special issues related to AI theories – as e.g. Special Issue on Explainable Artificial Intelligence from the Artificial Intelligence Journal, which encouraged submissions concerning the theoretical, philosophical and methodological field of Explainable AI [36];
- It is expected that journal editors may start to have preferences for certain AI theories, rather than importing theories from other fields (Sect. 4.2.). AI is becoming a hot research topic; therefore, the AI will start having a greater impact in its own theoretic field;
- Calls for papers, which expect to cover the theoretical and philosophical field – as e.g. "Conference on Philosophy and Theory of Artificial Intelligence" [37].

The systematic literature review has disclosed two artificial intelligence theories applied to service management, namely: Huang and Rust's [3] article entitled "Artificial intelligence in service" and Mbecke's [38] article entitled "Resolving the service delivery dilemma in South Africa through a cohesive service delivery theory" (Table 2).

The first theory has provided very relevant insights, covering fears associated to AI-technologies, that were already addressed in the conceptual background, such as the human job replacement by machines. Thus, Huang and Rust [3] developed a theory for understanding the nature of service work and how/why these technologies can substitute humans. According to the authors "the theory of AI job replacement provides a road map about how AI advances to take over tasks requiring different intelligences, how AI can and should be used to perform service tasks, and finally how workers can and should shift their skills to achieve a win-win between humans and machines" [3, p. 14]. The second theory makes an attempt to understand the phenomenon and create scenarios to solve several identified issues, in pursue of a sustainable service delivery system in South Africa. To that purpose, the author identified six relevant factors: (1) good governance; (2) public participation; e-governance; (3) planning, monitoring and evaluation; (4) government resources; and (5) laws, policies, regulations and practices [38, p. 274]. This theory is limited in two ways: first, it is not identified as an AI theory for service delivery systems, although it is developed using a Bayesian network system, an artificial intelligence tool that provided solutions to a specific issue related to public service management; second, this theory may not be applicable to other countries and contexts.

As we have mentioned, we may be experiencing an inflection point, given that AI theories are emerging in service management, albeit in a very small number. According to Schmenner et al. [39, p. 339] these theories can "explain facts and provide stories as to how phenomena work the way that they do. They can, and should, be used to make predictions", being very relevant for AI to assert itself as a field of knowledge.

Table 1. PRISMA Flow Chart

Databases:		Web of Science	Scopus
1st phase: identification			
Selected keywords	"Artificial Intelligence" AND "Theory" AND "Service" (all fields)	477 $\sum$ 166,610	166,133
2nd Phase: Screening			
Search	Title-abstract-keywords	50	1,823
Language	English	49	1,766
Source type	Journals	–	413
Document type	Articles	23	377
Major subject areas	Business, Management & Accounting; Decision Science (DS); Social Sciences (SS) Besides computer science, DS it is the only major subject on Scopus that integrates the AI category| SS is a discipline which typically includes management	3	85
3rd Phase: Eligibility			
Full-text articles assessed for eligibility	Exclusion criteria – repeated articles are excluded	–	2
4th Phase: Inclusion			
Journal articles to be included	Remaining articles	86	

Table 2. Newest (related) artificial intelligence theories in service management.

Author	Date	Theory
Huang and Rust	2018	*Theory of AI Job replacement*
Definition		
"The AI Job replacement theory addresses a double-edged impact. The theory specifies four intelligences required for service tasks – mechanical, analytical, intuitive, and empathetic – and lays out the way firms should decide between humans and machines for accomplishing those tasks" (p. 155)		
Relation with artificial intelligence		
"The theory asserts that AI job replacement occurs fundamentally at the task level, rather than the job level, and for "lower" (easier for AI) intelligence tasks first. AI first replaces some of a service job's tasks, a transition stage seen as augmentation, and then progresses to replace human labour entirely when it has the ability to take over all of a job's tasks" (p. 155)		
Relation with management		
"An important implication from our theory is that analytical skills will become less important, as AI takes over more analytical tasks, giving the "softer" intuitive and empathetic skills even more importance for service employees" (p. 155)		

(*continued*)

Table 2. (*continued*)

Author	Date	Related theory
Mbecke	2014	*Cohesive service delivery theory (CSDT)*

Definition
"The CSDT is an information, communication and technologies (ICTs)-based theory that facilitates the understanding, improvement and sustaining of service delivery system." (p. 266)

Relation with artificial intelligence
"The theory is developed using the Bayesian networks system, an artificial intelligence tool that facilitates the definition, quantification and combination of different factors contributing to service delivery" (p. 265)

Relation with management
The objective of this theory is to improve the service delivery system, thus the theory suggests different scenarios showing the contribution of established factors that influences the service delivery.

Table 3. Artificial intelligence theories which support service management research.

Author(s)	Date	AI-related theory
Nilashi et al.	2015	*Fuzzy Logic*

Definition
Fuzzy logic is an alternative to Boolean logic that determines the membership to a given class by either a 0 (no) or a 1 (yes), this is the basis of modern computing [40].

Supporting purpose
"It supported the research in the extent that reveal the real importance level of trust factors on customers' trust and decision making in selecting the appropriate trusted website. By considering interrelationships among the trust factors, analytic network process is employed for selecting the appropriate website for mobile commerce. Then, using fuzzy set theory, a Fuzzy Inferences System (FIS) is developed for revealing the real importance level of factors. The outcome of the proposed decision-making system helps shopping websites managers and service providers to ascertain the trust level of their websites and adequately allow them to improve the website quality." (p. 57)

Author(s)	Date	AI-related theory
Li et al.	2014	*Kernel-based Machine Learning Paradigm*

Definition
"Multi-theoretical kernel-based approach enables to map various theories into a uniform kernel form and convert the recommendation problem to a kernel-based machine learning problem" (p. 96)

Supporting purpose
"The rapid development of online social networking services provides an opportunity to explore social networks together with information used in traditional recommender systems. To tackle this social network-based recommendation problem, previous studies generally built trust models in light of the social influence theory. Li et al. [41] study inspected a spectrum of social network theories to systematically model the multiple facets of a social network and infer user preferences. In order to effectively make use of these heterogonous theories, the authors took a kernel-based machine learning paradigm, design and select kernels describing individual similarities according to social network theories, and employ a non-linear multiple kernel learning algorithm to combine the kernels into a unified model" (p. 95)

(*continued*)

Table 3. (*continued*)

Author(s)	Date	AI-related theory
Velaga et al.	2012	*Formal Argumentation Theory*

Definition

"Argumentation has come to be increasingly central as a core study within AI" [42, p. 619]. "It is, of course, the case that similar issues underpin one well-established and highly-developed theory: that of formal logic and mathematical proof. It is no coincidence that much of the formal computational treatment of argumentation has its roots in ideas developed from AI inspired contributions to logic and deductive reasoning. So one finds in mathematical proof theory core concepts such as: precisely defined means for expressing assertions (e.g. formulae in a given logical language); accepted bases on which to build theorems (e.g. collections of axioms); procedures prescribing the means by which further theorems may be derived from existing theorems and axioms (e.g. templates for inference rules); and precise concepts of termination (e.g. a sentential form is derivable as a theorem, "true"; or is logically invalid, "false")" [42, p. 620]

Supporting purpose

"Flexible transport systems (FTS) offer a promising approach to improving the efficiency and performance of passenger transportation services. Formal argumentation is a powerful technique borrowed from artificial intelligence, and in this context is used to weigh-up the conflicting choices available to both passengers and service providers" (p. 62)

Author(s)	Date	AI-related theory
Abad-Grau and Arias-Aranda	2006	*Bayesian Decision Theory*

Definition

Bayesian networks (BN) are widely accepted as an artificial intelligence tool for capturing uncertainty in problem solving [42]. "BN represent a general pattern of inference, which can be suited to particular studies" [44, p. 461]. By disaggregating an inference problem into smaller "problem modules" which are solved separately, it is possible to obtain solutions for the larger problem [45].

Supporting purpose

"Different learning algorithms based on BN were applied to extract relevant information about the relationship between operations strategy and flexibility in a sample of engineering consulting firms. Feature selection algorithms automatically were able to improve the accuracy of these classifiers" [44, p. 460].

Original author	Date	AI-related theory
Wu, R.	1997	*Neural Network Theory*

Definition

Consists on processing units or nodes that are interconnected in a network, similar to brain neurons. Neural networks adjust input weights to improve performance, while this capability to lean is essential for many of intelligent activities [46].

Supporting purpose

The author investigated the possibility of applying artificial intelligence to solve an audit decision problem faced by the public sector (Internal Revenue Service's – IRS) when targeting firms for further investigation. The author propose that the neural network will overcome problems faced by a direct knowledge acquisition method in building an expert system to preserve the expertise of senior auditors of the IRS. The results strongly suggest that neural networks can be used to identify firms requiring further auditing investigation.

4.2   Importing Theories from Other Fields

We have found that AI-related theories had several alternative purposes, such as developing: frameworks [47], theoretical prepositions [48], models [41, 49], decision-making systems [50], algorithms [51, 52] (...), or even used as research methodologies [53, 54]. Overall, the most relevant AI supporting theories are presented in Table 3, namely: Fuzzy logic [50]; Kernel-based machine learning paradigm [41]; Formal argumentation theory [55]; Bayesian decision theory [43]; and Neural network theory [45]. All the other 34 identified theories do not have any direct relation with AI (e.g. Grounded theory, Game theory).

All the selected studies (Table 3) are empirical, based on theoretical testing and not on theoretical building. We found a trend to investigating real-life phenomena (e.g. audit decision problem faced by the public sector), using AI tools from other areas of knowledge (e.g. Fuzzy logic – from computer science), to solve complex problems related to service management.

The rise of "automation is happening in a period of growing economic inequality, raising fears of mass technological unemployment and renewed call for policy efforts to address the consequences of technological change" [19, p. 6531]. To cope with the aforementioned argument, it is of paramount relevance to fill a gap in the literature that is related to the scarcity of AI theories in service management. In that regard, some researchers are already making some efforts to slowly develop new AI theories, consistent and appropriate to real life, moving away from existing theories from other fields. However, the truth is that the results showed a wide range of theories, but not all directly related to AI, so that the latter are smaller in number, making it difficult to define a clear pattern.

5   Concluding Remarks

The authors found the majority of the selected studies to be empirical, upon theory testing rather than theory building. The findings have suggested that AI in service management clearly needs to be developed from a theoretical perspective. Consequently, we strongly believe that there will be an increasing demand for theoretical studies in a near future, able to gather supporting evidence and guidelines for empirical research, which is being identified as the preferred method of conducting research by both academics and practitioners.

This study makes novel contributions by exploring the AI main theories that have been adopted by services management scholars, researchers and practitioners. In doing so, and by systematically analysing articles from the selected peer-reviewed databases, we have also identified some limitations. For instance, the literature review was confined to specific keywords, and thus some articles may be missing. We would also like to point out that this is an exploratory research that will be deepened in a next phase using the systematic literature review procedures of Kitchenham [56]. The need to deepen knowledge is due to the few theories found, but also because other keywords can be used; for example, it is possible to find articles that use specific AI techniques applied to the Service Management but that do not refer to it as "Theory".

Moreover, the selected databases are constantly being updated with new and relevant literature, therefore, our findings present a snapshot of a specific period of time. To mitigate some of the identified limitations we have adopted some strategies, as the inclusion of relevant articles besides the ones listed in the databases, for corroboration purposes. New avenues for future research almost necessarily lead to AI's theoretical advances in the field of service management, as a very few theories have been discovered in scientific journals articles indexed to major databases. Empirical articles on artificial intelligence in service management have focused on generic theories, which shows a great opportunity to publication on more focused theories to support the growing demand in the field of AI. If we consider one of the areas of greatest technological growth, such as the public services [57, 58], we found that, to analyse the impacts of the implementation of new technologies, there is theoretical support in relation to employment. However, there are other areas, as suggested by Reis et al. [17] that are largely poorly studied, such as the impact on political leadership or the quality of citizens' life, which lack adequate theoretical support.

References

1. JSM: AI and machine learning in service management. Spec. Issue Call Pap. J. Serv. Manag. Emerald J. http://www.emeraldgrouppublishing.com/products/journals/call_for_papers.htm?id=8053
2. Walker, H., Chicksand, D., Radnor, Z., Watson, G.: Theoretical perspectives in operations management: an analysis of the literature. Int. J. Oper. Prod. Manag. **35**(8), 1182–1206 (2015)
3. Huang, M., Rust, R.: Artificial intelligence in service. J. Serv. Res. **21**(2), 155–172 (2018)
4. Gershman, S., Horvitz, E., Tenenbaum, J.: Computational rationality: a converging paradigm for intelligence in brains, minds, and machines. Science **349**(6245), 273–278 (2015)
5. Spector, L.: Evolution of artificial intelligence. Artif. Intell. **170**(18), 1251–1253 (2006)
6. Simon, H.: Artificial intelligence: an empirical science. Artif. Intell. **77**, 95–127 (1995)
7. Desjardins-Proulx, P., Poisot, T., Gravel, D.: Scientific theories and artificial intelligence. *BioRxiv*, 161125 (2017)
8. Gioia, D., Pitre, E.: Multiparadigm perspective on theory building. Acad. Manag. Rev. **15**(4), 584–602 (1990)
9. Turing, A.: Computing machinery and intelligence. Mind **59**(236), 433–460 (1950)
10. Fox, J.: Expert systems and theories of knowledge. In: Artificial Intelligence. pp. 157–181. Academic Press (1996)
11. Clark, A.: Philosophical foundations. In: Artificial Intelligence, pp. 1–22. Academic Press (1996)
12. Silver, D., et al.: Mastering the game of Go with deep neural networks and tree search. Nature **529**(7587), 484–489 (2016)
13. Pan, Y.: Heading toward artificial intelligence 2.0. Engineering **2**(4), 409–413 (2016)
14. Silver, D., et al.: Mastering the game of Go without human knowledge. Nature **550**(7676), 354 (2017)
15. Jarrahi, M.: Artificial intelligence and the future of work: human-AI symbiosis in organizational decision making. Bus. Horiz. **61**(4), 577–586 (2018)
16. Petropoulos, G.: The impact of artificial intelligence on employment. Praise for Work in the Digital Age, p. 119 (2018)

17. Reis, J., Santo, P., Melão, N.: Impact of artificial intelligence on public administration: a systematic literature review. In: 2019 14th Iberian Conference on Information Systems and Technologies (CISTI), pp. 1–7. IEEE (2019) https://doi.org/10.23919/cisti.2019.8760893

18. Dhanabalan, T., Sathish, A.: Transforming Indian industries through artificial intelligence and robotics in industry 4.0. Int. J. Mech. Eng. Technol. **9**(10), 835–845 (2018)

19. Frank, M., et al.: Toward understanding the impact of artificial intelligence on labor. Proc. Nat. Acad. Sci. **116**(14), 6531–6539 (2019)

20. Pepito, J., Locsin, R.: Can nurses remain relevant in a technologically advanced future? Int. J. Nurs. Sci. **6**(1), 106–110 (2019)

21. Manyika, J., et al.: Jobs lost, jobs gained: workforce transitions in a time of automation. McKinsey Global Institute (2017)

22. Frey, C., Osborne, M.: The future of employment: how susceptible are jobs to computerisation? Technol. Forecast. Soc. Change **114**, 254–280 (2017)

23. Brones, F., Carvalho, M., Zancul, E.: Ecodesign in project management: a missing link for the integration of sustainability in product development? J. Cleaner Prod. **80**, 106–118 (2014)

24. Fink, A.: Conducting Research Literature Reviews: From the Internet to Paper, 3rd edn. Sage, London (2010)

25. Moher, D., Liberati, A., Tetzlaff, J., Altman, D.: Preferred reporting items for systematic reviews and meta-analysis: the PRISMA statement. Ann. Intern. Med. **151**(4), 264–269 (2009)

26. Liberati, A., et al.: The PRISMA statement for reporting systematic reviews and meta-analyses of studies that evaluate health care interventions: explanation and elaboration. PLoS Med. **6**(7), e1000100 (2009)

27. Dekker, R., Bekkers, V.: The contingency of governments' responsiveness to the virtual public sphere: a systematic literature review and meta-synthesis. Gov. Inf. Quart. **32**(4), 496–505 (2015)

28. Tursunbayeva, A., Franco, M., Pagliari, C.: Use of social media for e-Government in the public health sector: a systematic review of published studies. Gov. Inf. Quart. **34**(2), 270–282 (2017)

29. Mergel, I., Gong, Y., Bertot, J.: Agile government: systematic literature review and future research. Gov. Inf. Quar. **35**(2), 291–298 (2018)

30. Russell, S.: Rationality and intelligence. Artif. Intell. **94**(1–2), 57–77 (1997)

31. Li, L.: Evolving academic libraries in the future. In: Scholarly Information Discovery in the Networked Academic Learning Environment, pp 279–309, Elsevier (2014)

32. Feldman, J.: Artificial intelligence in cognitive science. In: International Encyclopedia of the Social & Behavioral Sciences, vol. 2., pp. 792–796, Elsevier Science (2001)

33. Müller, V.: Introduction: philosophy and theory of artificial intelligence. Minds Mach. **22**, 67–69 (2012)

34. Neuroth, M., MacConnell, P., Stronach, F., Vamplew, P.: Improved modelling and control of oil and gas transport facility operations using artificial intelligence. In: Ellis, R., Moulton, M., Coenen, F. (eds.) Applications and Innovations in Intelligent Systems VII, pp. 119–136). Springer, London (2000). https://doi.org/10.1007/978-1-4471-0465-0_8

35. Scopus, S.: Content coverage guide, pp. 1–28 (2017). https://www.elsevier.com/__data/assets/pdf_file/0007/69451/0597-Scopus-Content-Coverage-Guide-US-LETTER-v4-HI-singles-no-ticks.pdf

36. Artificial Intelligence: Special Issue on Explainable Artificial Intelligence. Elsevier. https://www.journals.elsevier.com/artificial-intelligence/call-for-papers/special-issue-on-explainable-artificial-intelligence. Accessed 16 Nov 2019

37. Philosophy & Theory of Artificial Intelligence. In: 3rd Conference on Philosophy and Theory of Artificial Intelligence. http://www.pt-ai.org/2017/. Accessed 16 Nov 2019

38. Mbecke, Z.: Resolving the service delivery dilemma in South Africa through a cohesive service delivery theory. Probl. Perspect. Manag. **12**(4-si) 265–275 (2014)
39. Schmenner, R., Van Wassenhove, L., Ketokivi, M., Heyl, J., Lusch, R.: Too much theory, not enough understanding. J. Oper. Manag. **27**(5), 339–343 (2009)
40. Thompson, J.A., Roecker, S., Grunwald, S., Owens, P.: Digital soil mapping: interactions with and applications for hydropedology. Hydropedology, pp. 665–709 (2012)
41. Li, X., Wang, M., Liang, T.: A multi-theoretical kernel-based approach to social network-based recommendation. Decis. Support Syst. **65**, 95–104 (2014)
42. Bench-Capon, T., Dunne, P.: Argumentation in artificial intelligence. Artif. Intell. **171**(10–15), 619–641 (2007)
43. Nadkarni, S., Shenoy, P.: A Bayesian network approach to make inferences in causal maps. Eur. J. Oper. Res. **128**, 479–498 (2001)
44. Abad-Grau, M., Arias-Aranda, D.: Operations strategy and flexibility: modeling with Bayesian classifiers. Indu. Manag. Data Syst. **106**(4), 460–484 (2006)
45. Pearl, J.: Probabilistic inference in intelligent systems. In: Networks of Plausible Inference. Morgan Kaufmann, San Mateo (1988)
46. Wu, R.: Neural network models: foundations and applications to an audit decision problem. Ann. Oper. Res. **75**, 291–301 (1997)
47. Katerna, O.: Conceptual framework for the formation of the integrated intelligent transport system in Ukraine. Економічний часопис-XXI **158**(3–4(2)), 31–34 (2016)
48. Chae, B.: A complexity theory approach to IT-enabled services (IESs) and service innovation: business analytics as an illustration of IES. Decis. Support Syst. **57**, 1–10 (2014)
49. Azadeh, A., Darivandi Shoushtari, K., Saberi, M., Teimoury, E.: an integrated artificial neural network and system dynamics approach in support of the viable system model to enhance industrial intelligence: the case of a large broiler industry. Syst. Res. Behav. Sci. **31**(2), 236–257 (2014)
50. Nilashi, M., Ibrahim, O., Mirabi, V., Ebrahimi, L., Zare, M.: The role of security, design and content factors on customer trust in mobile commerce. J. Retail. Consum. Serv. **26**, 57–69 (2015)
51. Hajipour, V., Farahani, R., Fattahi, P.: Bi-objective vibration damping optimization for congested location–pricing problem. Comput. Oper. Res. **70**, 87–100 (2016)
52. Liu, Z., Chu, D., Song, C., Xue, X., Lu, B.: Social learning optimization (SLO) algorithm paradigm and its application in QoS-aware cloud service composition. Inf. Sci. **326**, 315–333 (2016)
53. Schockaert, S., De Cock, M., Kerre, E.: Location approximation for local search services using natural language hints. Int. J. Geogr. Inf. Sci. **22**(3), 315–336 (2008)
54. Abubakar, A., Behravesh, E., Rezapouraghdam, H., Yildiz, S.: Applying artificial intelligence technique to predict knowledge hiding behavior. Int. J. Inf. Manag. **49**, 45–57 (2019)
55. Velaga, N., Rotstein, N., Oren, N., Nelson, J.D., Norman, T., Wright, S.: Development of an integrated flexible transport systems platform for rural areas using argumentation theory. Res. Transp. Bus. Manag. **3**, 62–70 (2012)
56. Kitchenham, B.: Procedures for Performing Systematic Reviews. vol. 33, pp. 1–26. Keele University, Keele (2004)
57. Reis, J., Amorim, M., Melão, N., Matos, P.: Digital transformation: a literature review and guidelines for future research. In: Rocha, Á., Adeli, H., Reis, L.P., Costanzo, S. (eds.) WorldCIST'18 2018. AISC, vol. 745, pp. 411–421. Springer, Cham (2018). https://doi.org/10.1007/978-3-319-77703-0_41
58. Reis, J., Amorim, M., Melão, N., Cohen, Y., Rodrigues, M.: Digitalization: a literature review and research agenda. In: Lecture Notes on Multidisciplinary Industrial Engineering (2020, forthcoming)

Conceptualizing the Role of Blockchain Technology in Digital Platform Business

Tim Schulze, Stefan Seebacher$^{(\boxtimes)}$, and Fabian Hunke

Institute of Information Systems and Marketing (IISM)
and Karlsruhe Service Research Institute (KSRI),
Karlsruhe Institute of Technology (KIT), Karlsruhe, Germany
tim.schulze@alumni.kit.edu,
{stefan.seebacher,fabian.hunke}@kit.edu

Abstract. Digital platforms have firmly changed the way we interact within the last decade and, thereby, created an entire platform economy. Services and offerings on such platforms are typically facilitated by intermediaries. While blockchain technology can serve as means for creating a digital platform, it makes these intermediaries obsolete by creating trust through the system itself. Yet, blockchain is often referred to as "innovative technology in search of use cases" – a statement which indicates that a real-world impact of blockchain is still missing. This research, therefore, intends to address this gap with a taxonomy of blockchain platforms. It thereby aims to provide researchers and practitioners alike with a structured and holistic approach for studying practical use cases to allow for a better understanding of blockchain-based platforms.

Keywords: Blockchain technology · Digital platform · Taxonomy

1 Introduction

Digitalization fostered the adoption of digital platforms as an "extensible, digital medium of exchange for products, information, and services" [1]. As of today, digital platforms have successfully disrupted a variety of industries, such as transportation, trading, or software development [2] and gave rise to some of the most valued companies, such as Apple, Facebook or Amazon [3]. This demonstrates the importance of digital platforms as a means of structuring and organizing interactions [4]. They provide the means for service-for-service exchange [5], an can, therefore, be regarded as nucleus for the development of a service ecosystem.

Digital platforms are typically owned or controlled by organizations that convey interactions between different parties. Such organizations can be viewed as third parties or intermediaries. Especially when transferring ownership rights or processing transactions in the contemporary world, intermediaries are needed. In turn, the involved parties need to trust the intermediaries to reliably act in their interests. While blockchain is a suitable technology for being adopted in the platform business, enabling the emerging type of blockchain-based platforms [1], it renders these intermediaries obsolete by creating trust through the system itself [6].

© Springer Nature Switzerland AG 2020
H. Nóvoa et al. (Eds.): IESS 2020, LNBIP 377, pp. 150–163, 2020.
https://doi.org/10.1007/978-3-030-38724-2_11

Blockchain is expected to disrupt existing business models, while potentially creating new opportunities for startups. Affected industries include the financial sector, supply chain management, insurance, digital knowledge management, e-business, e-commerce and the public sector [7, 8]. This reflects the potential blockchain applications have in parts of nearly every industry sector.

Yet, due to the uncertainty about the advantages and disadvantages and a lack of common knowledge of blockchain, a real-world impact of blockchain is still missing [9]. Therefore, it is often stated that "blockchain is an innovative technology in search of use cases" [7]. This phrase illustrates the great potential of blockchain, the current lack of suitable use cases and the need to identify such.

Therefore, a conceptualization of the role of blockchain technology in digital platform business is necessary. To achieve this, we seek to answer the following research question: *What are significant dimensions and characteristics to describe blockchain platforms?* In order to answer the research question, we develop a taxonomy of blockchain platforms in several iterations. The taxonomy development is based on a structured literature review and the analysis of practical blockchain applications.

This research aims at supporting researchers in the understanding of existing blockchain use cases while also enabling practitioners to develop new ideas for the application of this technology. It is conducive to the analysis of upcoming blockchain systems, facilitating their comparison as well as the identification of common characteristics among them. With the aid of a taxonomy, differences and similarities become clear and comprehensive statements about blockchain use cases can be made. Business leaders may use the findings of this research to determine whether the application of blockchain is reasonable in a future scenario.

The remainder of this paper is structured as follows: Sect. 2 provides foundational knowledge and definitions of digital platforms as well as blockchain technology. It further presents related work of both domains. Section 3 outlines the applied methodology and the research process of this work. Then, Sect. 4 presents the final taxonomy of blockchain platforms, while Sect. 5 demonstrates the practical applicability of the taxonomy by classifying the Libra blockchain platform. Section 6 discusses the implications on research and practice of this work. Lastly, Sect. 7 summarizes the findings and lays out avenues for future research.

2 Related Work

This research intends to contribute to the conceptualization of blockchain-based platforms. Therefore, we first provide foundations on digital platforms as well as related literature on taxonomies for such platforms. Second, we give an overview of the functionality of blockchain technology and extant taxonomies of blockchain systems. While all of them contribute to a different aspect of the technology, they miss out on addressing blockchain in context of digital platforms.

2.1 Digital Platforms

Digital platforms drastically change how human activities and value creation are structured and carried out [10]. A sound understanding of digital platforms helps to grasp the potential and importance of this technology. We follow the idea of Constantinides et al. [3] and define digital platforms as "a set of digital resources—including services and content—that enable value-creating interactions between external producers and consumers" [3]. These interactions can be of social and economic nature [10] and are an important task of digital platforms [11]. Consumers, producers, service providers, or third-party developers, among others, can interact on a platform and create added value through these interactions [3].

From a technical perspective, a digital platform can be divided into a stable core and variable components. The platform core is defined by Baldwin and Woodard [12] as "a set of stable components that supports variety and evolvability in a system by constraining the linkages among the other components". The stable components in the core create variety and evolvability by governing the relations between the external components with the help of constraints and design rules [12]. The components, also known as complements, exhibit a high variety and low reusability as they are created by the users of the platform during its evolution; the complements are therefore not part of the platform core itself [12, 13]. In conclusion, a platform can be seen as the unity of technical elements, organizational processes, and standards [11].

Various approaches have been made by researchers to structure and understand digital platforms. Amongst others, Blaschke et al. [14] develop a taxonomy for digital platforms from an architectural perspective. They identify high-level infrastructure and core characteristics of digital platforms in order to provide a better understanding and grouping of digital platforms into a coherent structure. By analyzing 100 practical use cases, Täuscher and Laudien [15] are able to develop a taxonomy for platform business models. The focus of their research lies on how value is created, delivered, and captured on digital and non-digital platforms. Gawer [16] develops a classification of technological platforms, thereby including digital platforms as well. She differentiates between internal platforms, supply-chain platforms, and industry platforms. Next to foundational architectural characteristics, other non-technical characteristics are addressed, such as types of agents and their coordination on a platform.

2.2 Blockchain-Based Platforms

Albeit platforms in general and digital platforms in specific are assessed from various directions through taxonomies, blockchain technology introduces unique characteristics and new perspectives that require a reassessment of blockchain in the context of digital platforms [11].

A blockchain is a distributed, transactional database [7], decentrally stored across a peer-to-peer network [17]. This distributed database consists of blocks containing the transactions created by the network participants [6, 18]. When a new block with new transactions is proposed to be appended to the existing chain [7], consensus about the validity of the new block has to be reached across the network participants. Once it is added to the blockchain [6] it cannot be changed anymore, which makes a blockchain immutable and tamper-resistant without the need for third-party supervision. Therefore, blockchain presents itself as novel technology with great potential. Through its unique features, it allows for decentralization and disintermediation [7]. Although, this might seem inconsistent with the design of previous digital platforms, Glaser et al. [1] underscore its suitability for today's digital platform economy. They further suggest research on blockchain platforms in a sociotechnical manner.

To conceptualize the unique properties of blockchain technology, extant research has provided a diverse set of taxonomies. In their research of decentralized consensus systems, Glaser and Bezzenberger [19] develop a taxonomy of such systems. They, however, do not assess the interrelation of blockchain and digital platforms. Rückeshäuser [20] investigated blockchain startups and corporations in order to create a typology of blockchain business models. She further argues for the simultaneous analysis of the distributed ledger and business model in future research. Beinke et al. [21] pursue a similar path and develop a taxonomy of blockchain startups with particular emphasis on the nature of the related business models. By doing so, the authors derive five dimensions to describe business models. The authors call for further research to include other industries and the analysis of further business model characteristics. Labazova et al. [22] identify important technical characteristics of blockchain applications, but do not relate their findings to non-technical, organizational aspects. However, the authors explain the limited success of blockchain-based systems to date as a result of a missing understanding of blockchain technology and its capabilities. They propose that future research fills this gap. By investigating 53 blockchain applications and 17 papers, Blossey et al. [23] derive high-level use case clusters for blockchain applications in SCM. They argue for similar studies in other sectors and point towards a lack of empirical research methods in this domain.

Although there are various endeavors, conceptualizing blockchain from different perspectives, research has not yet attached the appropriate importance towards blockchain technology in context of digital platforms. A holistic approach combining architectural design decisions and sociotechnical arrangements to comprehensively conceptualize blockchain platforms is still lacking. Therefore, the goal of this work is to transfer concepts of digital platforms to blockchain platforms and identify dimensions and characteristics for a taxonomy that helps to structure and understand blockchain platforms.

3   Methodology

This research aims to develop a taxonomy of blockchain platforms. Taxonomies structure and organize the knowledge of a domain [24]. They allow to generalize and differentiate previous research, foster a systematic description and thus provide a basis for a deeper theorizing process among researchers [25–27] —a contribution which is, so far, missing in the context of blockchain platforms.

This research follows the well-established procedure for developing taxonomies in IS proposed by Nickerson et al. [27]. As a first step in the taxonomy development, a meta-characteristic is formulated. It reflects the most comprehensive characteristic and aids in selecting appropriate characteristics in the development iterations. Therefore, each characteristic of the taxonomy should be chosen as a logical consequence of the meta-characteristic [27]. The targeted audience comprises scientific researchers as well as practitioners. Consequently, the following meta-characteristic is chosen: *Conceptualizing the role of blockchain technology in digital platform business.*

As the taxonomy development process is an iterative approach, ending conditions must be defined in order to terminate the process. If all ending conditions are fulfilled after an iteration, the taxonomy development process stops. Subjective and objective ending conditions are defined according to Nickerson et al. [27]. We conduct three iterations in total. The first iteration follows a conceptual-to-empirical (C2E) approach and is based on a structured literature review of blockchain publications that is further enriched by core concepts of digital platforms. The subsequent two iterations follow an empirical-to-conceptual (E2C) approach and each analyze a highly diverse set of blockchain platform use cases. Figure 1 illustrates the taxonomy development process. White boxes indicate that a dimension is newly added during an iteration while dark grey boxes are final dimensions.

3.1   First Iteration: Conceptual-to-Empirical

In order to build on prior research in the field of blockchain and digital platforms, the first iteration followed a C2E approach by conducting a structured and systematic literature review according to Webster and Watson [28].

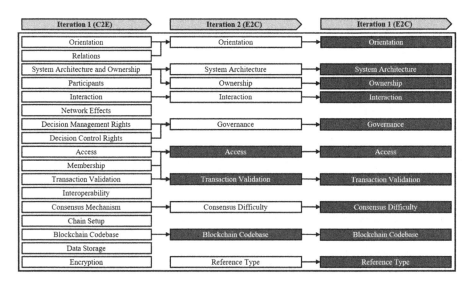

Fig. 1. Overview of the iterations and dimensions in the taxonomy development process

First, relevant literature in the blockchain domain is gathered from the AIS eLibrary, EBSCOhost, ScienceDirect, and JSTOR bibliographic databases. The keywords used for the database search are as follows: "Blockchain use case", "Blockchain application", "Blockchain systems", "Blockchain business networks", "Blockchain services", "Blockchain taxonomy", "Blockchain typology" and "Blockchain classification". For each search term, we use both the singular and plural form. After the initial search, the literature collection is enriched by a selection of highly cited publications on digital platforms from Google Scholar, which, in total, leads to an identification of 40 publications. They are conjointly analyzed with the help of a concept matrix, determining dimensions, which describe both blockchain and digital platforms. In a second step, we identify characteristics, which are related to the derived dimensions. The structured literature review thus provides a profound basis of knowledge to determine the first set of dimensions and characteristics - as depicted in Fig. 1.

At this early stage of the taxonomy development process, the ending conditions are not yet fulfilled. For instance, new dimensions were added in this iteration. Consequently, an additional taxonomy development iteration is necessary.

3.2 Second Iteration: Empirical-to-Conceptual

In order to enrich the initial findings with input from practical applications of blockchain platforms, this iteration analyzes a set of practical use cases. This E2C approach further aims at achieving practical applicability of the taxonomy.

To establish a set of information-rich and highly sophisticated use cases, the selection process focuses on blockchain success stories of leading software companies. We assume that customers looking for the implementation of blockchain technology reach out to these large, experienced software enterprises. The top three enterprises according to their total software revenue are selected with the help of the global software market analysis provided by Chitkara and McCaffrey [29]. By doing so, seven success stories from Microsoft, Oracle and IBM are identified and analyzed. As the seven success stories exhibit similar characteristics for the dimensions "Access" and "Transaction Validation", additional five success stories from the blockchains Ethereum, Hyperledger, and EOS are incorporated. This is done to create a more diverse set of objects to be analyzed in this iteration.

New insights and changes to the taxonomy result from the analysis of the success stories. For the dimension "Orientation", the characteristics are too narrow and not comprehensive. As the analyzed success stories suggest, a characteristic for platforms connecting enterprise and private participants is missing and is therefore added. Furthermore, the characteristics "Supply chain platform" and "Industry platform" are combined to a common characteristic "Inter-business" to help detach the focus from specific industries and branches. A new characteristic "Inter-individuals" is complementing this new focus. The characteristics "Internal platform" was removed, as no practical evidence could be identified among the success stories exhibiting this characteristic.

The success stories indicate a broad diversity in consensus mechanisms. To account for conciseness and explanatory power, the consensus mechanisms are grouped according to their difficulty to reach consensus in the dimension "Consensus Difficulty". Furthermore, the dimension "System Architecture and Ownership" is split into two separate dimensions "System Architecture" and "Ownership" to provide clear differentiation between the technical architecture and the ownership responsibility of a platform. One additional dimension "Reference Type" is added to capture the nature of the referenced assets on the blockchain, either being a purely digital asset or a representation of an asset outside the blockchain.

Lastly, the dimensions "Decision Management Rights" and "Decision Control Rights" are both too specific in light of the limited documentation of the success stories. Therefore, they are combined into a comprehensive dimension "Governance" to still allow for use cases to be analyzed regarding this concept.

Some dimensions had to be removed from the taxonomy due to a lack of available information about the success stories. This caused the following dimensions in this iteration to remain empty or to be only addressed by the minority of cases: "Network Effects", "Chain Setup", "Data Storage" and "Encryption". The reason for this information gap might originate form the deep technical orientation of those dimensions. The respective companies probably prefer to avoid disclosing their implementation details for the sake of secrecy. With regard to network effect, a different reason may cause the lack of information: Network effects are implicitly induced by the design and structure of the platform, they are not part of the platform's documentation. Moreover, the dimensions "Interoperability" and "Participants" are removed. Because all analyzed success stories exhibit the same characteristics regarding these dimensions, the explanatory power is lost. This leads to "null"-characteristics, as objects are related to only one of the characteristics within a dimension [27]. Therefore, the entire dimension is removed.

Since the dimensions and characteristics undergo several changes in this iteration, objective ending conditions are not met. The subjective ending conditions are considered to be partially fulfilled. For instance, the taxonomy accommodates ten dimensions and a maximum of three characteristics per dimension. As this represents a reasonable variety without overwhelming the researcher, the taxonomy is therefore deemed concise. However, so far, only a small sample of practical objects is incorporated in the taxonomy; thus, the taxonomy is not yet rated as robust.

3.3   Third Iteration: Empirical-to-Conceptual

Based on the analysis of additional practical objects, the goal is to augment the limited extent of success stories from the second iteration and enrich the taxonomy with input from the most valued and cutting-edge use cases, including young enterprises, startups and grown businesses alike. This wide sampling approach is chosen to achieve comprehensiveness and an overall robust taxonomy. The use cases are extracted from Crunchbase, an online database providing company data of public and private enterprises, with a focus on the "blockchain" category. Further, the results are sorted regarding their total funding amount. By doing so, we hope to identify the most promising and sophisticated blockchain use cases, as these are usually highly valued by investors. Furthermore, additional use cases are collected by sorting the database results with regard to their "Crunchbase rank". After an initial screening, a total of 22 use cases are added and analyzed, complementing the heretofore collected samples by the most prominent use cases in the Crunchbase database.

Only small changes were made in this iteration. In the dimension "Reference Type" an additional characteristic "Hybrid" is added to reflect that use cases may incorporate both endogenous and exogenous data on the blockchain. No further changes were made to the taxonomy, which consolidates the dimensions and characteristics.

Still, one objective ending condition is not met after this iteration because a new characteristic was added in the dimension "Reference Type". However, as this is considered to be only a minor change to the taxonomy, we deem the ending condition "No new dimensions or characteristics added" to be fulfilled. All other objective ending conditions are satisfied.

The subjective ending conditions are fulfilled as well. Nickerson et al. [27] argue that the number of dimensions and characteristics should be limited in order to achieve conciseness, as it may otherwise overstress the cognitive load of the researcher. The dimensions of the taxonomy are reduced from the initial 17 dimensions after the first iteration to the present ten dimensions. Nickerson et al. [27] suggest that a concise taxonomy consists of seven plus or minus two dimensions. Because the taxonomy covers aspects from both blockchain and digital platform research, we deem ten dimensions as reasonable and manageable, resulting in a concise taxonomy. Robustness of the taxonomy is achieved through two aspects. First, the taxonomy is based on an extensive body of knowledge from the scientific literature. Second, the taxonomy development builds upon two different samples of practical use cases. Consequently, the taxonomy internalizes knowledge from different perspectives on the domain and is therefore expected to be robust. We consider the taxonomy as comprehensive due to its ability to successfully classify the diverse set of analyzed practical blockchain platform use cases, abstracting their characteristics. We thus believe that the taxonomy strikes a reasonable balance between conciseness and comprehensiveness. Still, we regard the taxonomy as extensible, since, new dimensions and characteristics may be added in the future when necessary. Lastly, explanatory potential is achieved by abstracting from details and providing useful explanations of the objects' nature [27]. The taxonomy is considered to fulfill this condition as it provides an appropriate level of abstraction when compared to the many-faceted results in the first iteration of the taxonomy development process. Since all subjective and objective ending conditions are met, the development process is terminated.

4   Results

After finishing the taxonomy development cycle, the research question is answered through the development of a taxonomy for blockchain platforms. The final taxonomy consists of ten distinct dimensions and 26 characteristics and is depicted in Table 1. An explanation of the dimensions is provided in Table 2 to increase their explanatory power and provide a comprehensive understanding of the taxonomy.

Table 1. The final taxonomy of blockchain platforms

Dimension	Characteristics		
Orientation	Inter-business	Inter-individual	Heterogeneous
System Architecture	Proprietary	Semi-proprietary	Open Source
Ownership	Single Entity	Consortium	Public
Interaction	One-sided		Multi-sided
Governance	Hierarchical	Federated	Decentralized
Access	Public	Private	Variable
Transaction Validation	Permissioned		Permissionless
Consensus Difficulty	Limited		Unlimited
Blockchain Codebase	Existing		Independent
Reference Type	Endogenous	Exogenous	Hybrid

Table 2. Definition of the taxonomy's dimensions

Dimension	Underlying guiding question
Orientation	What is the operational scenario of the blockchain platform?
System Architecture	How are the rules implemented which govern the interactions? How open is the platform's foundation to externals?
Ownership	Who provides the blockchain platform core?
Interaction	How are the relations of the user groups structured?
Governance	Who has the decision rights on strategic questions, new features, parameters, or protocol changes?
Access	Which mode of transaction access is implemented?
Transaction Validation	Which mode of validating transactions is implemented?
Consensus Difficulty	How resource-intensive is finding a consensus?
Blockchain Codebase	Whereon is the blockchain developed?
Reference Type	Where does the referenced data on the blockchain come from?

5   Application to the Libra Case

Libra is planned as "a new decentralized blockchain, a low-volatility cryptocurrency, and a smart contract platform" [30]. Apart from Facebook [31], Libra is further backed by big names of the industry, such as Mastercard, PayPal, Uber or Spotify [30] and could, therefore, be available to billions of people [31]. Because of this large number of targeted users, we assume that Libra will be meticulously planned and implemented as a highly elaborate blockchain system. Documentation about Libra has already been published, making the practical application of the taxonomy to this use case and a discussion of the results fruitful. Table 3 depicts the classification of Libra according to the developed taxonomy of blockchain platforms.

Libra integrates a heterogeneous user base. It targets consumers, developers, and businesses. While external developers and organizations build the financial services on the Libra platform, these services, in turn, enable private users and businesses to make use of the Libra cryptocurrency [30]. The blockchain platform is designed as an open-source project [30], as the "technology behind the Libra Blockchain will be free for all to inspect, use, modify, and distribute, and this is understood by all" [32]. Libra is managed by the Libra association, an independent, non-profit membership organization [30]. The white paper states that the association facilitates the operation of the Libra blockchain and coordinates among its members to manage and expand the network [30]. Further, it "manage[s] the development of the technology" [32], which supports the classification of the ownership model as "Consortium". Multiple user groups come together on the Libra platform to interact with each other, such as individual end-users, businesses, providers of financial services. Libra further encourages external developers to contribute additional services, custom transactions and smart contracts to the platform, providing a smart contract language [33]. Libra follows a federated governance model, in which the decision rights reside with the Libra association; its members conjointly make decisions by voting [34]. The association oversees the development of the technology, structures the governance process and is responsible for the change management [32]. The central domains of the governance process are the alignment of the participants' work in the network based on a technical roadmap, the management of funding and the reserve of the Libra cryptocurrency [34].

Table 3. Classification of Libra Case

Dimension	Characteristics		
Orientation	Inter-business	Inter-individual	Heterogeneous
System Architecture	Proprietary	Semi-proprietary	Open Source
Ownership	Single Entity	Consortium	Public
Interaction	One-sided		Multi-sided
Governance	Hierarchical	Federated	Decentralized
Access	Public	Private	Variable
Transaction Validation	Permissioned		Permissionless
Consensus Difficulty	Limited		Unlimited
Blockchain Codebase	Existing		Independent
Reference Type	Endogenous	Exogenous	Hybrid

Libra employs an open access model allowing anyone to access and use the network. Externally developed applications contribute to a healthy competition on the platform [30]. Transactions on the Libra blockchain can only be validated in a permissioned setting by authorized nodes [30]. In the early stages of the Libra network, the members of the Libra association act as validators [34]. In the future, up to 100 validating nodes shall be appointed [30]. In order to agree on a certain state of the blockchain, the validating nodes apply the "LibraBFT" consensus mechanism, a variation of the byzantine fault tolerance (BFT) approach [30]. This process does not rely on extensive computing power [35] and is therefore characterized as having limited difficulty. According to the Libra white paper, the underlying blockchain of Libra is built from the ground up. It further combines a newly developed programming language for smart contracts and a new consensus mechanism, therefore resembling an independent codebase [30]. The asset traded on the Libra platform is the Libra cryptocurrency. Every "Libra" coin is backed by a reserve of deposits and securities in order to secure an underlying value [30]. Coins can only be created and destroyed by the Libra association in exchange for fiat money or traded on exchanges [30]. Since transactions of Libra coins can only be made on the Libra platform, it stands to reason that the asset exhibits an endogenous reference type.

In summary, at the current state, Libra resembles an industry blockchain platform, such as a supply chain blockchain, developed and managed by a few companies, with limited transparency and rights for external users. Ownership and control lie with the closed group of the Libra association members.

6 Discussion

Although some researchers from the blockchain domain have already started to include aspects of platform research into their projects [1, 36], blockchain research has not yet savored the merits of platform literature. Platform research has just started to recognize blockchain as a suitable technology for digital platforms (e.g., [11]). These two research streams are yet to be fully interwoven. Therefore, we follow several calls for research for investigating blockchain in the light of platformization [11, 37] employing a multidisciplinary approach [37, 38]. For instance, we illustrate that aspects such as the orientation, interaction structure or the openness and access to a platform also apply for blockchain systems and support the understanding of the blockchain phenomenon by leveraging these insights from existent platform research.

While existing conceptualizations of blockchain systems often neglect non-technical dimensions [38] or explicitly call for an assessment of such dimensions, e.g. governance [39], our taxonomy successfully integrates both technical and non-technical dimensions. Aspects often found in the literature, such as the consensus mechanism or access and transaction validation rights, are similarly often mentioned by the use cases. Governance was mentioned in many cases as well, however to the most varying intensity. To still account for concepts that are described to a limited extent, such as governance, they were reduced to their core characteristics; thereby still allowing the analysis of practical use cases in the light of these concepts.

Despite the hurdles discussed above, the developed taxonomy achieved comprehensiveness and proved practical applicability. The diverse sample of use cases chosen in the second and third iteration, from startups to incumbent companies, was crucial to cover the variety of blockchain applications of today's market. Further, by successfully classifying Libra with the help of the taxonomy, this work shows its value for researchers and practitioners alike as a structured and holistic approach to study blockchain platforms.

Researchers are given a common ground to further advance the research domain of blockchain platforms and base future findings on the taxonomy. In addition, this work makes a tool available to assist practitioners in the exploration and development of blockchain platforms. Through the taxonomy, managers and developers can grasp the variety of blockchain platforms without the need for a prior, deep scientific understanding of blockchain platforms. Our findings show that while all the use cases claim to make use of blockchain technology, they often do not offer a fully decentralized system. Some companies might only jump on the bandwagon of blockchain due to its recent hype. The taxonomy provides the means to carefully examine a given blockchain platform.

7 Conclusion

We set out to conceptualize the role of blockchain technology in digital platform business. Based on a structured literature review and the analysis of available vignettes of blockchain applications, we developed a taxonomy, integrating technical and non-technical aspects of blockchain platforms. Thereby, we provide means for understanding and comparing existing blockchain use cases while also enabling practitioners to develop their own blockchain platforms by illustrating their foundational elements. Business leaders may use the findings of this research to determine whether the application of blockchain is reasonable in a future scenario.

The research at hand is not without limitations. First, in order to achieve transparency and replicability of the literature review, a structured approach was followed. Although this was done with utmost caution, a subjective bias, e.g. regarding keyword or database selection, cannot be completely ruled out. Second, a limiting factor of the E2C iterations is the set of examined use cases. The composition and description of a use case depend on what and how much is published by a respective company or by a third party. Information about success stories might be prone to exaggerations and whitewashing. By selecting the most information-rich use cases, we tried to counteract this problem. Third, the examination of practical use cases always involves interpretation of the given information, mapping it to dimensions and characteristics of the taxonomy. By employing multiple researchers who individually assess the use cases and combine their findings, subjectivity may be reduced to an acceptable minimum.

Future research may identify relations and interdependencies between the dimensions and characteristics. It is of importance to unveil how design decisions and characteristics of blockchain platforms interrelate and influence each other. Therefore, a cluster analysis, applying the taxonomy, should be performed to identify archetypes of blockchain platforms. The derived archetypes could, in turn, guide practitioners when developing new blockchain-based platforms.

References

1. Glaser, F., Hawlitschek, F., Notheisen, B.: Blockchain as a platform. In: Treiblmaier, H., Beck, R. (eds.) Business Transformation through Blockchain, pp. 121–143. Springer, Cham (2019). https://doi.org/10.1007/978-3-319-98911-2_4
2. Asadullah, A., Faik, I., Kankanhalli, A.: Digital platforms: a review and future directions. In: Proceedings of the 22nd Pacific Asia Conference on Information Systems (2018)
3. Constantinides, P., Henfridsson, O., Parker, G.G.: Introduction—platforms and infrastructures in the digital age. Inf. Syst. Res. **29**, 1–20 (2018)

4. Zaki, M.: Digital transformation: harnessing digital technologies for the next generation of services. JSM **33**, 429–435 (2019)
5. Wieland, H., Polese, F., Vargo, S.L., Lusch, R.F.: Toward a service (Eco) systems perspective on value creation. Int. J. Serv. Sci. Manag. Eng. Technol. **3**, 12–25 (2012)
6. Nofer, M., Gomber, P., Hinz, O., Schiereck, D.: Blockchain. Bus. Inf. Syst. Eng. **59**, 183–187 (2017)
7. Glaser, F.: Pervasive decentralisation of digital infrastructures: a framework for blockchain enabled system and use case analysis. In: Proceedings of the 50th Hawaii International Conference on System Sciences (2017)
8. Friedlmaier, M., Tumasjan, A., Welpe, I.M.: Disrupting industries with blockchain: the industry, venture capital funding, and regional distribution of blockchain ventures. In: Proceedings of the 51st Hawaii International Conference on System Sciences (2018)
9. Lacity, M.: Addressing key challenges to making enterprise blockchain applications a reality. MIS Q. Exec. **17**, 201–222 (2018)
10. Kenney, M., Zysman, J.: The rise of the platform economy. Issues Sci. Technol. **32**, 61–69 (2016)
11. de Reuver, M., Sørensen, C., Basole, R.C.: The digital platform: a research agenda. J. Inf. Technol. **33**, 124–135 (2018)
12. Baldwin, C.Y., Woodard, C.J.: The architecture of platforms: a unified view. In: Gawer, A. (ed.) Platforms, Markets and Innovation. Edward Elgar, Cheltenham (2009)
13. Staykova, K., Damsgaard, J.: A typology of multi-sided platforms: the core and the periphery. In: Proceedings of the 23rd European Conference on Information Systems (2015)
14. Blaschke, M., Haki, K., Aier, S., Winter, R.: Taxonomy of digital platforms: a platform architecture perspective. In: Proceedings of the 14th International Conference on Wirtschaftsinformatik, pp. 572–586 (2019)
15. Täuscher, K., Laudien, S.M.: Understanding platform business models: a mixed methods study of marketplaces. Eur. Manag. J. **36**, 319–329 (2018)
16. Gawer, A.: Bridging differing perspectives on technological platforms: toward an integrative framework. Res. Policy **43**, 1239–1249 (2014)
17. Hileman, G., Rauchs, M.: Global Blockchain Benchmarking Study. Cambridge Centre for Alternative Finance, University of Cambridge (2017). https://doi.org/10.2139/ssrn.3040224
18. Notheisen, B., Hawlitschek, F., Weinhardt, C.: Breaking down the blockchain hype – towards a blockchain market engineering approach. In: Proceedings of the 25th European Conference on Information Systems, pp. 1062–1080 (2017)
19. Glaser, F., Bezzenberger, L.: Beyond cryptocurrencies - a taxonomy of decentralized consensus systems. In: Proceedings of the 23rd European Conference on Information Systems (2015)
20. Rückeshäuser, N.: Typology of distributed ledger based business models. In: Proceedings of the 25th European Conference on Information Systems (ECIS), pp. 2202–2217 (2017)
21. Beinke, J.H., Nguyen, D., Teuteberg, F.: Towards a business model taxonomy of startups in the finance sector using blockchain. In: Proceedings of the 39th International Conference on Information Systems (2018)
22. Labazova, O., Dehling, T., Sunyaev, A.: From hype to reality: a taxonomy of blockchain applications. In: Proceedings of the 52nd Hawaii International Conference on System Sciences (2019)
23. Blossey, G., Eisenhardt, J., Hahn, G.: Blockchain technology in supply chain management: an application perspective. In: Proceedings of the 52nd Hawaii International Conference on System Sciences (2019)

24. Eickhoff, M., Muntermann, J., Weinrich, T.: What do FinTechs actually do? A taxonomy of FinTech business models. In: Proceedings of the 38th International Conference on Information Systems (2017)
25. Glass, R.L., Vessey, I.: Contemporary application-domain taxonomies. IEEE Softw. **12**, 63–76 (1995)
26. Sabherwal, R., King, W.R.: An empirical taxonomy of the decision-making processes concerning strategic applications of information systems. J. Manag. Inf. Syst. **11**, 177–214 (1995)
27. Nickerson, R.C., Varshney, U., Muntermann, J.: A method for taxonomy development and its application in information systems. Eur. J. Inf. Syst. **22**, 1–24 (2012)
28. Webster, J., Watson, R.T.: Analyzing the past to prepare for the future: writing a literature review. MIS Q. **26**(2), xiii–xxiii (2002)
29. Chitkara, R., McCaffrey, M.: PwC global 100 software leaders. Digital intelligence conquers the world below and the cloud above (2016)
30. Libra Association: Libra White Paper. An Introduction to Libra. https://libra.org/en-US/white-paper/
31. Isaac, M., Popper, N.: Facebook Plans Global Financial System Based on Cryptocurrency. https://www.nytimes.com/2019/06/18/technology/facebook-cryptocurrency-libra.html
32. Libra Association: Libra Open Source. Libra Developer Documentation. https://developers.libra.org/docs/libra-open-source-paper
33. Blackshear, S., Cheng, E., Dill, D.L., Gao, V., Maurer, B., Nowacki, T., Pott, A., Qadeer, S., Rain, D.R., Sezer, S., Zakian, T.: Move: a language with programmable resources. https://developers.libra.org/docs/assets/papers/libra-move-a-language-with-programmable-resources.pdf
34. Libra Association: The Libra Association. https://libra.org/en-US/association-council-principles
35. Baudet, M., Ching, A., Chursin, A., Danezis, G., Garillot, F., Li, Z., Malkhi, D., Naor, O., Perelman, D., Sonnino, A.: State Machine Replication in the Libra Blockchain (2019). https://developers.libra.org/docs/assets/papers/libra-consensus-state-machine-replication-in-the-libra-blockchain.pdf
36. Subramanian, H.: Decentralized blockchain-based electronic marketplaces. Commun. ACM **61**, 78–84 (2017)
37. Risius, M., Spohrer, K.: A blockchain research framework. What we don't know, where we go from here, and how we will get there. Bus. Inf. Syst. Eng. **59**, 385–409 (2017)
38. de Rossi, L.M., Abbatemarco, N., Salviotti, G.: Towards a comprehensive blockchain architecture continuum. In: Proceedings of the 52nd Hawaii International Conference on System Sciences (2019)
39. Beck, R., Müller-Bloch, C., King, J.L.: Governance in the blockchain economy: a framework and research agenda. J. Assoc. Inf. Syst. **19**, 1020–1034 (2018)

Towards a Better Understanding of Smart Services - A Cross-Disciplinary Investigation

Ana Kuštrak Korper[1]([⊠]), Maren Purrmann[2], Kristina Heinonen[3], and Werner Kunz[4]

[1] Linköping University, Linköping, Sweden
ana.kustrak.korper@liu.se
[2] University of Paderborn, Paderborn, Germany
maren.purrmann@uni-paderborn.de
[3] Hanken School of Economics, Helsinki, Finland
kristina.heinonen@hanken.fi
[4] University of Massachusetts Boston, Boston, MA, USA
werner.kunz@umb.edu

Abstract. Leveraging new technology to advance smart service is a key service research priority. Such technologies enable smart interactions that address customers' needs in a more meaningful way. Thus, they have a profound effect not only on customer experience but also on streamlining future service offerings, business models and service ecosystems. However, while research is emerging, the concept of smart service is still vague, complicating its successful integration in business practice. This paper aims to characterize smart service through a comprehensive analysis of its theoretical and conceptual building blocks. A systematic literature review is conducted to reconcile the existing understanding of diverse disciplines. A comprehensive keyword-based search approach across 33 academic and three practitioner-oriented journals yielded 157 relevant articles (out of 13.022 articles). The contribution of this investigation is the cross-disciplinary overview of smart service with implications to a broader understanding of the role of smart service in individuals' everyday lives.

Keywords: Smart services · Smartness · Artificial intelligence · Mobile · Digital · AI

1 Introduction

Leveraging new technology such as artificial intelligence (AI) and the internet of things (IoT) to advance smart service has been named one of the top service research priorities [1–4]. Such technologies can enable smart interactions addressing customers' needs in a more meaningful way than ever before [5]. Thus, they can have a profound effect not only on customer experience but also on streamlining future service offerings, innovating business models and transforming relationships that organizations have with different actors in the service system [6].

Understanding smart service becomes even more important as technology infusion transforms digital, physical, and social dimensions of service [7]. So far, the service

© Springer Nature Switzerland AG 2020
H. Nóvoa et al. (Eds.): IESS 2020, LNBIP 377, pp. 164–173, 2020.
https://doi.org/10.1007/978-3-030-38724-2_12

literature distinguished smart service from other technology-enabled services like the one entailing some form of real-time, data-driven and embedded interactions supported by a connected object with learning capability [2, 8, 9]. However, the concept of smart service is still rather vague. While "service" usually implies some form of interactive and value-creating activity [10], the term "smart" mostly has buzz-worthy qualities or is tied to specific technologies in non-service-related disciplines. All this hinders the conceptual and theoretical clarity of the "smart service" phenomenon.

Thus, the purpose of this paper is to characterize the main conceptual building blocks of smart service. By using the existing literature to delineate the concept of smart service, a systematic, cross-disciplinary literature review is conducted to provide an integrative insight and set the research directions. The study is positioned in service research but takes a broad approach to smart service literature. The contribution of the study is the cross-disciplinary overview of smart service and the implication it can have for a borad understanding of its role in individuals' everyday lives.

2 Smart Service in Academic Literature

As a response to companies' imminent need to adopt and integrate service technologies in different processes, research on smart service is emerging in many fields. Smart service has been defined as "individual, highly dynamic and quality-based service solutions that are convenient for the customer, realized with field intelligence and analyses of technology, environment and social context data (partially in real-time), resulting in co-creating value between the customer and the provider in all phases from the strategic development to the improvement of a smart service" [10]. Characteristics of smart service include intelligent, sensing, anticipatory, and adaptable use of data and technology [7, 9]. It is not only the device itself that is intelligent, but the "smartness" is contained in its ability to connect to other technologies [11]. Research is providing a broad array of different perspectives on smart service, both with a conceptual and empirical basis, in consumer and industrial contexts [8, 9, 11]. While it seems that researchers and practitioners may have different views on prioritizing service tech-nologies, there is an emerging consensus that more research is needed in the area, especially in understanding transformative power of smart service and managerial responses related to strategy, design, and operations [1, 10].

Recently, the research of smart service has been reviewed to provide a compre-hensive research agenda for future development. In their review, Dyer et al. [10] approached smart service from the perspective of the service lifecycle categorizing existing research into five groups. Kabadayi et al. [7], in turn, explored the applications of smart service in hospitality and tourism describing the dimensions of customers' smart service experience. While both of these studies are methodologically based on a literature review of smart service their contributions are framed within a specific context or concept such as the life cycle of smart service. However, these literature reviews do not account for different development, conceptualizations, or theoretical and empirical focus areas of smart service across different disciplines. A need for expansion and reconciliation of the diverse landscape of smart service can serve as a basis for the cross-disciplinary exploration of its transformative role. This study aims to address this research challenge.

3   Method

A systematic literature search was conducted to analyze and reconcile the understanding of smart service. This method has increasingly been used in different disciplines to provide a comprehensive body of knowledge [3]. The review follows general recommendations for systematic reviews in business research [13] and the method has also been used to explore smart service [3, 5]. While recent research with the same approach has been conducted [7, 10], this study differs from these in two respect. First, the review in this paper is not specifically focused on any empirical context other than "service," and secondly, it uses a broader array of keywords to identify relevant studies.

We used 16 search words (plus multiple synonyms of them) across 33 academic (peer-reviewed) and three practitioner-oriented journals both service-specific and non-service-specific. In addition to keywords with abbreviations of "smart" and "digital," we used tech-related keywords such as "big data," data-driven", "innovative" and "value-added." While the use of a broad array of keywords resulted in an extensive number of hits, we ensured that the paper would provide a diverse perspective on smart service. While we did not use a temporal limitation for the data collection, the relevant time frame was defined by the data. In other words, the initial data search resulted in articles from the 1920s but was subsequently excluded in the screening process.

The considered time frame for the publication year was therefore from 1983 to 2019. The author team used two-coder screening of the articles. In case of a discrepancy, the article was screened by a third author to resolve the conflictual view. Articles were mainly excluded because of lack of relevance to the topic. The search yielded 157 relevant results with a total hits N = 13.022. 21% of the relevant results came from practitioner-oriented journals.

We conducted a structural analysis of the papers by developing an analysis framework that enabled conceptual structuring [14]. The analysis framework was based on general themes that described the content of the article, as well as more specific themes related to the technology focus. New themes emerged along the analysis process.

4   Findings

The data collection was conducted in the first part of 2019, with the first iteration of data analysis completed. In the following sub-sections we present the findings by focusing on theoretical and empirical approaches to smart service in extant literature.

4.1   General Theoretical Focus and Applicability

The findings indicate that research on smart services is still in its conceptual and theoretical infancy which might be challenging for defining it and subsequently producing relevant contributions. In other words, while many studies are being conducted, the theoretical basis used is diverse and sometimes even not thoroughly presented. Furthermore, some theoretical perspectives tend to rely on disciplinary boundaries. Research in marketing domain would typically be framed within various psychology theories [12, 13], general management in organizational learning theories and technology diffusion [14, 15], and informations systems in technology acceptance model (TAM) and organizational networks [16]. On the other hand the research in the service domain is quite disperse with theoretical focus varying from learning theories in the frontline context [17], customer collaboration and TAM [6] and interdisciplinary frameworks and methods [18, 19]. This creates conceptual ambiguity in terms of what exactly smart service includes, in comparison to traditional service offerings, as well as what is the scope of different types of smart service. The terminology used for different types of smart service technology is very broad, further complicating the generalizability and transferability across technologies. To resolve this conceptual vagueness, it is appropriate that conceptual research is also appearing, mainly such research that provides an overview of the topic. However, while research papers and literature reviews can provide an overview of smart service and identify key research areas, more systematic exploration is needed to develop a conceptual framework for smart service.

4.2   Different Disciplinary Perspectives

Smart service is researched in various academic and practical disciplines and across various industries such as healthcare [20, 21], tourism [22, 23], retailing [24, 25] household security [26], electricity management [27, 28], industrial service processes [6, 29], transportation [30–32], as well as public and governmental services [33]. The levels of empirical inquiry vary from micro (service offerings) [34, 35], meso (service system) [28, 36] to macro levels (i.e. smart cities) [31, 33].

The service-related journals used for this analysis generally cover the service, marketing, management and information systems domains. While all of them indicate the imminent need for academic research on smart services, they did so from different perspectives. The service and marketing domains focus primarily on the customer perspective on smart service, such as on customer adoption and experience of smart technology such as wearables and robotics [4, 55, 58]. The research in this domain also explores the repercussions that smart services bring in the context of a wider service system, especially healthcare and the implementation of IoT technology or service robots. "Smartness" is also discussed concerning intelligent agents enhancing the customer experience [37], the effect of personalization possibilities [38] and general possibilities for more informed and data-driven decision making [39]. Some studies in particular investigate perceived risks and benefits of smart systems from the customer perspective as well as how the perceived level of control impacts technology adoption in a specific context [35, 40]. Also, as service research is often dealing with the relationship between the provider and the customer, the smart service research in this domain investigates changing dynamics of this relationship due to smart technology.

While several researchers in the management domain (i.e. general management and operations management) also investigate how smart technology affects customers and their behavior [25, 41], they predominantly take a company perspective. The primary research focus is on understanding the smartness of the IT processing system and how data can be used to increase productivity in manufacturing and supply chains [31, 32]. Thus, big data is often discussed in this domain focusing on how they effect strategic decision-making and business model innovation [42]. How smart technology and data can be used to innovate supply chains is also discussed in the information systems domain [43], although this domain primarily investigates how information systems, interfaces and technologies need to be designed for enabling smart services [27, 44].

Despite the variety of industries, disciplines, and empirical levels of inquiry, the findings, however, indicate similarities in smart service conceptualization. For instance, smart service is usually defined through the presence of an autonomous system or object. Management domains predominantly define smart products in relation to its autonomy and consider the level of autonomy customer-related. Smart, intelligent products become aware of their local context, have knowledge and reasoning capabilities for performing tasks autonomously [32]. The information system domain considers smart services as autonomous, interactive, intelligent and knowledge-based systems that are capable of interacting with the environment and self-regulate [19, 45].

Further, the findings suggest that the majority of research on smart service in service-related journals focuses on B2C markets, revealing important opportunities for further research on smart service in B2B markets. Current B2B-related smart service research is primarily conducted in the context of logistics and transportation [31, 32].

Generally, existing research on smart service takes a rather broad technology focus by conceptualizing smart service through multiple technologies or smart technology in general, instead of specific type of technology.

4.3   Development of Literature over Time

The data reveals how the research of smart service has changed over time. This change was predominantly facilitated by the technology development and progress that enabled higher degrees of objects, people and systems being interconnected and embedded. The research focus has thus shifted from the need to understand internet-based systems and their effect on operations management or workplace management [46], through the usage of smart cards to reap the benefits of advance selling [47], towards smart meters [48], tracking technology [32], IoT [49] and virtual assistants [37]. The progression of "smartness," although driven by technology, is related to the capacity for connectedness, autonomy, and the interaction level that it allows between users and other systems. In the research from the 1980s and 1990s, smart technology had a clear function of serving as a promise for solving front-line problems more efficiently or increasing productivity processes in manufacturing. Examples include expert-system technology for hotels [50] or intelligent purchasing agents for e-commerce [51]. In the 2000s, smart technology is further investigated within specific contexts such as e-commerce [52], but the focus is put increasingly on understanding smart technology as a supporting tool in prediction or decision-making. Its transformative effects in customer experience, relationship management, organizational capability building, workplace management, and new forms of value creation became a focus in the past decade. The transformational potential of smart service is emphasized due to both digital and physical aspects of smart technologies that are interwoven and can profoundly affect personal as well as societal behaviors and practices. This explains the drastic increase of smart service research in recent years. About two-third of all relevant articles for the analysis of this research were published since the year 2010 with an increasing focus on autonomous and adaptive smart technologies.

4.4 Application of Smart Services

In contrast to the conceptual development, the empirical application and exploration of smart service has been developing very quickly in recent years. Smart service has been explored in a diverse set of empirical and contextual applications, while conceptual papers are predominant in service-related journals. Three general themes appeared when reviewing the data from an empirical perspective. A key theme of smart service research is focused on the applications of technology within businesses and has been explored in frontline and back-office service processes [17]. Many studies are exploring how technology is changing and transforming the everyday activities of individuals – in their roles as consumers, employees or citizens [37]. While smart technology is mainly seen as a facilitator and enabler of improved and more efficient activities and experiences, research is also pointing to the potential downside of connected and autonomous technology [6, 59] with key concerns related mainly to the balance between human touch and technology interactions. The key sectors that have been addressed are healthcare, tourism, retailing, household security, electricity management, industrial service processes, transportation, as well as public and governmental services. The early research on smart service was mainly focused on individual technology applications whereas a recent shift toward smart service systems is emerging [19].

5 Discussion and Future Research

Our literature review indicated that research in smart service is still mainly very unidisciplinary, although different disciplines are in unison in positioning "smartness" as a multifaceted and complex phenomenon that can benefit from a multidisciplinary research lens. Disciplinary foci are, therefore, mostly consistent with the expectations: marketing and service domains revolve around usage, experience, and behavioral challenges, management domains around capability building and organizational learning, business models, and strategic challenges, and information system domains around process support, and optimization challenges. However, technology as an important element of any smart service takes different forms. In some studies, it is specified and defined very narrow (e.g. RFID technology) [53], while in others it is discussed on a more general level (e.g. IoT) with its application varying from single to multiple systems or even contexts [54, 55]. Thus, different technology-focus is not discipline-specific. Interestingly, as technology focus related to smart service varies both between and within disciplines, practitioners' journals offer a valuable overview of case studies exploring more recent technologies offering more specific managerial guidance [56, 57].

By using existing literature in service-related journals to delineate the concept of smart service, this systematic literature review provides an integrative insight and scope of the future research directions. To our best knowledge this is the first systematic cross-disciplinary investigation of the smart service phenomenon. This paper reveals that researchers are approaching smart services from different perspectives and we contribute by providing insight into them. The understanding of varying theoretical and disciplinary perspectives delineated in the results helps in understanding the current status and trends of research in smart service. This shows that there is a common ground of approaching the phenomenon across disciplines indicating that an integrated cross-disciplinary perspective is particularly fruitful for further knowledge building. Finally, implications of our study are important for research in the service domain because it reveals challenges but also research opportunities. Smart service is researched mostly in the conceptual realm, setting the research agendas, priorities, or arguing for a specific perspective, technology or a phenomenon to be addressed in future studies. Although the conceptual work brings forward useful frameworks to understand the landscape of smart service, there is still a need for extensive empirical research. The current status of empirical research might indicate the difficulty in grasping both the complexity and the inherent transformative power related to smart service. Thus, the agenda for applying a multidisciplinary lens might prove as a more relevant empirical approach.

The theoretical and empirical contributions of the study involve reconciling the diverse understanding of smart service studies in disciplines including service, marketing, management, and information systems. This research used structural analysis that enabled descriptive systematization of the results, an important initial step in understanding cross-disciplinary positioning of smart service. The study delineates the diverse and emerging challenges related to research on smart service. It should serve as a reference for future service research and practice. Future conceptual work in the area should focus on developing an integrative definition of smart service, and relating it to other conceptual phenomenon and empirical contexts. In the next phase further research should include quantitative approaches for more nuanced and robust analysis that can move the field forward by providing a comprehensive research framework.

References

1. Kunz, W.H., Heinonen, K., Lemmink, J.G.A.M.: Future service technologies: is service research on track with business reality? J. Serv. Mark. 33(4), 479–487 (2019)
2. Ostrom, A.L., et al.: Service research priorities in a rapidly changing context. J. Serv. Res. 18, 127–159 (2015)
3. Wünderlich, N.V., et al.: "Futurizing" smart service: implications for service researchers and managers. J. Serv. Mark. 29(6/7), 442–447 (2015)
4. Wirtz, J., et al.: Brave new world: service robots in the frontline. Print. J. Serv. Manag. 29, 1–25 (2018)
5. Beverungen, D., et al.: Conceptualizing smart service systems. Electron. Mark. 29(1), 7–18 (2019)

6. Wünderlich, N.V., Wangenheim, F.V., Bitner, M.J.: High tech and high touch: a framework for understanding user attitudes and behaviors related to smart interactive services. J. Serv. Res. **16**(1), 3–20 (2013)
7. Kabadayi, S., et al.: Smart service experience in hospitality and tourism services: a conceptualization and future research agenda. J. Serv. Manag. **30**(3), 326–348 (2019)
8. Allmendinger, G., Lombreglia, R.: Four strategies for the age of smart services. Harv. Bus. Rev. **83**(10), 1–10 (2005)
9. MSI: Research Priorities 2018–2020. Marketing Science Institute Report (2018)
10. Dreyer, S., et al.: Focusing the customer through smart services: a literature review. Electron. Mark. **29**(1), 55–78 (2019)
11. Porter, M.E., Heppelmann, J.E.: How smart, connected products are transforming competition. Harv. Bus. Rev. **92**(11), 64–88 (2014)
12. Johnson, D.S., Lowe, B.: Emotional support, perceived corporate ownership and skepticism toward out-groups in virtual communities. J. Interact. Mark. **29**(C), 1–10 (2015)
13. Köhler, C.F., et al.: Return on interactivity: the impact of online agents on newcomer adjustment. J. Mark. **75**(2), 93 (2011)
14. Plouffe, C.R., Vandenbosch, M., Hulland, J.: Intermediating technologies and multi-group adoption: a comparison of consumer and merchant adoption intentions toward a new electronic payment system. J. Prod. Innov. Manag. **18**(2), 65–81 (2001)
15. Töytäri, P., et al.: Aligning the mindset and capabilities within a business network for successful adoption of smart services. J. Prod. Innov. Manag. **35**(5), 763–779 (2018)
16. Planko, J., et al.: Managing strategic system-building networks in emerging business fields: a case study of the Dutch smart grid sector. Ind. Mark. Manag. **67**, 37–51 (2017)
17. Marinova, D., et al.: Getting smart: learning from technology-empowered frontline interactions. J. Serv. Res. **20**(1), 29–42 (2017)
18. Grenha Teixeira, J., et al.: The MINDS method: integrating management and interaction design perspectives for service design. J. Serv. Res. **20**(3), 240–258 (2017)
19. Lim, C., Maglio, P.P.: Data-driven understanding of smart service systems through text mining. Serv. Sci. **10**(2), 154–180 (2018)
20. Lowe, B., Fraser, I., Souza-Monteiro, D.M.: A change for the better? Digital health technologies and changing food consumption behaviors. Psychol. Mark. **32**(5), 585–600 (2015)
21. Zahedi, F.M., Walia, N., Jain, H.: Augmented virtual doctor office. J. Manag. Inf. Syst. **33**(3), 776–808 (2016)
22. Femenia-Serra, F., Neuhofer, B., Ivars-Baidal, J.A.: Towards a conceptualisation of smart tourists and their role within the smart destination scenario. Serv. Ind. J. **39**(2), 109–133 (2019)
23. Hartley, N., Green, T.: Consumer construal of separation in virtual services. J. Serv. Theory Pract. **27**(2), 358–383 (2017)
24. Hilken, T., et al.: Augmenting the eye of the beholder. J. Acad. Mark. Sci. **45**(6), 884–905 (2017)
25. Montgomery, A.L., et al.: Designing a better shopbot. Manag. Sci. **50**(2), 189 (2004)
26. Jankowski, S.: The sectors where the internet of things really matters. Harv. Bus. Rev. Digit. 2–4 (2014)
27. Loock, C.-M., Staake, T., Thiesse, F.: Motivating energy-efficient behavior with green IS: an investigation of goal setting and the role of defaults. MIS Q. **37**(4), 1313–1332 (2013)
28. Nordin, F., et al.: Network management in emergent high-tech business contexts. Ind. Mark. Manage. **74**, 89–101 (2018)
29. Coreynen, W., Matthyssens, P., van Bockhaven, W.: Boosting servitization through digitization. Ind. Mark. Manag. **60**, 42–53 (2017)

30. Alberti-Alhtaybat, L., Al-Htaybat, K., Hutaibat, K.: A knowledge management and sharing business model for dealing with disruption. J. Bus. Res. **94**, 400–407 (2019)

31. Mehmood, R., et al.: Exploring the influence of big data on city transport operations. Int. J. Oper. Prod. Manag. **37**(1), 75 (2017)

32. Meyer, G.G., et al.: Intelligent products for enhancing the utilization of tracking technology in transportation. Int. J. Oper. Prod. Manag. **34**(4), 422 (2014)

33. Ben Letaifa, S.: How to strategize smart cities. J. Bus. Res. **68**(7), 1414–1419 (2015)

34. Kim, S., Chen, R.P., Zhang, K.: Anthropomorphized helpers undermine autonomy and enjoyment in computer games. J. Consum. Res. **43**(2), 282–302 (2016)

35. Rijsdijk, S.A., Hultink, E.J.: "Honey, have you seen our hamster?" Consumer evaluations of autonomous domestic products. J. Prod. Innov. Manag. **20**(3), 204–216 (2003)

36. Lim, C., et al.: Using data to advance service. J. Serv. Theory Pract. **28**(1), 99–128 (2018)

37. Bolton, R.N., et al.: Customer experience challenges: bringing together digital, physical and social realms. J. Serv. Manag. **29**(5), 776–808 (2018)

38. Chung, T.S., Rust, R.T., Wedel, M.: My mobile music. Mark. Sci. **28**(1), 52–68 (2009)

39. Martínez-López, F.J., Casillas, J.: Artificial intelligence-based systems applied in industrial marketing. Ind. Mark. Manag. **42**(4), 489–495 (2013)

40. Paluch, S., Wünderlich, N.V.: Contrasting risk perceptions of technology-based service innovations in inter-organizational settings. J. Bus. Res. **69**(7), 2424–2431 (2016)

41. Aloysius, J.A., Hoehle, H., Venkatesh, V.: Exploiting big data for customer and retailer benefits. Int. J. Oper. Prod. Manag. **36**(4), 467 (2016)

42. Sorescu, A.: Data-driven business model innovation. J. Prod. Innov. Manag. **34**(5), 691–696 (2017)

43. Nissen, M.E., Sengupta, K.: Incorporating software agents into supply chains. MIS Q. **30**(1), 145 (2006)

44. Bichler, M., Gupta, A., Ketter, W.: Designing smart markets. Inf. Syst. Res. **21**(4), 688–699 (2010)

45. Deng, P.-S., Chaudhury, A.: A conceptual model of adaptive knowledge-based systems. Inf. Syst. Res. **3**(2), 127–149 (1992)

46. Haley, M.: The economic dynamics of work. Strateg. Manag. J. **7**(5), 459–472 (1986)

47. Xie, J., Shugan, S.M.: Electronic tickets, smart cards, and online prepayments. Mark. Sci. **20**(3), 219–243 (2001)

48. Zhang, T., Nuttall, W.J.: Evaluating government's policies on promoting smart metering diffusion in retail electricity markets via agent-based simulation. J. Prod. Innov. Manag. **28**(2), 169–186 (2011)

49. Ng, I.C.L., Wakenshaw, S.Y.L.: The internet-of-things. Int. J. Res. Mark. **34**(1), 3–21 (2017)

50. Cho, W., Sumichrast, R.T., Olsen, M.D.: Expert-system technology for hotels. Cornell Hosp. Q. **37**(1), 54–60 (1996)

51. Maes, P.: Smart commerce. J. Interact. Mark. **13**(3), 66–76 (1999)

52. Li, H., Daugherty, T., Biocca, F.: Characteristics of virtual experience in electronic commerce. J. Interact. Mark. **15**(3), 13–30 (2001)

53. Mukherjee, A., Smith, R.J., Turri, A.M.: The smartness paradox: the moderating effect of brand quality reputation on consumers' reactions to RFID-based smart fitting rooms. J. Bus. Res. **92**, 290–299 (2018)

54. Mani, Z., Chouk, I.: Consumer resistance to innovation in services: challenges and barriers in the internet of things era. J. Prod. Innov. Manag. **35**(5), 780–807 (2018)

55. Verhoef, P.C., et al.: Consumer connectivity in a complex, technology-enabled, and mobile-oriented world with smart products. J. Interact. Mark. **40**, 1–8 (2017)

56. Almirall, E., et al.: Smart cities at the crossroads: new tensions in city transformation. Calif. Manag. Rev. **59**(1), 141–152 (2016)

57. Porter, M.E., Heppelmann, J.E.: Managing organizations how smart, connected products are transforming companies the operations and organizational structure of firms are being radically reshaped by products' evolution into intelligent, connected devices. Harv. Bus. Rev. **10**(93), 96 (2015)
58. Rauschnabel, P.A., He, J., Ro, Y.K.: Antecedents to the adoption of augmented reality smart glasses: a closer look at privacy risks. J. Bus. Res. **92**, 374–384 (2018)
59. Wood, S.C., Brown, G.S.: Commercializing nascent technology: the case of laser diodes at Sony. J. Prod. Innov. Manag. **15**(2), 167–183 (1998)

Service Robots in the Hospitality Industry: An Exploratory Literature Review

Ana Rosete, Barbara Soares, Juliana Salvadorinho, João Reis[(⌗)], and Marlene Amorim

Department of Economics, Management Industrial Engineering and Tourism,
GOVCOPP, Aveiro University, Aveiro, Portugal
{anarosete, barbaramsoares, juliana.salvadorinho,
reis.joao, mamorim}@ua.pt

Abstract. The service sector is changing drastically due the use of robotics and other technologies, such as Artificial Intelligence (AI), Internet of things (IoT), Big Data and Biometrics. Consequently, further research opportunities in the service industry domain are also expected. In light of the above, the purpose of this paper is to explore the potentialities and limitations of service robots in the hospitality industry. To this end, this paper uses a conceptual approach based on a literature review. As a result, we found that in contexts of high customer contact, service robots should be considered to perform standardized tasks due to social/emotional and cognitive/analytical complexity. The hospitality industry is therefore considered closely related to empathic intelligence, as the integration of service robots has not yet reached the desired stage of service delivery. In a seemingly far-fetched context of our reality, organizations will have to decide whether the AI will allow the complete replacement of humans with robots capable of performing the necessary cognitive and emotional tasks. Or investing in balanced capacities by integrating robot-human systems that seems a reasonable option these days.

Keywords: Digital transformation · Service robots · Artificial intelligence · Big data · Hospitality industry · Service industry

1 Introduction

The emerging domain of service robots encompasses a broad spectrum of advanced technologies and holds the potential to outperform industrial robots in both scope and diversification [1]. Service robots are becoming more common in various activities of daily life [2], such as healthcare [3], mobility [4], and so on. Hospitality services are no exception, as several researchers have been investigating the effects of robotic services on hotel brand experiences [5, 6].

It has also been advanced in the literature that service robots and artificial intelligence are being used to provide services to humans, and are gaining increasing attention from hotel and tourism businesses [7]. As a result, we have found that leading companies are combining AI-based technologies with complementary technologies (i.e., robots) with the intent of enhancing or even modernising their service delivery

© Springer Nature Switzerland AG 2020
H. Nóvoa et al. (Eds.): IESS 2020, LNBIP 377, pp. 174–186, 2020.
https://doi.org/10.1007/978-3-030-38724-2_13

systems (SDS's). Despite the growing interest of academics in the service robots' arena, little academic attention has been devoted to this theme, especially in the hospitality frontline services. Recent developments have been labelled by practitioners, who quite dominate the AI field of knowledge [8]. It is worth noting that AI is a classical domain of Computer Science, with significant contributions since the 1960s. The technological phenomenon mention above, especially AI, also deserves the opportunity to be further explored in other fields, such as Business Economics [9, 10], and therefore in the hospitality industry. While we are convinced that current technological developments are being achieved by practitioners, the latter are challenging academics to keep pace.

So, while this article is based on service robot theories, it also tries to shed some light on academia, by reducing the gaps in the literature. In order to conduct scientific research on the shoulders of giants, we selected an article entitled: "Brave new world: service robots in the frontline" by Wirtz *et al.* [11]. The selected article allowed us to study the service robots in the hospitality industry through the lens of the main dimensions of frontline service delivery by robots, which brought stimulating results. Similar research in the literature focused on scenarios and comparative analysis, notably examples are provided by Bazzano and Lamberti [2], who considered solutions and new implementations that use interfaces to request and give directions (e.g., voice) and/or various embodiments (e.g., physical robot, virtual agents). Our research is different in that it analyses a real-life scenario of a hotel that used service robots in its service systems without any human intervention. This unique case unit allowed us to analyse the positive and negative outcomes of dealing with service robots in frontline services.

We have structured the paper into five sections: it beings with an overview of the topic; followed by a discussion of the most relevant terms and a section that explains the methodological process; we also discuss the results of a case study in the hospitality industry, considering the main dimensions of robot delivery; finally, we explore the implications for practitioners and academics, as well as some suggestions for future research.

2   Conceptual Background

Because this paper focus on frontline services in the hospitality industry, it first defines service robots and provides a brief overview of AI; following is a definition of the main dimensions of customer acceptance regarding the introduction of service robots in the frontline services, a categorization of service robots by type of service and it possible roles.

It is generally acceptable that service robots are creating new forms of customer interaction and service experience in the hotel industry [12]. Service robots can be defined as "robotic systems that function as smart, programmable tools, that can sense, think, and to benefit or enable humans" [1, p. 31], which soon covered a broad spectrum of advanced technologies and have the potential to surpass industrial robots in both scope and diversity [1]. Notable example of this is the hospitality industry – the Henn-na hotel in Japan. This hotel, which opened in July 2015, held 80 robots, including luggage-carrying arm robots, in-house customer service desk robots, cleaning robots, etc. [13]. Henn-na is known as a fully automated hotel, so guests cannot find any staff [14]. Recent AI advances have also spurred robot development even further, while robots are navigating more complex scenarios due to improved image recognition and processing techniques, and having sophisticated interactions with humans due to increased processing capabilities of natural language [15]. Thus, AI has developed cognitive skills, enhancing human capabilities or strengthening SDS when associated with other technologies, for example robotics.

Wirtz *et al.* [11] defined three main dimensions of customer acceptance of service robots in the frontline – functional; social-emotional and relational.

Functional dimensions are linked to the customer intention of using new technologies, which depends on the customer cognitive assessment of their utility and ease of use [16]. Regarding the *social-emotional dimension* [17], it is reported that the presence of robots is best accepted by customers when interlinked with elements such as: (a) perceived humanness, which refers to the indistinguishability of robots from humans [18], examples can be identified as anthropomorphism levels; (b) Social interactivity, where robot design does not have to be necessary as a human to be seen as a social competitor [20], that is, robot behaviour can mimic the human performance; (c) Social presence, identified as a situation where robots are "taking care" and customers feel that they are in the presence of another social being [19]. The *relational dimension* [20] is closely linked to aspects such as the feeling of trust that robots can convey to customers [21], and the relationship that is characterized as the customer's perception of a pleasant interaction with a service robot, as well as a personal connection between customer and robot [11].

Wirtz *et al.* [11] also categorize the service robots by service type, namely: task type and service recipient; emotional-social and cognitive complexity; physical task functionality and workload.

Task-type and recipient of service were extensively studied by Lovelock's [22] who distinguished whether a service is directed at people or owned by them and whether these services were tangible or intangible in nature. *Emotional-social and cognitive complexity* was recently studied by Huang and Rust [24], who defined four level of analytical intelligence: mechanical (i.e. routine and repeated tasks), analytical (i.e. information processing), intuitive (i.e. ability to process complex information), and emphatic intelligence (i.e. ability to read, understand and respond to customers' emotions). According to Wirtz *et al.* [11], there is a consensus in the literature that the first three levels of AI will develop at a higher level and robots will become dominant delivery mechanisms, but there is considerable debate about whether robots will be effective in providing emotional and social services at a human level, as we will see later in the discussion section. With regard to *physical task functionality and service volume*, it is expectable that in a near future, service employees will work side by side with robots, which must deal with heavy work or hazardous tasks.

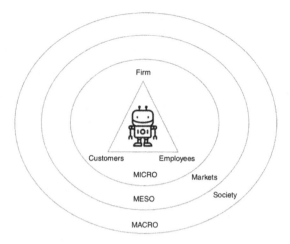

Fig. 1. The impact of service robots on key stakeholders [11]

Finally, Wirtz *et al.* [11] explore the potential role of service robots, in particular, focus on the impact of service robots on three main levels (Fig. 1): (a) *micro level*, is identified as the individual customer experience, emphasizing privacy/customer security [24, 25], dehumanization/depersonalization [26] and social deprivation [11]; it is also associated with service training and learning, which is an important ability for employees to provide consistent services; (b) *meso level*, focus on markets of a particular service and market prices, due to fall in the cost of the service robots, this technology is increasing its viability and range on service industry contexts [27]; (c) and *macro level*, which refers to societal issues related to employment [23] and inequality [28] within and across societies [11]. All three of these levels will be useful for our analysis.

3 Methodology

This article follows a qualitative analysis of samples retrieved from Scopus.com, which has been used as a bibliographic database of peer-reviewed literature. Selecting this database is justified by Scopus's ability to provide a coverage of articles, as it indexes a broad range of AI journals when compared to other known databases, such as Web of Science from Thomson Reuters [8]. A truly comprehensive approach would lead us to select more than one digital repository, however, given that our priority is transparency and easy reproduction of results, this choice may be acceptable when comparing the pros and cons [29].

The data search was conducted on March 8th, 2019, and we started with the inclusion criteria using "service robots" in topic (article title, abstract and keywords), and "Henn-na" in all search fields (Table 1).

Table 1. Literature review process

Scopus database				
Selected keywords	"Service robots"		"Henn-na"	
Search	Title-abstract-keywords	4,048	All fields	19
Language	English	3,848	English	17
Source type	Journals	1,049		
Document type	Articles	935		
Years	2019–2017	242		

To cover a large number of international publication, we have selected journal articles in English language; furthermore, to avoid misinterpretations, articles in other languages were excluded. In addition, to ensure quality, the study included indexed scientific research from over a three-year period covering the current state-of-the-art. Regarding the second keyword, i.e. Henn-na, we did not apply any filters other than the language, as it would further restrict the few results obtained. In order to finding corroboration, we included additional articles, especially conference proceedings, as it was difficult to find corroboration among researchers. This is therefore a hot topic that, in our understanding, needs to be further explored by scholars.

Regarding the content analysis, the fourth author manually coded the manuscripts, hierarchizing all categories and subcategories in order to identify emerging ideas. To ensure high quality, unbiased review and reliability, the remaining authors independently read and coded the data with a qualitative analysis software package (i.e., NVIVO 11) for the purpose of making adjustments, until redundancies and contradictions were clarified. The software allowed the researchers to handle a large volume of data at relative fast pace to cross-check the first results. The differences identified between manual analysis and software coding were discussed until a consensus was reached among researchers, allowing the results to be refined and, consequently, to find evidence.

The results of this research have been limited due to the methodological options. The first restriction is related to Scopus database as it is constantly being updated with newly peer-reviewed research. Our sample mainly consisted of journal articles based on the assumption that these manuscripts were of better quality due the rigorous revision process, although we recognize that eventually some relevant articles from other sources have been left behind. A related restriction is associated to the selected keywords; as another similar keyword might yield different results. Recognizing the limitations mentioned, we still believe the literature review has its value, as it synthesizes the existing body knowledge in few pages and provides an exploratory overview of the phenomenon. Last but not least, we have just reviewed a human-robot hotel due to space limitations of the paper, although we believe that the results are similar to equivalent hotels, so more research is needed.

4   Findings and Discussion

4.1   Is the Hospitality Industry Moving to Full Range Automation?

Despite valuable research from the past, there are relatively few studies on guest rating studies on the hotel brand experience that compare service delivery by a human staff and/or a robot [5]; therefore, no study to date has examined the performance of frontline service robots through the lens of the key dimensions of frontline service delivered by robot. But, why is this relevant? As some authors argue, "it is likely to have technological capability of delivering service with few or no human is a possibility in the field of hospitality" [30, p. 2] our opinion is that further research is needed in this regard.

The need to deepen the theoretical research is quite obvious, as we find contradictory studies in the literature. For instance, Nakanishi *et al.* [31] conducted a research to investigate whether humanoid robots engage in heart-warming interaction service in hotels. From Nakanishi *et al.* [31] customers' impression is that "humanoid robot's potential for heart-warming interaction service enhances the customer's satisfaction of the whole service" (p. 45); at the same time, the authors reinforce the fact that "technology of the human-robot social interaction will hopefully permeate other hospitality services or emotional labour from now on" (p. 52). Although somewhat more sceptical, Osawa *et al.* [13] do not exclude the aforementioned possibility, and they argue that "although human emotional labour seems difficult to simply replace with robot technologies, there is a possibility that human emotional labour can be replaced in a very complicated way" (p. 223). The sceptical positions, above all, point out that the technological gains of robot services to perform complex social-emotional tasks are still a vision. In that regard, Can and Tung [5] state that "among all levels of hotels, robotic service decreased affective experience, a possible reason for this result is the limitations of current robotic technology as service robots cannot imitate humans to the same extent, in terms of emotions" (p. 466). Consequently, it is common to find in the literature articles that aim to make non-humanoid mobile robots more enjoyable with the intention of imitating human social behaviour [32]. The Henn-na hotel, mentioned earlier, which included in its reception services realistic humanoid robots [33] or anthropomorphic-human like, also required human intervention in its service operations [34]. Accordingly, human-robot interactions (HRI) have attracted considerable attention in the robotic research community [35]. The point is that the Henn-na hotel recently removed more than half of its robotic workforce, as these robots were not advanced enough to perform social activities (e.g. in-room robots failed to answer customer questions) and guests were frustrated [36]. Overall, many Henn-na hotel guest suffered disruption as they were unable to communicate with robots due to language barriers, while another Henn-na guest felt the technology was "not there yet" because it could not help clients where needed [37].

Given the above, Lu *et al.* [7] suggest that the emotional/hedonic and utilitarian aspects (e.g., performance efficacy) of AI-based technology and service robots are critical determinants of their integration into service delivery.

4.2   Service Robots' vs Humans: Semi-automated Systems

Within the ecosystem framework, we acknowledge other domains in which service robots have been evolved, from one-function automation to intelligent systems of versatile features, sharing the same space and tasks with humans [38]. Lu *et al.* [7] state that their article sheds light on the hedonic and utilitarian nature of AI technology infusion in hospitality services, as the service industry pioneered the adaptation to new technologies to enhance the user experience. Ivanov *et al.* [14] have made some progress in this regard as they argued that robotics and AI can become a viable alternative to human employees in travel, tourism and hospitality companies. However, the aforementioned authors believe that not all service processes can and should be automated or performed by robots – "at the end of the day it is the economic efficiency, customer experience, company's competitiveness and other factors that will determine whether the automate and robotise the service delivery process" (p. 1512). Despite the advancements made on AI technologies in the context of hospitality services, scholars have not yet extended sufficient advances in the field of service robots [7]. In addition, we can find some empirical developments on the application of AI-enabled technologies that are not consensual in other SDS's. For instance, Pezzullo's *et al.* [39] surveyed radiologists who expressed dissatisfaction with AI technologies, feeling frustration that results from erroneous radiology reports were higher compared to traditional transcription services. On the other hand, Gartner *et al.* [40] found that AI methods and programming-based features increased the hospital's contribution margin.

So what solutions can be considered, in addition to the use of AI-enabled technologies, that can enhance the service robot capabilities in the hospitality industry? In light with Wirtz *et al.* [11], which mapped the customer needs to robot capabilities, it is quite evident that hospitality, in a first instance, can be conceived as a task of simple analytical-cognitive complexity and high emotional complexity; for this reason, given the limitations presented by robots in a first stage of service, their assignment will tend to be more in favour of humans. Therefore, robot interactions in frontline hospitality services should be considered insufficient because they usually do not meet the dimensions of human-oriented perception, demonstrating high levels of impersonality – "these services will by and large continue to be delivered by people" [11, p. 919].

It should be noted that in the hospitality industry, a preponderant part of the service generates emotions relevant to customer satisfaction and, therefore, to service loyalty. Service robots have been used in highly standardized service operations [5], without a high level of interpersonal demand [7, 15], making the service a unique act. However, in order for the service to become highly personalized, providers need to have complex cognitive and emotional social skills (e.g. healthcare services), which are categorized as highly customized services with often uncertain outcomes [11]. Wirtz *et al.* [11] argues that it seems unlikely that robots possess the social intelligence and communications skills to deal with complex emotional issues. In the light of the above, we understand that frontline hotel services can be considered an experience of cognitive and emotional complexity, let's see if a customer wants to hail a cab, or ask for directions – simple instructions that can be partially resolved by AI; but complex issues, such as mediating service complaints, may require demanding skills from frontline employees who are likely to be more comfortable acting with the help of AI or robot support – that is, service delivered by teams of robot and humans. Similarly, Reis *et al.* [41] argue that companies are adopting strategies to solve complex service failures because frontline employees are playing important roles in performing service recovery activities, while in some cases these employees complement their activities with offline technologies. Tung and Law [42] go further as they uncover unexplored possibilities such as "human operators to be telepresent on screen from copresent robots" (p. 17), which could be a reliable option when dealing with complex service failures that surpasses the capabilities of frontline employees. Special attention to be given to issues related to anthropomorphic-human like – which are critical components to customers' acceptance, particularly hospitality services contexts [15]. Although, Tung and Au [37] findings suggest that experienced discrepancies at the Henn-na hotel between anthropomorphic robots still fails to perform human tasks, resulting in negative guest experience and deterrence against robot services as they are considered "just a gimmick" or marketing ploy. At the end of the line, advances in robotics and AI technologies incorporate broad research opportunities in the hospitality segment, while practitioners are advancing ideas that can inspire academics to find new avenues for research [42, 43].

4.3   Human-Robot Systems: Micro, Meso and Macro Levels

The discussion in this section continues to provide information about hospitality services and, consequently, about the customer experience from a micro, meso, and macro perspective (Table 2). The frontline services can be characterized as highly cognitive and social-emotional, so the use of human-robot systems is likely to be considered. In light of the above option, these systems can add the best features of humans and robots.

Overall, the literature considers that frontline employees should provide services that require a high degree of personalization, heterogeneous and strong emotional relation; on the other hand, high consistency services and high analytical-cognitive complexity can be performed by service robots.

Table 2 shows that Henn-na Hotel employees pooled 5 positive/3 negative items against service robots which have collected 4 positive/4 negative items. This analysis reinforces the reasons why robots in the frontline services were replaced by humans four years after the opening of Henn-na [45].

Micro Level: Henn-na hotel staff has engaged in out-of-box thinking, bringing together creative problem-solving solutions, seeking answers that are tailored to customer needs. For instance, at Henn-na, human work not only provides customers with answers to questions, but employees were required to experience the difficulties visitors may face – robots could not perform these tasks [44]. On the other hand, employee integration is demanding from a training angle, unlike service robots. From customers' perception, employees can provide personalized service, while robots have the ability to deliver consistent, high-performance services. However, front-desk robot's ability to communicate was very limited, especially in the absence of any complementary AI technology. These limitations can cause user frustration and disappointment, especially if guests face the same challenge multiple times; therefore, additional staff training may be required to communicate and explain to guests the type of services that robots may or may not perform [37].

Table 2. Human-robot systems: micro, meso and macro levels (Adapted from Wirtz *et al.* [11])

Frontline services			Employees		Robots		Henn-na hotel
Micro	Employees training	+	Can engage in out-of-box thinking and creative problem solving	−	Limited out-of-box thinking		Reduced customer relationship [34]
		−	Need training	+	Upgradable, system-wide		At Henn-na hotel, education is provided so that any employee can handle such work [44, p. 242]
		+	Understanding of their customers	−	Pattern recognition		Employees are required to sense difficulties that visitors may be facing, and visitors expect that employees will try to ascertain their reactions, the reception robot cannot perform such tasks [44]
	Customer experience	+	Customization and personalization depend on employee skill and effort	+	Customization and personalization can be delivered on scale at consistent quality and performance		The Japanese considers cleaning guest rooms as a human work, while robots cleans common facilities such as corridors and lobby's [44, p. 241]
		+	Can engage in deep acting	−	Can engage in surface acting		The front desk robot's ability to communicate is low or very limited [13, 37, 44]
Meso	Markets	+	Service employees can be a source of competitive advantage	−	Service robots are unlikely to be a source of competitive advantage		While hotels are investing in service robots in the frontline services, Henn-na, which opened in 2015 have replaced many of its robots by humans just four years after opening [45]

(continued)

Table 2. (*continued*)

Frontline services			Employees		Robots	Henn-na hotel
Macro	Society	–	Many service employees work in unattractive jobs	+	Mundane and unattractive service jobs can be robot-delivered	In some western countries cleaning activities are considered a routine and unattractive job
		–	Important services are expensive and scarce if delivered by service employees	+	Cost savings of robot-delivered services will be competed away, leading to lower prices, and increased consumption and higher standards of living	The efficiency of activities performed by humans is measured by the time needed to execute them; time spent on robot labour is less expensive than paying humans [44, p. 242]. In some cases, robots can reduce the human workload [46]

Meso Level: Service robots are unlikely to become an essential source of competitive advantage, at least in the medium to long term [11]. For example, while hotels are investing in frontline robot services, and while there is no consensus in the literature, our view is that humans are still a source of competitive advantage. However, for companies operating in innovative markets that wat to take risks and where the "winners take everything" service robots and advanced AI technologies can become an interesting asset.

Macro Level: At the Henn-na hotel, the efficiency of human activities is measured by the time required to perform them [44] and their relevance to customers. While the nature of hospitality work has created an unparalleled number of part-time jobs, especially seasonal jobs [47], robots are seen as a solution that can help hotels to deal with seasonal employment permanently [12]. Moreover, the time spent working with robots is less expensive than paying to humans [44] and thus more permanent. The above evidence may reinforce social concern about the risk of unemployment caused by robots [48] and/or artificial intelligence [23].

5   Concluding Remarks

Although a significant number of hospitality companies are trying to incorporate service robots into their frontline services, only a few have succeeded. The point is that empirical results differ from different service sectors and, despite the academic efforts made so far, there is no consensus in the literature. Therefore, in our view, managers should be very cautious when deciding to integrate service robots into their service delivery systems. To the best of our knowledge, in the context of the hospitality industry, frontline services are considered closely related to empathic intelligence, as the integration of service robots has not yet reached the desired stage of service delivery. In a seemingly far-fetched context, organizations will have to decide whether AI will allow the complete replacement of humans by robots capable of performing the required emotional tasks or investing in balanced capacities by integrating robot-human systems. This study contributes to hospitality management by highlighting current frontline service practices, i.e. using service robots and AI technologies; If used well, it can improve future costumer experiences. More research is needed to find appropriate, manageable and replicable solutions to service delivery contexts, especially at the frontline of SDS, where involvement of these service technologies requires complex cognitive and emotional tasks.

References

1. Engelhardt, K.: Service robotics and artificial intelligence: current research and future directions. ISA Trans. **29**(1), 31–40 (1990)
2. Bazzano, F., Lamberti, F.: Human-robot interfaces for interactive receptionist systems and wayfinding applications. Robotics **7**(3), 56 (2018)
3. Ramoly, N., Bouzeghoub, A., Finance, B.: A framework for service robots in smart home: an efficient solution for domestic healthcare. IRBM (2018). https://doi.org/10.1016/j.irbm. 2018.10.010
4. Ding, J., Lin, R., Lin, Z.: Service robot system with integration of wearable Myo armband for specialized hand gesture human–computer interfaces for people with disabilities with mobility problems. Comput. Electr. Eng. **69**, 815–827 (2018)
5. Chan, A., Tung, V.: Examine the effects of robotic service on brand experience: the moderating role of hotel segment. J. Travel Tourism Mark. **36**(4), 458–468 (2019)
6. Choi, S., Liu, S., Mattila, A.: "How may I help you?" Says a robot: examining language styles in the service encounter. Int. J. Hosp. Manag. **82**, 32–38 (2019). https://doi.org/10. 1016/j.ijhm.2019.03.026
7. Lu, L., Cai, R., Gursoy, D.: Developing and validating a service robot integration willingness scale. Int. J. Hosp. Manag. **80**, 36–51 (2019). https://doi.org/10.1016/j.ijhm. 2019.01.005
8. Reis, J., Santo, P., Melão, N.: Impacts of artificial intelligence on public administration: a systematic literature review. In: 14th Iberian Conference on Information Systems and Technologies (CISTI). IEEE (2019)
9. Jarrahi, M.: Artificial intelligence and the future of work: human-AI symbiosis in organizational decision making. Bus. Horiz. **61**(4), 577–586 (2018)
10. Joerss, M., Schröder, J., Neuhaus, F., Klink, C., Mann, F.: Parcel Delivery: The Future of Last Mile. McKinsey & Company (2016)
11. Wirtz, J., et al.: Brave new world: service robots in the frontline. J. Serv. Manag. **29**(5), 907–931 (2018)
12. Kuo, C., Chen, C., Tseng, C.: Investigating and innovative service with hospitality robots. Int. J. Contemp. Hosp. Manag. **29**(5), 1305–1321 (2017)
13. Osawa, H., et al.: Analysis of robot hotel: reconstruction of works with robots. In: International Symposium on Robot and Human Interactive Communication, pp. 219–223. IEEE (2017)

14. Ivanov, S., Webster, C., Berezina, K.: Adoption of robots and service automation by tourism and hospitality companies. Revista Turismo e Desenvolvimento **27/28**, 1501–1517 (2017)

15. Murphy, J., Gretzel, U., Pesonen, J.: Marketing robot services in hospitality and tourism: the role of anthropomorphism. J. Travel Tourism Mark. **36**(7), 784–795 (2019). https://doi.org/10.1080/10548408.2019.1571983

16. Davis, F.: Perceived usefulness, perceived ease of use, and user acceptance of information technology. MIS Q. **13**(3), 318–340 (1989)

17. van Doorn, J., et al.: Domo Arigato Mr. Roboto: emergence of automated social presence in organizational frontlines and customers' service experiences. J. Serv. Res. **20**(1), 43–58 (2017)

18. Wünderlich, N., Paluch, S.: A nice and friendly chat with a bot: user perceptions of AI-based service agents. In: Proceedings of the 38th International Conference on Information Systems (2017). https://aisel.aisnet.org/icis2017/ServiceScience/Presentations/11/. Accessed 20 Apr 2019

19. Breazeal, C.: Towards sociable robots. Robot. Auton. Syst. **42**(3–4), 167–175 (2003)

20. Nomura, T., Kanda, T.: Rapport–expectation with a robot scale. Int. J. Soc. Robot. **8**(1), 21–30 (2016)

21. Doney, P., Cannon, J.: An examination of the nature of trust in buyer-seller relationships. J. Mark. **61**(2), 35–51 (1997)

22. Lovelock, C.: Classifying services to gain strategic marketing insights. J. Mark. **47**(3), 9–20 (1983)

23. Huang, M.-H., Rust, R.: Artificial intelligence in service. J. Serv. Res. **21**(2), 155–172 (2018)

24. Saravanan, S., Ramakrishnan, B.: Preserving privacy in the context of location based services through location hider in mobile-tourism. Inf. Technol. Tourism **16**(2), 229–248 (2016)

25. Broadbent, E.: Interactions with robots: the truths we reveal about ourselves. Annu. Rev. Psychol. **68**, 627–652 (2017)

26. Forman, A., Sriram, V.: The depersonalization of retailing: its impact on the 'lonely' consumer. J. Retail. **67**(2), 226–243 (1991)

27. Dyrkolbotn, S.: A typology of liability rules for robot harms. In: Ferreira, M.I.A., Silva Sequeira, J., Tokhi, M.O., Kadar, E.E., Virk, G.S. (eds.) A World with Robots. ISCASE, vol. 84, pp. 119–133. Springer, Cham (2017). https://doi.org/10.1007/978-3-319-46667-5_9

28. Freeman, R.: Who owns the robots rules the world (2015). https://wol.iza.org/articles/who-owns-the-robots-rules-the-world/long. Accessed 20 Apr 2019

29. Reis, J., Amorim, M., Melão, N., Matos, P.: Digital transformation: a literature review and guidelines for future research. In: Rocha, Á., Adeli, H., Reis, L.P., Costanzo, S. (eds.) WorldCIST'18 2018. AISC, vol. 745, pp. 411–421. Springer, Cham (2018). https://doi.org/10.1007/978-3-319-77703-0_41

30. Ivanov, S., Webster, C., Garenko, A.: Young Russian adults' attitudes towards the potential use of robots in hotels. Technol. Soc. **55**, 24–32 (2018). https://doi.org/10.1016/j.techsoc.2018.06.004

31. Nakanishi, J., Kuramoto, I., Babam, J., Kohei, O., Yoshikawa, K., Ishiguro, H.: Can a humanoid robot engage in heartwarming interaction service at a hotel? In: Proceedings of the 6th International Conference on Human-Agent Interaction, pp. 45–53. ACM (2018). https://doi.org/10.1145/3284432.3284448

32. Kaiser, F., Glatte, K., Lauckner, M.: How to make nonhumanoid mobile robots more likable: employing kinesic courtesy cues to promote appreciation. Appl. Ergon. **78**, 70–75 (2019). https://doi.org/10.1016/j.apergo.2019.02.004

33. Wisskirchen, G., et al.: Artificial intelligence and robotics and their impact on the workplace. IBA Global Employment Institute (2017)
34. Rajesh, M.: Inside Japan's first robot-staffed hotel. The Guardian (2015). https://www.theguardian.com/travel/2015/aug/14/japan-henn-na-hotel-staffed-by-robots. Accessed 20 Apr 2019
35. Stock, R., Merkle, M.: A service robot acceptance model: user acceptance of humanoid robots during service encounters. In: IEEE International Conference on Pervasive Computing and Communications Workshops, pp. 339–344. IEEE (2017)
36. Shead, S.: World's first robot hotel fires half of its robots. Forbes (2019). https://www.forbes.com/sites/samshead/2019/01/16/worlds-first-robot-hotel-fires-half-of-its-robots/#fc2fe3ce1b1d. Accessed 20 Apr 2019
37. Tung, V., Au, N.: Exploring customer experiences with robotics in hospitality. Int. J. Contemp. Hosp. Manag. 30(7), 2680–2697 (2018)
38. Savela, N., Turja, T., Oksanen, A.: Social acceptance of robots in different occupational fields: a systematic literature review. Int. J. Soc. Robot. 10(4), 493–502 (2018)
39. Pezzullo, J., Tung, G., Rogg, J., Davis, L., Brody, J., Mayo-Smith, W.: Voice recognition dictation: radiologist as transcriptionist. J. Digit. Imaging 21(4), 384–389 (2008)
40. Gartner, D., Kolisch, R., Neill, D., Padman, R.: Machine learning approaches for early DRG classification and resource allocation. Informs J. Comput. 27(4), 718–734 (2015)
41. Reis, J., Amorim, M., Melão, N.: Multichannel service failure and recovery in a O2O era: a qualitative multi-method research in the banking services industry. Int. J. Prod. Econ. (2018). https://doi.org/10.1016/j.ijpe.2018.07.001
42. Tung, V., Law, R.: The potential for tourism and hospitality experience research in human-robot interactions. Int. J. Contemp. Hosp. Manag. 29(10), 2498–2513 (2017)
43. Ivanov, S., Gretzel, U., Berezina, K., Sigala, M., Webster, C.: Progress on robotics in hospitality and tourism: a review of the literature. J. Hosp. Tourism Technol. (2019, forthcoming)
44. Osawa, H., et al.: What is real risk and benefit on work with robots? From the analysis of a robot hotel. In: Proceedings of the Companion of the 2017 ACM/IEEE International Conference on Human-Robot Interaction, pp. 241–242. ACM (2017)
45. Gulliver Travels: Why the world's first robot hotel was a disaster. The economist (2019). https://www.economist.com/gulliver/2019/03/27/why-the-worlds-first-robot-hotel-was-a-disaster. Accessed 26 Apr 2019
46. Hochschild, A.: The Managed Heart: Commercialization of Human Feeling. University of California Press, Berkeley (2012)
47. Baum, T.: Human Resource Management for Tourism, Hospitality and Leisure: An International Perspective. Thomson, London (2006)
48. Marchant, G.E., et al.: International governance of autonomous military robots. In: Valavanis, K.P., Vachtsevanos, G.J. (eds.) Handbook of Unmanned Aerial Vehicles, pp. 2879–2910. Springer, Dordrecht (2015). https://doi.org/10.1007/978-90-481-9707-1_102

Understanding FinTech Ecosystem Evolution Through Service Innovation and Socio-technical System Perspective

Paola Castro[✉], José Pedro Rodrigues, and Jorge Grenha Teixeira

INESC TEC, and Faculty of Engineering, University of Porto, Porto, Portugal
paolalbuquerque@gmail.com, {jpcr,jteixeira}@fe.up.pt

Abstract. Although interest in FinTech businesses has been growing, research about these companies is still scarce. To address this gap, this paper aims to understand the evolution of the FinTechs ecosystem, through a socio-technical system theory and service innovation lense. A case study research methodology was used, in which 6 Brazilian and 5 Portuguese FinTechs were analyzed. Primary data was collected using semi-structured interviews with managers and employees of the startups, while secondary data was obtained through the analysis of reports from consulting firms and public relations materials of the startups. Results show the evolution of FinTech ecosystems from the perspective of socio-technical system theory and service innovation. From the socio-technical system perspective it was possible to understand the roles of social, technological and organizational actors in the evolution of these ecosystems. From the service innovation perspective, it was possible to understand the dynamics of the evolution of the FinTech ecosystems and its results.

Keywords: FinTech · FinTech ecosystem · Service innovation · Process innovation · Social-technical theory

1 Introduction

The word FinTech is a combination of the terms "financial" and "technology", and is intended to denote the use of technology to deliver a financial solution [1]. Interest in FinTech business has been growing. According to Gagliardi, Dickerson [2] the value of global investment in FinTech grew 75% in 2015, and is now equivalent to US$ 22.3 billion. Despite the considerable growth, academic research about FinTechs is still scarce [3], which motivates this work.

This study has the purpose of understanding the dynamics of creating a FinTech startup and its evolution. Moreover, it is intended to emphasize the interaction between FinTechs evolution, and the development of the services offered by these startups. In order to understand FinTechs evolution dynamics, it is necessary to analyze their ecosystem [4]. According to Lee and Shin [4], the FinTech ecosystem is composed by five elements: FinTech Startup, Government, Financial Customer, Technology Developers, and Traditional Financial Institutions.

© Springer Nature Switzerland AG 2020
H. Nóvoa et al. (Eds.): IESS 2020, LNBIP 377, pp. 187–201, 2020.
https://doi.org/10.1007/978-3-030-38724-2_14

Socio-technical system theory was found appropriate to better understand the relation between the aforementioned elements of the ecosystem and their roles in the FinTech evolution. The socio-technical theory proposes a structure to analyze the interactions between social and technical dimensions of organizations [5]. By applying the socio-technical systems perspective it is, possible to identify the structure of the FinTech ecosystem and new stakeholders that are part of the ecosystem. Moreover, by applying service innovation perspective it is possible to understand the dynamics of FinTech's ecosystem evolution and its results, since this perspective provides the tools to understand the co-creation of value between the startup and its clients. Accordingly, this study had the goal of understanding FinTech ecosystems evolution, namely by understanding:

- How are the various actors for the FinTechs ecosystem characterized?
- How do these actors influence the evolution of the FinTechs ecosystem?
- How do the perspectives of service innovation and socio-technical systems theory support the understanding of FinTech ecosystems' evolution?

To achieve this goal, an empirical study was developed following case research [6]. Data was collected through semi-structured interviews with C-levels managers and employees of Portuguese and Brazilian FinTechs. In order to validate data, other sources of evidence were also used, such as reports from large consultancy companies, archives of the FinTech, among others. This study consists of four main sections: literature review, empirical study, analysis of results and conclusion.

2   Literature Review

2.1   FinTech

The word FinTech is a combination of the terms "Financial" and "Technology", and is intended to denote the use of technology to deliver a financial solution [1]. According to Leong, Tan [7], a FinTech provides a financial solution developing a technology-based product and/or service. Regarded as the most significant innovation in the financial sector, FinTechs promise to reshape the industry by cutting costs and growing quality of service delivery [4]. FinTech business models are developed to be affordable and cost-effective, and, therefore, stand out from traditional financial service providers [2]. FinTechs offer financial solutions such as payment services (payment through cryptocurrency, blockchain technology), financing and loans (crowdsourcing and crowdfunding), insurance (usage-based insurance), and interaction with customers (personal finance management) [8].

The global financial crisis had a significant negative impact on the confidence in the banking system [9]. The financial sector experienced a series of ruptures in its operations and processes [10]. Therefore, after the crisis, new regulatory issues emerged and focused on greater transparency from financial agents and greater protection for consumers [8]. Although the financial sector regulations became stricter after the crisis, some initiatives in some countries reduced FinTech entry barriers for new companies [1]. This led to the emergence of the FinTech in 2008, along with technological development for mobile devices, changes in financial customer behavior and e-finance development [4, 10]. The spread of mobile devices and digital financial services have enabled customers to gain access to their financial information anytime and anywhere [8].

Since then, the sector has been gaining prominence among investors. According to Gagliardi, Dickerson [2], the value of global investments in FinTech grew 75% in 2015, which is equivalent to US$ 22.3 billion. Gagliardi, Dickerson [2], emphasize that the expansion of FinTech startups is due to low bureaucratic boundaries, great knowledge about customer needs and highly qualified dynamic teams. However, although one can find research about the digitalization of the financial services industry, the literature about FinTech startups has begun to develop recently and is still scarce [1, 3].

2.2 FinTech Ecosystem

FinTech services can influence financial institutions, regulators, customers and retailers in a wide range of industries [7]. New trends and needs related to the financial services delivery have been driving the development of an entirely new ecosystem, affecting both FinTechs and non-financial firms [1]. From a Service-Dominant Logic perspective. the service ecosystem can be understood as a structure in which the actors co-create value and exchange service [11]. This ecosystem is of paramount importance to ensure that technological innovation is created to make financial services more efficient and to improve customer experience [12].

According to Lee and Shin [4], the ecosystem of a FinTech comprises five elements: FinTech startups; government; traditional financial institutions; financial customers; and, technology developers. A well-developed ecosystem can stimulate the local economy and generate opportunities for growth [12]. In order to understand the dynamic of a FinTech and how it develops, it is necessary to comprehend in detail this ecosystem [4].

The FinTech are characterized as follows:

- FinTech Startups: FinTech startups are new technology-based companies that offer innovative solutions in the financial industry [3]. These companies are responsible for the innovation leaps in that industry, so they should be considered the central piece of the ecosystem [4]. Fintech startups' businesses have many elements that directly impact the consumer in the financial value chain, such as the use of digital channels [3] and prioritizing meeting niche market needs by offering customized services [4]. FinTech startups adopt a customer-centric strategy, although such strategy does not remove uncertainties about long-term profits and success rates [3].
- Government: The global financial sector has experienced significant changes in regulations due to digital technologies and their disruptive effect [7]. Governments and regulatory agencies can positively impact different dimensions of the ecosystem, e.g., by simplifying trade regulations or by reducing taxes and duties. However, they can also have a negative impact, namely by creating more rigid and bureaucratic regulations [12]. After the global economic crisis of 2008, governments and regulation agencies further developed regulations, prioritizing

transparency in an attempt to reduce fraudulent behavior and protect consumers [8]. At the same time, some countries started initiatives that promote the emergence of FinTechs [1].

- Traditional Financial Institutions: Traditional financial institutions (TFIs) are essential for FinTech ecosystems [12]. After the first impact of FinTechs emergence in the financial sector, TFIs have been reviewing their business models and developing new strategies to innovate, particularly by using technology [4]. Although at the beginning TFIs faced FinTech startups as a threat, recently they have started to work in collaboration with those new companies [4], through acquisitions and the creation of in-house incubators seeking to create new services with lower operational costs and more competitive prices [13]. According to Diemers [12], the relationship of TFIs with FinTech startups may stimulate innovation within the former and further strengthen their competitive position.

- Financial Clients: One of FinTech startups' key feature is the ability to identify customer needs [3]. FinTech startups focus on offering services that meet the needs of market niches by delivering high quality and personalized services [4]. This approach is extremely important to acquire of new clients, since customers evaluate the benefits and risks of the services of a FinTech before using them [14]. Customer satisfaction is of paramount importance for FinTech startups because word-of-mouth recommendations can be crucial for business success in such a highly competitive industry [4]. Moreover, high standard and value-added services attract customers who have been served by TFIs, since those institutions do not offer services that meet the specific customers' needs [15].

- Technology Developers: With the advance of information technology, technology developers deliver digital technologies, such as big data, cloud computing, social media, and artificial intelligence (AI), that are one of the factors responsible for FinTech startups success [4, 8]. The internet of things (IoT), cloud computing, big data, social computing, among other technologies, enable startup companies to automatize their business processes and offer unparallel services and products within the financial sector [1].

2.3 Service Innovation and Social-Technical System Theory

Customers have been changing the way they perceive the purchasing process, and therefore, companies realized the need to redesign their businesses to improve customer experience, by using service innovation [16, 17]. According to Snyder, Witell [17], service innovation focuses on co-creating value from service actors' experiences since its primary focuses should be value creation and customer experience. Service actors are clients, organizations or other individuals that may be related to the process or offer of services, and who participate in the process of value creation [18].

Service innovation fosters the development of service sectors and is an engine of society renewal [17]. Service innovation involves value creation for clients, organizations or other actors, through the development of new processes or services offers [18]. Gallouj and Weinstein [19] identify different modes of service innovation: radical, which involves the creation of an entirely new service, or incremental, which involves changing some characteristics of the service without changing the general offer. In addition to the modes of service innovation, Gallouj and Savona [20] define service innovation from three approaches: (1) assimilation or technologist approach, (2) service-based or differentiation approach, and (3) synthesis approach. Socio-technical systems (STS) theory analyses the relationship between social and technology dimensions and identifies whatever may emerge from this interaction [5]. The STS is considered appropriate for the analysis of emerging technology-based areas, as is the case of financial services, since the industry seeks a model that is able to explain the social and technological changes in the sector [5]. In order to understand the complex relationship between technology and workers within an organization, it is essential to analyze the organization and its context as a STS [5, 21]. The STS consists of two subsystems: the social subsystem, which comprises people (workers) and structures, and the technical subsystem that encompasses technologies, processes, procedures, and the physical environment [5].

Furthermore, according to Fuenfschilling and Truffer [22], a socio-technical transition (change) involves interactions between three pillars: actors, institutions, and technologies. STS only work due to the interactions between human actors (companies, consumers, industries, citizens, public authority, social groups), technologies (radical innovations and incremental improvements), and institutions (legislation and formal rules) [22–24]. In this context, technologies define the organizational environment in which human agents act [24]. A socio-technical transition means a change from one socio-technical regime to another, with the interaction between actors, institutions (organizations), and technologies [22]. This transition is a moment when new products, services, organizations, business models, or regulations that may complement or replace the previous offer emerge [24, 25].

3   Method

The method used in this research was case research. Case research is appropriate to understand complex social phenomena, such as organizational processes and management, and the maturation process of industries [6]. Yin [6] defines case studies as particular situations that have great empirical detail, and which are based on several data sources. Multiple case studies may increase the external validity and help prevent researcher bias [26, 27].

Interviews were the main source of data used in case research [26]. Semi-structured interviews were used and secondary data that included materials of public relations, such as press releases, newspaper articles, information on the website, among others, was also used. Between March and April of 2019, a total of 11 FinTech startups have accepted to be part of this study. These companies have been categorized as FinTech in consulting reports such as Portugal FinTech Report 2018 and Finnovista FinTech Radar Brazil 2018 [28]. The FinTech startups are presented in Table 1.

Table 1. Characterization of the FinTechs startups interviewees.

FinTech startup	Business type	FinTech size (employees)	Year of foundation	Meeting type	Country	Interviewee
FinTech BR1	Wealth management	51–200	2014	Web call	Brazil	Co-founder
FinTech BR2	Payment	11–50	2014	Web call	Brazil	Head of business development
FinTech BR3	Payment	201–500	2016	Web call	Brazil	Public relationship
FinTech BR4	Capital Market	201–500	2010	Web call	Brazil	Founder
FinTech BR5	Capital Market	11–50	2017	Web call	Brazil	Marketing manager
FinTech BR6	Crowdfunding	11–50	2010	Web call	Brazil	CEO
FinTech PT1	Payment	11–50	2014	In person	Portugal	Co-founder
FinTech PT2	Crowdfunding	2–10	2011	Web call	Portugal	Co-founder
FinTech PT3	Financing	2–10	2018	In person	Portugal	CEO
FinTech PT4	RegTech	11–50	2017	Web call	Portugal/Switzerland	Product owner
FinTech PT5	InsurTech	11–50	2015	In person	Portugal	CEO

4   Results

Considering the analysis of the findings, based on coding techniques, it was possible to identify two main categories, and several other subcategories.

The "FinTech Ecosystem" category, characterizes the relationship between the stakeholders of the startups under study and their respective business models. This analysis was performed based on the diagram of the FinTech ecosystem developed by Lee and Shin [4]. In addition to the stakeholders identified by Lee and Shin [4], investors were also found to be important stakeholders. Table 2 illustrates the relationships found between each stakeholder and FinTech startup.

The key customers of the startups under study were categorized as other companies and individuals. Most of the FinTechs studied have other companies as the target, being characterized as B2B (business to business). Other FinTechs target individuals and are defined as B2C (business to consumer). There is also a small group of FinTechs that serves both targets, which were classified as B2B and B2C. They also can be identified as actors from the S-D logic perspective, which swapping service in an A2A network [11]. However, in these results, it will be used the Innovation literature for roles of actors, as B2B and B2C.

Characteristics of the FinTech startups also were identified, related to the founders, capital investments, and value proposal, as shown in the Table 2.

Among the companies studied, several behaviors regarding the relationship with the government have been identified. No single pattern emerged from the data concerning the relationship between Financial Regulators and FinTech startups. Instead, five types of different behaviors were uncovered: a closer relationship to Financial regulators, a nonexistent relationship, a loophole relation between the company and the government, a proactive relationship, and an indirect relationship to Financial regulators.

Table 2. Coding tree – category: FinTech ecosystem

Category	Subcategory 1° level	Subcategory 2° level	Sources
FinTech ecosystem			11
	Financial customer		11
		B2B	5
		B2C	4
		B2B and B2C	3
	FinTech startup		11
		Foundation	11
		Investment capital	11
		Value proposition	11
	Government		11
		Financial regulators and legislature	11
	Investor		11
		Pivotal	7
		Irrelevant	3
		Neutral	3
	Technology developers		11
		In-house	8
		Outsource	4
	Traditional financial institution		10
		Banks	9

Different relationships between FinTech startups and investors were also identified. Pivotal, refering to FinTechs that consider their relationship with investors essential for the business. Neutral relationships with investors, given that startups only needed the third-party capital when the company was already at a more mature stage, and its operations were more structured. And, irrelevant relationship, i.e., startups with no connection with investors.

Due to the importance of technology for a FinTech, in every interview, this element was identified as essential for the business. Furthermore, the relationship between technology developers and startups were divided into two groups, as seen in Table 2: companies working with either outsourced or in-house team of technology developers.

The FinTech startup relationship with traditional financial institutions is directly related to the startup business type. In most of the interviews, banks were considered the most important financial institution for the business. However, in the case of Financing and InsurTech startups, the insurers and credit brokers were considered key institutions.

The "Innovation Process" category, describes the stages of the startups' innovation processes according to the service innovation literature, from idea generation, idea selection, development and dissemination to product and service delivery [29]. In

Table 3. Coding Tree – category: innovation process

Category	Subcategory 1° level	Subcategory 2° level	Sources	Quotes
Innovation process			11	
	Idea generation		11	
		Market opportunity	9	"At the time, there was no crowdfunding platform in Portugal. So, we decided to replicate what existed, for example, in the USA, the Kickstarter, adapting it to the Portuguese context"
		Solving Own problems	2	"The company was created out of the founder dissatisfaction with the bank he used"
	Business definition		6	
		Headquarters definition	2	"We decided to go to Porto Alegre, because São Paulo were very expensive (…)"
		Organization process	2	"We had to open up a wealth manager and go through the whole process. And since everything was 100% digital, (…)we were fight with the regulator of the financial market"
		Teambuilding	2	"One of the fundamental factors and 100% vital for this business to get off the ground, is that the 3 founders have completely different backgrounds"
	Development		11	
		Product process	11	"Product development for sure. We have a very strong focus on UX, and the further we go in this direction, the more updated, the better it will be"
		Improvement driver	8	"We do a lot of usability testing with the customer"
		Radical evolution	6	"We created a simple investment course and started delivering education online. In

(*continued*)

Table 3. (*continued*)

Category	Subcategory 1° level	Subcategory 2° level	Sources	Quotes
				2010 there wasn't much talk of FinTech, and this combination of technology and education is very powerful"
		Incremental adaptation	5	"At the beginning, we had to adapt our reality, the means of payment are not the same as those of the USA, [(used as a reference for the business creation)]so we had to have local means of payment"

Table 3, it is possible to observe these three stages of the innovation process based on the FinTechs studied. The first stage of the innovation process was the Idea Generation, where the real reasons for the emergence of the idea that generated the FinTech were collected. Two main reasons were identified: a market opportunity and the solution to a personal problem. Most of the FinTechs indicated that the idea of their business arose due to an opportunity the founders identified in the market. The second stage of the innovation process was the Definition of the FinTech business. At this stage it was possible to understand some decisions that the startups needed to make in order to build the business. Some decisions were considered particularly important by most of the startups, such as: definition of organizational processes and team building with different backgrounds. The third stage of the innovation process was the Development of startups. In this phase subcategories were identified that show the development of the business in several areas of the startup, from the processes of validation of the service or product offered, to issues related with the evolution and improvement of the startup and the product or service it provides (Table 3).

A crucial element identified in every case study was the product development process. During the interviews, it was possible to identify some characteristics of the main products that changed and some practical use cases that induced change. As seen on Table 3, during the business development stage, adjustments and improvements to the products and services provided were classified as Radical Evolution or Incremental Adaptation. Among these radical changes or improvements that startups have been through, the drivers responsible for those changes were identified. In most cases, the consumer was considered the main reason for these adaptations, which can be related to user experience design (UX) or the mode of providing the service. Regulatory agents were the second driver identified. In some cases, these are considered responsible for changes in FinTech startups. The last driver identified were other companies that the startups consider as benchmarks.

5   Discussion and Conclusion

5.1   FinTech Ecosystem Evolution - Socio-technical System Perspective

From the results, and using the socio-technical system perspective, it is possible to propose a new configuration for Lee and Shin [4] diagram that includes a new element identified, the investors. The FinTech ecosystem diagram proposed illustrates the relationship between a FinTech startup and each of the stakeholders that influence its evolution. This new diagram is not intended to substitute the framework proposed by Lee and Shin [4] since it was developed with the purpose of understanding the evolution of the FinTech and its ecosystem, while the Lee and Shin [4] diagram had the purpose of characterizing such ecosystem.

According to the findings from the case studies, relationships between startups and stakeholders occur on a two-way direction. FinTechs startup are considered as the central element of the ecosystem, as the focus of this research is the evolution of the FinTech. Adapting the Lee and Shin framework, allows to more clearly identify how do the different actors in the ecosystem influence the evolution of the FinTech and its ecosystem, while still being able to characterize such actors, based on the framework proposed by Lee and Shin [4]. Based on the results of the case studies, the following paragraphs characterize the actors of the ecosystem, and their influence in the evolution of the FinTech startup and the ecosystem.

FinTech Startups: The FinTech startup was considered the key and central element in the diagram since it interacts with every other stakeholder of the ecosystem. During this study, it was found that FinTechs are responsible for significant structural and cultural changes both in Brazil and Portugal. The structural changes emerge from the use of new technologies to provide improved services, the implementation of new regulation within the context of financial services that affects and is affected by the FinTechs, and the increasingly closer relationship between big financial institutions and FinTechs startups. Cultural changes are related to the new form of consumer behavior, as customers are now used to have omnipresent access to their financial data [8] that creates new requirements for the services provided, such as further accessibility, convenience and customized products [4].

Government: Government and regulatory agencies are some of the main actors that realize the relevance of the development of FinTechs, and for this reason, they are increasingly searching for solutions to facilitate the development of these startups. Results from this study show that the relationship between FinTech startups and the government can be characterized as close, indirect, loophole, non-existent and proactive. In the Brazilian context there is a predominance of close relationships, since there is a strong demand for new regulatory issues that strongly influence the new business models of FinTechs startups. In Portuguese context, there is a predominance of indirect relationships, since startups recognize the relevance of regulatory issues for the ecosystem, however, changes in regulation do not impact directly the way in which startup businesses have been developed.

Traditional Financial Institutions: Over the years, FinTech startups started to prove the potential of their business, which led traditional financial institutions to start showing interest in these startups. Despite that rapprochement, according to the results of this case research, the relationship between FinTech startups and traditional financial institutions, especially banks, is still difficult, because the processes involving banks are still perceived as bureaucratic and time-consuming. The relationship between FinTech startups and traditional financial institutions was found to be characterized as a relationship of dependence, neutral or partnership, with no relevant predominance of any of these emerging from the data.

Technology Developers: The relationship between technology developers and Fin-Techs is of paramount importance and was found to be defined as outsource and in-house. Most of the startups studied have internalized the processes related to technology development (to protect their intellectual property) since these can be considered an important element for their competitive advantage. However, some of the currently used technologies (such as cloud computing and mobile services), despite being essential for the startups, have become commodities. Therefore, technology developers also need to be aware of trending and promising technological innovations, such as Artificial Intelligence, automation of robotic processes (RPA), blockchain, and IoT [30] to be lead providers if they also become commodities for these startups.

Customers: Most businesses emerged from a market opportunity. Therefore, the role of customers in the FinTech ecosystem is of high relevance to the business, being considered a key stakeholder to guide the development of new services. In addition, customers also have an important role in the adaptation and improvement of existing services. Those improvements can be incremental adaptations, or sometimes can result in the radical evolution of a service. The relationship of FinTechs startups with consumers can be defined as business to business (B2B), business to consumer (B2C), or B2B and B2C. In a global analysis, B2B FinTechs are increasing [30], and B2C FinTechs typically focus on the Millennium generation since these consumers search for better and more innovative experiences using digital channels [31].

Investors: Although investors are not part of the FinTech ecosystem proposed by Lee and Shin [4], from the findings of the case research this stakeholder is very important for FinTech startups. These stakeholders were considered as key for the creation or growth of the startups. The relationship between investors and startups can begin during the development of the business, through venture capital or business angel capital, or in a more mature phase of the startup, thought other sources of funding, such as investment funds, private equity, debt financing, IPO or acquisitions. The relationship of FinTechs startups with investors was found to be characterized as pivotal (in most cases), neutral (when the investor has little influence on the startup) or irrelevant (mostly when there is self-financing).

5.2   FinTech Ecosystem from a Service Innovation Perspective

Based on the information collected, and using a service innovation perspective, it was possible to identify the stakeholders with more influence on each stage of the innovation process and their respective roles in those stages. Although innovation processes are typically not linear, they are presented as continuous and linear processes for clarity purposes. As the objectives of this study are to characterize the FinTech ecosystem and the stages of their innovation process, and not the flow of the innovation process, the non-linearity of that process was not explored.

Stakeholders of the FinTech ecosystem influence every stages of the innovation process of creating a new FinTech startup, and consequently, directly influence the innovation of the services provided by FinTechs startups. However, apart from the startup, which is the object of creation and evolution and therefore is the center of the whole process, some stakeholders were found to have a stronger influence in some stages of the process, which is explain in the following paragraphs.

The first stage of the innovation process is the generation of ideas. The most influent stakeholders in this stage are customers that drive the creation of the FinTech through their needs, the government that influence the context in which the FinTech is created, namely by defining regulation to comply with, and the investor, when the startup needs external capital. In most cases, the origin of business idea is a combination of customer needs, regulatory issues and market opportunities.

The second stage of the innovation process is the business definition, where organizational processes are defined and teams are structured. At this stage, the key stakeholders are investors that might bring their experience from working with other startups, and the government since it is necessary to validate the service within the existing regulation.

The third stage of the innovation process is the business development. At this stage, customers, technology developers, government, traditional financial institutions are the most important stakeholders. They are directly or indirectly related to the radical evolutions and incremental adaptations that FinTech startups implement in the services they provide and that enable the growth of the startup's business. Customers guide the development of new offers, and technology developers are key to provide the technology that startups need to improve their services, i.e., FinTechs depend on them to innovate. Then, the government might still be relevant since regulation needs to be complied with (usually contributing to incremental adaptations of the services). Traditional financial institutions might be relevant partners to be able to provide the new services as they might be more efficient in providing some parts of the service, and investors are still an important source of experience and, naturally, are also very important to finance the innovation.

As mentioned above, government and regulatory agencies have a significant influence on most of the cases studied. However, regulatory issues contribute specifically for incremental adaptations of FinTech startups, since, after the validation of the business, it is only necessary to adjust it to operate within the regulation established. On the other hand, traditional financial institutions influence some incremental adaptations of the startups, especially when there is an interdependence between them and the FinTech startup. Investors also have an influence on incremental adaptations and radical evolutions, as they are identified as a key element for startup growth.

5.3 Contributions to Theory and Management

In this study, a deeper understanding about the evolution of the FinTech ecosystem is provided, based on the perspectives of service innovation and STS theory.

The FinTechs ecosystem framework developed by Lee and Shin [4] fits into the STS theory, and thus presents a structure that allows understanding the evolution and changes that exist together with the social, technological and organizational actors. Therefore, it was possible to identify that just as FinTech startups go through changes, the other agents of their ecosystems also develop and adapt to the new reality created by the change of FinTech. This adaptation of other actors was particularly evident in regulatory issues, which have changed due to FinTech emergence, and in traditional financial institutions, which changed their perceptions about FinTech as they matured.

The services innovation perspective was particularly important to identify how FinTech startups search for the appropriate business models to provide the services that they are offering. During this innovation process, it was observed how stakeholders of the FinTech ecosystem impact the evolution of the startup and of the services it provides. The innovations observed were characterized as radical or incremental. Customers and regulatory issues are an example, and particularly important stakeholders, of how actors in the ecosystem can drive a change that would initially be an incremental innovation, but eventually may become a radical innovation. These two stakeholders are particularly important in this process because customer needs and regulation are the most important drivers of the creation of new FinTech startups, and they have a broad impact in the innovation process, with important presence in almost all its stages.

5.4 Research Limitations and Future Research

As with any research, this study has some limitations, which are listed below, in addition to some suggestions for future research. The first limitation of the study is related to the limited number of FinTech startups studied, due to the limited time for the completion of this study. Future research may consider a more extensive sample of FinTech startups from Portugal, Brazil, and other countries, to extend the findings to other countries as well. In this context, it would be interesting to analyze startups operating the same types of business in the different countries, and compare the evolution of the FinTech ecosystems in the different countries. A quantitative study may also be carried out in order to validate these findings on a larger sample and measure which stakeholders have greater importance for the evolution of the business.

References

1. Puschmann, T.: Fintech. Bus. Inf. Syst. Eng. **59**(1), 69–76 (2017)
2. Gagliardi, L., Dickerson, J., Skan, J.: FinTech and the evolving landscape: landing points for the industry. Accenture, pp. 1–12 (2016)
3. Gimpel, H., Rau, D., Roglinger, M.: Understanding FinTech start-ups - a taxonomy of consumer-oriented service offerings. Electron. Markets **28**(3), 245–264 (2018)

4. Lee, I., Shin, Y.J.: Fintech: ecosystem, business models, investment decisions, and challenges. Bus. Horiz. **61**(1), 35–46 (2018)

5. Durkin, M., Mulholland, G., McCartan, A.: A socio-technical perspective on social media adoption: a case from retail banking. Int. J. Bank Mark. **33**(7), 944–962 (2015)

6. Yin, R.K.: Case Study Research: Design and Methods. Sage (2003)

7. Leong, C., et al.: Nurturing a FinTech ecosystem: the case of a youth microloan startup in China. Int. J. Inf. Manag. **37**(2), 92–97 (2017)

8. Alt, R., Beck, R., Smits, M.T.: FinTech and the transformation of the financial industry. Electron. Mark. **28**(3), 235–243 (2018)

9. Dietz, M., Olanrewaju, T., Khanna, S., Rajgopal, K.: Cutting Through the FinTech Noise: Markers of Success, Imperatives For Banks. McKinsey (2016)

10. Alt, R., Puschmann, T.: The rise of customer-oriented banking - electronic markets are paving the way for change in the financial industry. Electron. Markets **22**(4), 203–215 (2012)

11. Lusch, R.F., Nambisan, S.: Service innovation: a service-dominant logic perspective. MIS Q. **39**(1), 155–175 (2015)

12. Diemers, D.L., Salamat, J., Steffens, T.: Developing a FinTech ecosystem in the GCC: let's get ready for take off. PWC: Strategy (2015)

13. Dany, O.G., Schwarz, J., Berg, P.V.D., Scortecci, A., Baben, S.: FinTechs May Be Corporate Banks' Best "Frenemies". The Boston Consulting Group (2016)

14. Ryu, H.S.: What makes users willing or hesitant to use Fintech?: the moderating effect of user type. Ind. Manag. Data Syst. **118**(3), 541–569 (2018)

15. Gomber, P., Koch, J.A., Siering, M.: Digital finance and FinTech: current research and future research directions. J. Bus. Econ. **87**(5), 537–580 (2017)

16. Ostrom, A.L., et al.: Service research priorities in a rapidly changing context. J. Serv. Res. **18**(2), 127–159 (2015)

17. Snyder, H., et al.: Identifying categories of service innovation: a review and synthesis of the literature. J. Bus. Res. **69**(7), 2401–2408 (2016)

18. Patricio, L., Gustafsson, A., Fisk, R.: Upframing service design and innovation for research impact. J. Serv. Res. **21**(1), 3–16 (2018)

19. Gallouj, F., Weinstein, O.: Innovation in services. Res. Policy **26**(4–5), 537–556 (1997)

20. Gallouj, F., Savona, M.: Innovation in services: a review of the debate and a research agenda. J. Evol. Econ. **19**(2), 149–172 (2009)

21. O'Hara, M.T., Watson, R.T., Kavan, C.B.: Managing the three levels of change. Inf. Syst. Manag. **16**(3), 63–70 (1999)

22. Fuenfschilling, L., Truffer, B.: The interplay of institutions, actors and technologies in socio-technical systems - an analysis of transformations in the Australian urban water sector. Technol. Forecast. Soc. Chang. **103**, 298–312 (2016)

23. Geels, F.W.: From sectoral systems of innovation to socio-technical systems - insights about dynamics and change from sociology and institutional theory. Res. Policy **33**(6–7), 897–920 (2004)

24. Zhang, H.P., Tang, Z.W., Jayakar, K.: A socio-technical analysis of China's cybersecurity policy: Towards delivering trusted e-government services. Telecommun. Policy **42**(5), 409–420 (2018)

25. Markard, J., Suter, M., Ingold, K.: Socio-technical transitions and policy change - advocacy coalitions in Swiss energy policy. Environ. Innov. Soc. Transitions **18**, 215–237 (2016)

26. Voss, C., Tsikriktsis, N., Frohlich, M.: Case research in operations management. Int. J. Oper. Produ. Manag. **22**(2), 195–219 (2002)

27. Eisenhardt, K.M., Graebner, M.E.: Theory building from cases: opportunities and challenges. Acad. Manag. J. **50**(1), 25–32 (2007)

28. Finnovista FinTech Radar Brazil 2018 (2018). https://www.finnovista.com/fintech-radar-brazil-may2018/?lang=en. Accessed 02 Feb 2019
29. Salerno, M.S., et al.: Innovation processes: which process for which project? Technovation **35**, 59–70 (2015)
30. Pollari, I.R.: The Pulse of Fintech 2018 - Biannual global analysis of investment in fintech. KPMG (2018)
31. Taiar, A.R., Neves, R.: Brazil FinTech Deep Dive 2018. PricewaterhouseCoopers Brasil (2018)

Understanding the Impact of Artificial Intelligence on Services

Pedro Ferreira[1], Jorge Grenha Teixeira[1,2(✉)], and Luís F. Teixeira[1,2]

[1] Faculty of Engineering, University of Porto, Porto, Portugal
`jteixeira@fe.up.pt`
[2] INESC TEC, Porto, Portugal

Abstract. Services are the backbone of modern economies and are increasingly supported by technology. Meanwhile, there is an accelerated growth of new technologies that are able to learn from themselves, providing more and more relevant results, i.e. Artificial Intelligence (AI). While there have been significant advances on the capabilities of AI, the impacts of this technology on service provision are still unknown. Conceptual research claims that AI offers a way to augment human capabilities or position it as a threat to human jobs. The objective of this study is to better understand the impact of AI on service, namely by understanding current trends in AI, and how they are, and will, impact service provision. To achieve this, a qualitative study, following Grounded Theory methodology was performed, with ten Artificial Intelligence experts selected from industry and academia.

Keywords: Service science · Artificial Intelligence · Grounded Theory

1 Introduction

Artificial intelligence (AI) has been defined as "a computational method that attempts to mimic, in a very simplistic way, the human cognition capability to solve engineering problems that have defied solution using conventional computational techniques" [1]. Russell and Norvig [2] claim that systems that think like humans, act like humans, think rationally and act rationally can be related with AI. Hence, AI is and will be, increasingly more prevalent in our lives. Google or Amazon products are being integrated into our daily life, and many more companies are providing smart services in their own products. When observing AI evolution, it is possible to posit that AI will shape the service field, as new service channels arise and are reinvented and jobs and tasks are replaced by AI [3]. Thus, robotics and Artificial Intelligence will create changes in economics and business, bringing new ways of living and sociological effects [4].

Concurrently, there has been a growing academic interest in services, with more disciplines rethinking their curricula, due to the growth of services [5]. In this way, many companies are facing new demands and intense competition. Therefore, they aim to apply services to differentiate themselves from the competition [6].

Lovelock and Wirtz [7] define services as economic activities that are offered by one party to another, applying time-based performances to bring expected results in

© Springer Nature Switzerland AG 2020
H. Nóvoa et al. (Eds.): IESS 2020, LNBIP 377, pp. 202–213, 2020.
https://doi.org/10.1007/978-3-030-38724-2_15

recipients themselves or in objects or other assets that purchasers have liability. This definition has since evolved towards an understanding of service as the process of reciprocal application of resources for others' benefit [8]. Furthermore, service customers are expected to obtain value from accessing goods, labor, skills, facilities, networks and systems but not taking ownership of any physical elements involved, in exchange for money, time and effort [7]. Following a more technology-oriented perspective, service science has emerged as "the study of service systems, which are dynamic value co-creation configurations of resources (people, technology, organizations and shared information)" [9].

The interest in technology from a service perspective has also led to a growing interest on smart services [10]. According to Wuenderlich et al. [10], smart services are powerful, allowing real-time data collection, continuous communication and interactive feedback. However, research on these topics is still in its early stages, despite the accelerating technological development [10]. Due to the accelerated evolution in technology, it is important to incorporate the technological change to improve service strategy [11]. Thus, technology in services with a smart oversight has a huge potential, but it requires further research to bring success to organizations and customers [10]. Advances in technology are aiming to develop new and transforming services, changing how customers behave [6]. Smart interactive services can offer several research lines [12], and it is crucial and far-reaching to understand how will AI alter and transform service delivery.

Some authors claim that AI will reshape jobs, threatening human service jobs [4, 13, 14]. Others say that it will ease the daily life, constituting a major source of innovation [3]. Accordingly to Makridakis [14], "the AI revolution aims to substitute, supplement and amplify practically all tasks currently performed by humans, becoming in effect, for the first time, a serious competitor to them". Thus, since services are increasingly reliant on technology, the application of AI has a disruptive effect that will transform the traditional way we provide services.

This research paper has the purpose of exploring the possible impacts of AI in service provision. A qualitative study based on Grounded Theory methodology [15, 16] was performed to achieve this objective, involving semi-structured interviews with ten AI experts from academia and practice. This exploratory approach enables a rich and in-depth understanding of what services will be most probably impacted by AI and how. Before introducing the research methodology and the results of the study, the next section includes relevant literature on AI.

2 Literature Review

The first recognized AI work was done by Warren McCulloch and Walter Pitts in 1943, with a model of artificial neurons characterized by "on" and "off". Marvin Minsky and Dean Edmonds, built the first neural network computer in 1950. Alan Turing's work was the most notable with the introduction of the Turing Test in the 50's [2, 17].

From 1956 until the early 80's, new hypotheses were formulated, such as AI programming languages and limited domains known as microworlds. On the other hand, these developments suffered from the lack of significant technology and early scientific knowledge. From 1980 to the present, investment from the industry towards understanding and building AI systems started to become more relevant. Neural Networks started to be again a trending subject, as well as Data Mining and concepts such as the Human-Level AI, Artificial General Intelligence - computers as smart as humans - and Big-Data [2, 18]. State-of-the-art examples of AI technologies such as robotic vehicles, speech recognition, autonomous planning, machine translation and skin cancer classification through image recognition are expected to be seen in the near future. In the service sector, chatbots, robotic service providers and intelligent agents are also emerging technologies [2, 11, 14]. Furthermore, it is expected that computers pass the Turing test successfully, indicating intelligence indistinguishable from humans, by the end of this decade [13]. Further in the future, and according to Miller [19], technological singularity is "a threshold of time at which AI's that are at least as smart as humans, and/or augmented human intelligence, radically remake civilization". According to Kurzweil [13], a superintelligence will thrive leading to a technological singularity [20], after the nanotechnology settles in the society, virtual reality will become a common feature within the nervous system, and the human society will have the possibility to be a different person, physically and emotionally. Hence, Kurzweil predicted that computers will reach human intelligence around 2029 while Singularity will come by 2045 [13].

While far from the futuristic scenarios drawn by these authors, nowadays, technology can be found everywhere in service and it is the main origin for innovation, empowering firms to leverage its benefits based to their strategic position. Thus, it can aid firms to standardize, customize, transactionalize and relationalize service [11]. However, technology needs to be adequately deployed to support service innovation in order to provide seamless customer experiences [21]. As such, further exploration of the effects of smart services have on organizations should be properly addressed [10]. To effectively leverage technology and enable a seamless customer experience across interfaces and systems, the design of the service and its implementation must be carefully managed [21]. However, technology in services brings challenges, especially when they are implemented in existent business models. Organizations require adaptations in their business models and their services offerings need to hold new smart business models. Thus, collaborations between researchers and managers are beneficial for the development of smart service offerings [10].

In a service encounter, aside from the fact that this represents a cross-disciplinary process integrating technical knowledge with psychological knowledge of interaction dynamics, it has been verified that when social robots show innovative service behaviors during a service encounter, they normally exceed customer expectations and deliver exceptional experiences to customers [22]. Being AI at the cutting edge of smart technologies used in services, it can help to provide better service over time [11]. However, the study of AI in service is still in an initial stage.

Huang, Ming Hui, Rust [3], have proposed a theory of AI job replacement in service, based on the premise that there are four types of intelligence which will be mastered by AI chronologically – mechanical, analytical, intuitive and empathetic – being the first one the easiest to master and the last one the harder to overcome. Furthermore, portraying what humans do nowadays and defining what can be augmented by machines is known as *Augmentation* [23]. That is why Huang, Ming Hui,

Rust [3] claim in their article that the job replacement will happen at the task level, from the easier and lower-skilled intelligent tasks (mechanical) to the higher intelligence tasks.

According to Ulrich [24], mechanical intelligence is a possibility "to imbue a mechanism with the ability to respond and react to the environment without guidance from a controller". The tasks that normally embed this type of intelligence include the ones which do not require further education or skills, like call-center agents or waiters/waitresses [3]. An everyday use that we give to one of the most known tools which are Google, Bing or any other type of search engine platform is considered a type of AI, included in the category of Mechanical Intelligence, because it retrieves the most relevant information, accordingly with previous searches. A further profound change could happen with the self-driving cars, which can be considered a very good example of a mechanical intelligence AI [25].

Whenever there is information that needs to be processed, analyzed, evaluated, judged and compared, the need of analytical intelligence is adequate [26]. Thus, problem-solving, logical and mathematics are primary characteristics that define analytical intelligence, normally acquired by training and refined by cognitive thinking, used by "computer and technology-related workers, data scientists, mathematicians, accountants, financial analysts, auto service technicians, and engineers" [3]. The collection of data and inter-machine communication are the two main and most important features of AI that it is known in the service field [3].

Intuitive intelligence requires a higher state of creativity and problem-solving abilities, like management consultants, lawyers, doctors and marketing managers have [3]. This type of intelligence is distinguishable from the analytical since it is able to analyze several data, understand the subject and the content of a certain fact. As Del Prado [25] defends, if a certain system could read all pages of millions of books that are related to a specific subject and combine them all in a single answer, then we would have evolved to a new level of intelligence. Tasks that are creative, chaotic, things that most skilled humans in sports or entertainment do, require intuitive intelligence for providing the best service [3].

Seddon and Biasutti [27] presents an example in music of how Empathising is intrinsically connected with interpersonal social skills and requiring collaboration towards creativity. From a service provision perspective, the way an employee shows their emotions while fulfilling service tasks to a certain customer, defines how good the service was, from the customer point-of-view. An example is a flight attendant or a psychiatric doctor [3]. Empathetic intelligence is different from emotional or cognitive intelligence, being the connection between thinking and feeling, being inspired by intuition and reflective practice, requiring recognition on a complex system between culture and human responsiveness [28]. Being directly connected with people and with other's feelings, it is a type of intelligence normally required by psychologists or customer-related personnel [3]. This is the most difficult stage of intelligence to simulate in a machine, due to the fact that "emotion is considered a biological reaction" [3]. As Fabrega [29] argues, the cognitive-emotional space of consciousness was always embedded in humans, from pre-historical times to what we are today. In Turing, 1950, the AI pioneer Alan Turing starts his book with the following statement: "Can machines think?". The Turing test determines if a computer can think like a human based on their behaviour (human or computer) observed by an outsider, no mattering how they achieved that level. Reaching this type of intelligence is very hard on AI machinery, and some are often designed not to look humanoids, for ethical purposes. Some examples are *Han* and his evolution and more intelligent version – *Sophia* – who was recently awarded a Saudi Arabia citizenship [3].

Current research shows that AI will have a very significant impact on service provision over time however it does not further characterize this impact. Based on a qualitative study, this paper aims to contribute towards closing this research gap.

3   Methodology

This research aims to understand the impact of AI on service. Qualitative research helps to address these exploratory questions, enabling researchers to address "why" and "how" questions, as opposed to quantitative research that is focused on measuring and validation [30]. Therefore, qualitative research is suitable to address the aims of this research. As such, this study follows a qualitative approach based on Grounded Theory [15] involving semi-structured interviews with experts from the AI field, in academia and industry. A total of ten interviews were made, five to university professors, four to experts working in industry, and one to a full-time researcher. The experts from the industry work in areas related to software development and data science, while university-related experts perform their research on areas such as Data Mining, Data Science and Machine Learning.

Data for this study was collected between the 20th of March and the 4th of May of 2018. Firstly, a selection of AI experts was made, based on their professional activity. All the interviewees were contacted by e-mail to briefly explain the study's objectives and, if agreed upon, to schedule the interviews. All the interviews were audio-recorded and the interviews were given and signed an informed consent. This informed consent included guarantees of confidentiality and anonymity of the interviewee. Interviewees were asked questions related to their background and qualifications, their opinion on the trends and challenges for AI on the short term, what impacts will AI have on society, what new services could arise using AI and what will be the impact on the existing ones. Finally, interviewees were asked about the long-term (10 years or more) trends and challenges for AI. All gathered data was literally transcribed and then coded and analysed using a Computer-Assisted Qualitative Data Analysis Software (NVIVO11®).

4   Results

In this section, the results of the qualitative study are presented. The codes from the interview's information were aggregated according to different services and shown in Table 1: healthcare, customer service, transportation and logistics, education, environment-related, financial services and retail. These different service contexts emerged directly from the data analysis. The positive and negative impacts of AI on each of these different kinds of services are specified, as well as challenges and barriers to develop these services and possible practical applications. Some quotes from the respondents are also provided to add further detail. This study also uncovered cross-cutting topics that impact all the identified services. These topics are discussed at the end of this section. The results do not distinguish between respondents from academia, or from industry, as responses were not significantly different. It is also important to notice that some interviewees who are on academia also, in some moment of their life, worked in industry.

First and foremost, the impact on healthcare was mentioned by all interviewees. The interviewed experts emphasize how AI has a great potential to bring about new drugs and improve early diagnosis of disease. While interviewees voiced concerns about human care being replaced, others emphasize that AI will never replace the need of a human opinion to validate the results, meaning that AI cannot work independently in these highly skilled services. However, AI has great potential to help healthcare professionals giving better care to their patients.

Customer services was also one of the most mentioned types of service. Most of the respondents suggested applications mainly for Call Centers, with AI helping call center automation with, for example, chatbots. However, only one source mentioned the complete disappearance of call centers. From a positive impact perspective, five respondents mentioned that the customization and personalization of customer services. This means that the service will be adapted to the needs of each customer, and therefore be more valuable for customers.

The impact on transportation and logistics was also considered as very relevant as AI will enable non-stop autonomous transportation with increased safety, eliminating the error factor that comes from the human side. Furthermore, having a controlled environment where autonomous cars communicate and learn from themselves will also increase safety, but will require a great investment. Also, autonomous driving raises important ethical issues that remain unaddressed by regulatory authorities.

Dealing with AI will require that education services create new courses and programs dedicated to AI. However, this will be problematic since AI specialists are on short supply and mostly working in industry.

From an environmental point-of-view, AI can support the development of entirely new services dedicated to the forecast of natural disasters, including smart fire detections, pollution measurement systems, atmospheric sensors and better waste management systems. However, AI might bring great environmental impacts because it requires great amounts of electricity.

For financial services, AI will enable improved security and fraud detection on credit card transactions, cloning, tax evasion and bank accounts. It will also enable more convenient security measures, fostering the disappearance of passwords. Ensuring the security of all these systems will be a significant challenge.

Experts also highlighted the impact of AI on retail that comes mostly from the replacement of jobs from humans to robots. Physical stores will still exist in the future but more oriented towards customer care. AI can also power smarter recommendation systems for advertisement.

Table 1. Impact of AI on services

Name	Sources (n = 10)	Positive impacts	Negative impacts	Barriers and challenges	Application examples	Quotes
Impact on healthcare	10	Creation of new drugs Improved diagnostics	Human care being replaced by AI systems	Assurance of robust algorithms that work for all possible patients	AI in diagnostics AI in preventive medicine AI as second opinion systems	"Technology that gets health signals from our body and alerts for a disease development, even before our symptoms"
Impact on customer services	9	Customization of services	Replacement of humans	Ensure that humans still have a place in customer service	Call centers and chatbots Automated supermarket	"Everything will be more customized"
Impact on transportation and logistics	9	Non-stop autonomous transportation. Reduce car crashes. Learning capabilities of autonomous vehicles	Autonomous cars will reduce the driving pleasure	Ethical issues. Creation of exclusive roadways for autonomous cars. Lack of regulation	Drone delivery to remote areas	"Autonomous cars will increase the comfort of not having to drive. Consequently, it will increase safety, since there will be less accidents"
Impact on education	6	New courses in school	–	Lack of AI experts	–	"I think that AI should be taught, at least the programming part, right from elementary school"
Impact on environment-related services	4	New service opportunities for societal well-being	High electricity consumption from AI technologies	No coverage from the current energetic production	Forecast of natural disasters Pollutants measurement	"On an ecologic level, preventive measurements could be made – studying the forecast of forest fires"
Impact on financial services	4	Fraud detection. Disappearance of passwords for facial or digital print recognition	Lack of credibility of cryptocurrencies	Ensure cybersecurity	Avoid credit card cloning, tax evasion	"AI will help to reduce the number of frauds"
Impact on retail	4	Physical stores will still exist in the future for customer care	Replacement of jobs	–	Shopping recommendation systems. Sensors in clothes	"The capacity of having sensors in every clothing article"

Finally, some crosscutting topics that impact all of the above services were also uncovered. These are shown in Table 2. The impact on privacy is specially concerning with AI possibly leading to mass surveillance and isolation among humans. The impact on labor was also emphasized as AI will influence the way companies operate and how they are managed, leading to an unavoidable shift in employment. This might lead to increased unemployment or a massive redirection of current jobs. Legal aspect also crosscut all other topics, and concern the lack of regulation, along with the political issues that are inherent to these regulations. As AI works as a black-box, the explanation of its actions will be hard to phantom, leading to legal issues as well.

Table 2. Crosscutting impacts on service.

Name	Sources (n = 10)	Positive impacts	Negative impacts	Barriers and challenges	Application examples	Quotes
Impact on privacy	10	–	Loss of privacy Isolation	Distrust in AI	Mass surveillance	"There will be one challenge, which is fight the first fear of people using these systems"
Impact on labour	10	Smart redirection of Human Resources on society Repetitive tasks automation	Unemployment Replacement of jobs	–	–	"[There is] no turning back on the reconversion of jobs"
Impact on legal aspects	8	Reduce the number of ID falsifications	No regulation of technologies Political issues Failure of explanation	The right of explanation from a machine	GDPR and connection with AI	"There is a set of regulations, a domain that needs to be improved, in order to build a safer path [in the future]"

5   Discussion

The study generated a detailed set of results that sparked interesting questions for discussing how AI impacts service. First, it is important to notice that, some of the observed trends can be related to the overall digitalization of services, such as job replacement. These trends are expected to be reinforced with the AI coming-of-age.

Other challenges, such as the ethical-related ones, are mostly new and arise with the introduction of AI in a service context.

Job replacement is probably one of the feared consequences, as it was also verified in the results. Since jobs are already migrating from manufacturing to services, and the service sector is expanding along with AI, it is important to assess how AI can replace low-skilled service jobs. Empathetic and intuitive skills are not yet mimicked by AI, which means that people who work daily in high-skilled service jobs, like healthcare workers, will not be as much impacted like those who perform tasks that are possible to automatize [3], which goes in agreement with the study results. The automation of repetitive tasks is already common for precise and well-defined procedures [31]. According to Goux-Baudiment [32], 47% of today's jobs will be automated by 2033. Furthermore, with state-of-the-art sensors, robots are capable of producing goods with higher quality that humans [33]. Results support this position for some services, like customer services. Chatbots are a tool that will – disruptively, for some respondents, or progressively and supporting, for others - change customer support. This is supported by literature as call-center employees are considered a type of jobs that does not require high skilled labour people, being easily replaceable by AI [3]. Stock and Merkle [22] performed a study with humanoid service robots and they concluded that, if they are able to express behavioural and innovative notions during the service encounter, they could exceed customer expectations and deliver noteworthy experience to customers. However, in services that require a stronger human interaction with empathetic characteristics, humans will be harder to replace. Services that require social and interpersonal skills, which Huang and Rust [3] also called "soft" people skills, like healthcare, will increasingly become the most important components for employability. People with these features will have more employment opportunities within an AI world [14]. Also, occupations that require a high degree of creative intelligence are unlike to be automated [33].

Regarding transportation and logistics, the case of the autonomous vehicle running over a pedestrian was referred to illustrate the serious questions facing regulatory agencies and society. What would happen if a car must decide between a person or a group of people? Automated vehicles are not considered "drivers" when it concerns speed limit regulation and enforcement. However, in case of an accident, the person responsible for that automated vehicle could be liable. There is a big lack of knowledge and guidelines when we talk about liability in this type of technology [2]. As such, the creation and development of autonomous transportation services might be limited while lawmakers establish the required boundaries and liabilities for the use of autonomous transportation.

Crosscutting all of these topics, the discussion on data security and privacy is also an important subject [4]. Some of our respondents mentioned a trade-off concerning privacy and security. To have more security, privacy will be lost. Hence, AI has the potential to mass-produce surveillance, but this colossal computerization leads to a loss of privacy [2]. Kurzweil [13] argues that the balance between privacy and protection will be one of the massive challenges to overcome with AI. As such, when designing and developing services that are supported by AI, additional measures should be taken to ensure the privacy of their users. For example, interfaces and software should be primarily designed with privacy guidelines through encryption and data access control

[18]. Public personalities like Bill Gates and Stephen Hawking also mentioned that humanity should have a very safeguarding and conservative view of AI, due to the harm that it can cause [18]. The creation of services that can humanize AI open an opportunity to converge technologies with customers.

AI will also enable freer and more quality time for humans. According to Makridakis [14], AI will shrink the level of work left for people and increase their free time, making people able to pursue their interests. This might open the possibility to develop services related to leisure and entertainment. Tasks requiring creativity, strategic thinking, entrepreneurship are not expected to be algorithmically accomplished, relying strongly on exclusive human skills.

6 Conclusion and Future Research

Technology innovation is rapidly transforming service experiences [21]. AI is a breakthrough technological innovation with yet unforeseen consequences for service. Through this qualitative study with AI experts, this article aims to start understanding the impacts that AI can bring to services, by understanding the perspective of those who are developing these systems. The findings of this article empirically corroborate many of the impacts conceptually described in the literature, and complements this literature by focusing on the impact on services.

As such, while on one hand some of the employment in services will suffer with AI, where mechanical and repetitive actions will mostly be replaced by AI engines, on the other hand, it was possible to understand that new tasks will arise with these emerging technologies, making either possible a smart redirection of employees within companies or in some cases a job disintegration. The most known example in customer service is the chatbot, that companies are increasingly implementing to answer most of the repetitive and usual inquiries.

This study has some limitations that can be overcome by future research efforts. A deeper study, involving multidisciplinary efforts, for each type of service identified is necessary. Further studies can explore the different impacts individually, depending on the evolution of complex abilities from AI. A relation between the identified impacts of AI can be established with other relevant service research topics, such as the service digitalization. Finally, this study can be complemented with a quantitative study in order to quantify and validate these impacts in a wider sample.

Although there is plenty to explore in this topic, it is possible to claim that this study offers an first understanding of the impacts of AI on services, and can help to prepare service providers to the sweeping changes that will arrive with the rise of AI.

References

1. Shahin, M.A.: State-of-the-art review of some artificial intelligence applications in pile foundations. Geosci. Front 33–44 (2016)
2. Russell, S., Norvig, P.: Artificial Intelligence: A Modern Approach, 3rd, illustr edn, 1151 p. Pearson/Prentice Hall, Upper Saddle River (2013)

3. Huang, M.H., Rust, R.T.: Artificial intelligence in service. J. Serv. Res. **21**(2), 155–172 (2018)
4. Dirican, C.: The impacts of robotics, artificial intelligence on business and economics. Procedia – Soc. Behav. Sci. **195**, 564–573 (2015)
5. Spohrer, J., Maglio, P.P.: Production and operations management the emergence of service science: toward systematic service innovations to accelerate co-creation of value. Prod. Oper. Manag. **17**(3), 238–246 (2008)
6. Ostrom, A.L., Parasuraman, A., Bowen, D.E., Patrício, L., Voss, C.A.: Service research priorities in a rapidly changing context. J. Serv. Res. **18**(2), 127–159 (2015)
7. Lovelock, C.H., Wirtz, J.: Services Marketing: People, Technology, Strategy, 6th illustr edn, 648 p. Pearson/Prentice Hall, Upper Saddle River (2007)
8. Vargo, S.L., Lusch, R.F.: Service-dominant logic: continuing the evolution. J. Acad. Mark. Sci. **36**(1), 1–10 (2008)
9. Maglio, P.P., Spohrer, J.: Fundamentals of service science. J. Acad. Mark. Sci. **36**(1), 18–20 (2008)
10. Wuenderlich, N.V., Heinonen, K., Ostrom, A.L., Patricio, L., Sousa, R., Voss, C., et al.: "Futurizing" smart service: implications for service researchers and managers. J. Serv. Mark. **29**(6/7), 442–447 (2015)
11. Huang, M.H., Rust, R.T.: Technology-driven service strategy. J. Acad. Mark. Sci. **45**(6), 906–924 (2017)
12. Wünderlich, N.V., Wangenheim, F.V., Bitner, M.J.: High tech and high touch: a framework for understanding user attitudes and behaviors related to smart interactive services. J. Serv. Res. **16**(1), 3–20 (2013)
13. Kurzweil, R.: The Singularity is Near: When Humans Transcend Biology, 672 p. Viking Press, New York (2005)
14. Makridakis, S.: The forthcoming Artificial Intelligence (AI) revolution: its impact on society and firms. Futures **90**, 46–60 (2017)
15. Charmaz, K.: Constructing Grounded Theory: A Practical Guide Through Qualitative Analysis. Illustrate, vol. 10, 208 p. Sage Publications, London (2006)
16. Corbin, J., Strauss, A.: Grounded Theory Research: Procedures, Canons and Evaluative Criteria, vol. 19, pp. 418–427. Zeitschrift fur Sociologie (1990)
17. Turing, A.M.: Computing Machinery and Intelligence: Parsing the Turing Test, vol. 59, pp. 433–460. Springer, Heidelberg (1950)
18. Gurkaynak, G., Yilmaz, I., Haksever, G.: Stifling artificial intelligence: human perils. Comput. Law Secur. Rev. **32**(5), 749–758 (2016)
19. Miller, J.D.: Singularity Rising: Surviving and Thriving in a Smarter, Richer, and More Dangerous World, 262 p. Benbella Books, Inc., Dallas (2012)
20. Čerka, P., Grigienė, J., Sirbikytė, G.: Is it possible to grant legal personality to artificial intelligence software systems? Comput. Law Secur. Rev. **33**(5), 685–699 (2017)
21. Grenha Teixeira, J., Patrício, L., Huang, K.H., Fisk, R.P., Nóbrega, L., Constantine, L.: The MINDS method: integrating management and interaction design perspectives for service design. J. Serv. Res. **20**(3), 240–258 (2017)
22. Stock, R.M., Merkle, M.: Can humanoid service robots perform better than service employees? A comparison of innovative behavior cues 2. Theoretical background & hypotheses. In: Proceedings of the 51st Hawaii International Conference on System Sciences, pp. 1056–1065 (2018)
23. Davenport, T.H., Kirby, J.: Beyond automation: strategies for remaining gainfully employed in an era of very smart machines. Harv. Bus. Rev. **93**(6), 58–65 (2015)
24. Ulrich, N.T.T.: Grasping with mechanical intelligence. Tech reports, p. 846, December 1989

25. Del Prado, G.M.: AI robots don't need to be conscious to become human enemies - Business Insider [Internet], 5 August 2015. http://www.businessinsider.com/artificial-intelligence-machine-consciousness-expert-stuart-russell-future-ai-2015-7?IR=T. Accessed 1 Feb 2018
26. Sternberg, R.J.: The theory of successful intelligence. Interam J Psychol. **39**(2), 189–202 (2005)
27. Seddon, F.A., Biasutti, M.: Modes of communication between members of a string quartet. Small Group Res. **40**(2), 115–137 (2009)
28. Arnold, R.: Empathic Intelligence: Teaching, Learning, Relating. University of New South Wales, 238 p. (2005)
29. Fabrega, H.: The feeling of what happens: body and emotion in the making of consciousness. Psychiatr. Serv. **51**(12), 1579 (2000)
30. Neuman, W.L.: Social Research Methods: Qualitative and Quantitative Approaches: Relevance of Social Research, 7th edn, 608 p. Pearson Education, Limited, Edinburgh (2014)
31. Autor, D.H., Dorn, D.: The growth of low skill service jobs and the polarization of the U.S. labor market. Am. Econ. Rev. **103**(5), 1553–1597 (2013)
32. Goux-Baudiment, F.: Sharing our humanity with robots. World Future Rev. **6**(4), 412–425 (2014)
33. Frey, C.B., Osborne, M.A.: The future of employment: How susceptible are jobs to computerisation? Technol. Forecast Soc. Change **114**, 254–280 (2017)

Service Design and Innovation

Igniting the Spark: Overcoming Organizational Change Resistance to Advance Innovation Adoption – The Case of Data-Driven Services

Tobias Enders[✉], Dominik Martin, Garish Gagan Sehgal, and Ronny Schüritz

Institute of Information Systems and Marketing (IISM) and Karlsruhe Service Research Institute (KSRI), Karlsruhe Institute of Technology (KIT), Karlsruhe, Germany
{tobias.enders,dominik.martin,ronny.schueritz}@kit.edu,
garish.sehgal@alumni.kit.edu

Abstract. The launch of innovative products and services is accompanied by customer resistance towards innovation. While this phenomenon is well understood in research, little is known about resistance towards innovation on an organizational level. Servitization and digitalization have paved the way for innovative B2B offerings such as data-driven services to augment traditional value propositions. Being perceived as innovation, it is relevant to understand if customer change resistance also applies to an organizational level. We conduct a set of expert interviews with organizations that face a situation of slow data-driven service adoption. We find that service adoption is strongly connected to factors of customer change resistance: routine seeking, cognitive rigidity, emotional reaction, and short-term focus. Our study contributes to the body of knowledge by adapting a construct of change resistance from an individual to an organizational level and by offering guidance to practitioners on how to overcome customer change resistance.

Keywords: Data-driven services · Change resistance · Innovation adoption · Data monetization

1 Introduction

Despite customers being generally open towards trying and evaluating new products, the launch of innovative products and services is accompanied by resistance to accept innovation [1]. Product failure rates of around 50% underscore that customers do not act based upon the expected paradigm [2,3]. Driven by cognitive and behavioral barriers to accept innovation-induced change, consumers aim for a "psychological equilibrium" that gives them a feeling of safety and control [4]. The change resistance of customers in accepting innovative offerings has put a burden on companies to find ways to ease the adoption process in order to establish sustainable business models.

© Springer Nature Switzerland AG 2020
H. Nóvoa et al. (Eds.): IESS 2020, LNBIP 377, pp. 217–230, 2020.
https://doi.org/10.1007/978-3-030-38724-2_16

Servitization and digitalization have triggered a new wave of service offerings being released to the market. Through servitization, "an organization's capabilities and processes to shift from selling products to selling integrated products and services that deliver value in use" [5], companies have broadened their offering portfolio to better serve their customer's needs [6]. Literature on servitization has attempted to provide guidance on how to implement servitization in an organization (e.g. [7,8]), however, customer acceptance of emerging services has largely been neglected. There is a shift in our understanding of value creation and service activity as the substance of economic exchange as described in Vargo and Lusch's [9] seminal work on service-dominant logic (SDL). By offering product-service systems, companies have found ways to co-create value with their customers and amend their value propositions accordingly. The process of servitization is augmented by the digitalization of our society. An ever-increasing amount of data is being created that enables new opportunities of joint value creation between the customer and provider. A special form of new services arising are data-driven services. As such, "Data-driven services use data and analytics to support the decision-making process of the customer through data and analytics-based features and experiences in the form of a stand-alone offering or bundled with an existing product or service" [10]. Examples of data-driven B2B services are manifold and include monitoring services for the status of an elevator, usage reporting for vehicles, and predictive maintenance for machinery.

Fueled by digitalization activities, data-driven services are launched as innovative offerings. With innovation resistance being observed on an individual level, it is relevant to understand if this broader phenomenon also exists in a B2B setting. Our research objective is therefore framed as follows: We want to explore if resistance towards innovation-induced change also exists in an organizational context, how this potentially relates to customer change resistance, and how to overcome it.

We conducted a series of expert interviews with organizations that are offering data-driven services and face a situation where business customers show a slow adoption of data-driven services. We find that the slow adoption is strongly connected to customer change resistance. We therefore adapt a change resistance construct from an individual to an organizational level to better understand business customer behavior. By this, we contribute to the body of knowledge on change resistance and broaden its application to an organizational level. We further provide guidance to providers of data-driven services in understanding the reasons that slow-down the adoption process and offer a set of practices to overcome them.

The remainder of the paper is structured as follows: Sect. 2 outlines fundamentals and related work on change resistance and data-driven services. Section 3 illustrates how we collected and analyzed our data. The results of our study are summarized in Sect. 4. We conclude the paper in Sect. 5 by summarizing the findings and contributions, pointing out limitations and future research.

2   Fundamentals and Related Work

This section introduces fundamentals and related work on resistance to change followed by a brief introduction of data-driven services as our application domain.

2.1   Resistance to Change

Resistance to change is – despite its negative connotation – a common human dispositional inclination [2,11–13]. This innate resistance behavior is an individual's intrinsic desire to maintain a 'psychological equilibrium' that gives people a feeling of security and control [4]. Thus, this desired balance might be disturbed by behavioral change, which is why individuals usually shy away from actively initiating or adopting change.

For instance, this behavior can be observed in consumers' acceptance of new products and innovations in the market [14]. In particular, partially high rates of failed product launches implies that—although customers could potentially benefit—some kind of innovation resistance exists, which is typically expressed through different behavioral traits: adoption shift, rejection, and active opposition [2,3,12,15].

However, the reasons for resistance are either of active or passive nature: Active resistance is characterized by conscious rejection, which is motivated by contradictions with beliefs or traditions or by functional issues. In contrast, passive resistance already occurs before the actual evaluation of an innovation and is caused by a general aversion to behavioral or attitude changes [1,2,16]. This general aversion, also called 'dispositional resistance to change', has, according to Oreg [13] four determinants: routine seeking, cognitive rigidity, emotional reaction to imposed change and short-term focus (e.g. [17–19]).

Routine seeking is anchored in human anxiety to lose the emotional control in changing life situations. The inclination to adopt routines therefore causes a sense of stability in these situations [13]. Cognitive rigidity describes the frequency and ease with which people change their minds and a general unwillingness to give up habits [18]. The emotional reaction to change is defined as the ability of humans to cope with stress and uneasiness induced by change [13]. Finally, the trait short-term focus describes the human tendency to perceive long-term gains as less important compared to short-term gains [20].

Yet, all forms and causes of resistance to change described in this section are directly related to behavioral aspects of individuals who seek to perceive their interests, regardless of the organizational structures in which they reside. However, the present work combines these individual characteristics with observations in an organizational context.

2.2   Data-Driven Services

With the age of digitalization and digitization in full swing, businesses have various opportunities to create new services—either as stand-alone or by enriching a core product or core service. Wixom and Ross [21] refer to the latter as 'wrapping', which describes, for instance, the monitoring service of an elevator or the tracking of a parcel. Especially the availability of large amounts of data from sources such as the Internet of Things (IoT), web or user-created content, enables businesses to offer services that build on data and to create mutual value through value co-creation with customers. With increasing data availability, the opportunities for large-scale analytics applications, from traditional, mainly internal decision-support tools to complex analytics solutions, which enable businesses creating additional value for customers by enriching products and services through the exploitation of data, appear almost unlimited [22,23].

Thus, new data-driven business models arise through the generation, aggregation and analysis of data [24]. Offerings based on data as their key resource and the utilization of analytics are referred to as data-driven services [23]. Despite data-driven services becoming more widespread, organizations find it difficult to define suitable revenue models to capture the value of their services [25].

One example for a data-driven service wrapped around a certain core product would be a predictive maintenance service. Here, for instance, a machine is additionally equipped with a service that allows to monitor the entire asset and its health condition. In addition, a warning alert is emitted when a particular part needs to be replaced in the near future to avoid a complete shutdown of the machine. An exemplary stand-alone service could be, for example, a software that intelligently calculates a demand-oriented staff planning by taking into account external factors such as weather forecast and holidays as well as internal factors as historical data and experience to calculate the optimal number of employees to meet the demand.

3   Methodology

In line with our research objective to develop a better understanding of the reasons that slow down the adoption process of data-driven services in B2B markets, we conducted an explorative study based on a series of expert interviews. Our data collection and analysis approach follows that of a qualitative content analysis based on Kuckartz [26].

3.1   Data Sources

In order to develop a better understanding of the reasons that trigger change resistance and ways to overcome them, we need to collect data from providers offering such services. Talking to service providers only can be a good source of information for understanding change resistance at the customer side if interviewees have direct customer contact [27].

We ensured that all of the organizations that participated in our interview study had launched data-driven services in the industrial B2B market. For the sampling of the interviewees, we followed a criterion-i purposeful sampling approach [28]. The criteria were defined that only provider representatives that are directly involved in customer interaction on data-driven services qualified for the interviews. We therefore approached sales representatives, product managers, and account executives in our personal network. The persons approached were either available themselves for an interview or forwarded the request to a colleague. This resulted in a total of eight interview conversations with nine experts within five different companies. Based on availability, the interviews were either conducted in-person or over the phone between July and August 2018. The interviewees hold various positions in their organization such as product, sales and project managers. An overview of the companies interviewed in the course of this study is provided in Table 1.

Table 1. Overview of companies interviewed

Company	Revenue in Bn € (2018)	Employees	Active countries	Interview partners
Automotive I	78.0	400,000	60	3
Automotive II	40.0	50,000	>150	1
Manufacturing I	3.3	24,000	50	2
Manufacturing II	3.2	21,000	61	1
Manufacturing III	3.1	14,000	13	2

We chose a semi-structured interview approach to guide the conversation and at the same time to allow the interviewees to emphasize topics that are important to them. Each interview was structed into two parts: First, we collected general information on the interviewees' tasks and responsibilities before focusing on data-driven services. The questions in part two of the interview revolve around the themes of the company's experience in introducing data-driven services into the market and how customers reacted to it. Furthermore, interviewees were asked how the company reacted to customer hesitations, which solution approaches were chosen and why. All interviews were recorded and transcribed for further analysis.

3.2 Data Analysis

We analyze the empirical data collected using a structured qualitative content analysis [26], which finds its origin in grounded theory research [29]. This approach is the most suitable since it helps to develop theories in areas where there is not much guidance in the literature [30].

Interviews were analyzed in an iterative manner by performing two coding cycles [31]. To account for the explorative nature of the study, the cycle uses an open coding approach in which interviews are labelled according to our research objective. The aim is to identify reasons that explain slow adoption of data-driven services and approaches that companies have successfully applied to overcome them. Initially, 27 codes have been identified. For the second iteration, we perform axial coding to resemble codes identified in the first iteration and to relate categories and sub-categories accordingly. This step ensures that codes are sharpened and have the best fit towards the research objective [29].

To ensure a high quality of the results, an independent second researcher codes all interviews again based on the coding structures derived from second cycle coding. Discrepancies are discussed until a mutual agreement is reached. To facilitate the coding process, the MAXQDA software is used. It is worth noting that we did see a saturation in the data in the final interviews.

4   Results

Service providers suffer from the slow adoption of data-driven services in the market and customer's hesitations to pay for them. Understanding the reasons that cause slow adoption is critical to understand and develop approaches to address these reasons. We find that the reasons for slow adoption of data-driven services go back to factors that are generally associated with change resistance. We link these insights to determinants of change resistance and offer solution approaches in the subsequent section.

4.1   Cognitive and Behavioral Barriers

The reasons for customers to show resistance in the adoption process of data-driven services are manifold. We find evidence that the causes for the slow adoption have strong parallels with determinants found in human change resistance constructs. Since our analysis has taken place on an organizational rather than an individual level, we adopt the change resistance construct from an individual to an organizational level and thereby redefine its boundary conditions. We show how determinants of change resistance, namely *routine seeking, cognitive rigidity, emotional reaction to imposed change, and short-term focus*, map onto an organizational level.

Routine Seeking. Analogous to human behavior, organizations seek for stability in their routines. This is especially true for historically grown companies that rely on proven processes to keep their core business running. We find that organizations reject the introduction of data-driven services by pointing to smoothly running processes of their legacy business while at the same time disregarding changing market conditions and customer demands. An introduction of a data-driven service requires new processes to be implemented and therefore a change in the daily routine across multiple departments. This deviation from a routine fuels the fear of losing (emotional) control on an individual and organizational level. Our evidence further shows that organizations fear of losing financial control: data-driven services are oftentimes billed on a subscription or per use basis, which is uncommon in most manufacturing industries. The risk of giving up the routine of predictable one-time purchases for e.g. material or one-off maintenance service requests in favor of perceived unpredictable expenses for data-driven services inhibits the willingness to adopt.

Cognitive Rigidity. The sum of habits defines the way that an organization runs and operates. We find that there is a narrow-mindedness across all the organizations that we interviewed caused by rigid structures that subsequently inhibit change. For instance, employees prefer to apply the skill set that they possess over and over again. The introduction of data-driven services, however, may require re-skilling part of the workforce to support new processes and applications. Evidence suggests that the threat of needing to learn new skills creates a negative atmosphere and spirit, which ultimately leads to rejection of the imposed change. Furthermore, we find that customers are having a difficult time grasping the value and economical benefit that a data-driven service provides. "The benefit is, or the positive effect is, not always directly tangible for the customer" reports a strategic sales manager of automotive company I. Some even deny that data-driven services add any value to their business. This trait of dogmatism therefore inhibits service providers' efforts to roll-out the service and subsequently establish new sustainable revenue streams. Even if the value from data-driven services is recognized, supporting functions such as procurement departments may slow down the process of adoption. We find that legacy processes oftentimes do not allow for service contracts to be charged by e.g. on a pay-per-use model. The introduction of new processes or approval of exceptions hence delays the timely ordering and introduction of the services.

Emotional Reaction to Imposed Change. While emotional reaction or stress clearly is associated with human behavior, we find that organizations show a similar form of uneasiness. The introduction of data-driven services impacts the operations of multiple departments, causing significant change and therefore inducing stress. We find that this elevated stress level stems primarily from the fear of the potentially negative consequences of induced change. For instance, there is a fear of the lock-in effect with one service provider. Customers worry that once a data-driven service has been embedded into their operations and IT systems, it is almost impossible to disentangle the processes without causing significant interruptions to their business. The sales manager of automotive company I describes this situation: "I [the data-driven service customer] have to convert my complete systems, I have to redesign my interfaces, I have to refer my [end] customers to another address and so on. So, there is a relatively high change barrier created by such a service partner, which also deeply intervenes in the organization." This finding is consistent with customers' demand of keeping their flexibility to switch providers easily in case of a disappointing service experience.

With data being a valuable asset, organizations raise concerns about data privacy and security. Data-driven services oftentimes require the transfer of data between the customer and provider for further processing. Data leaving the company's boundaries induces stress due to fear of theft and misappropriation. We find that this concern by itself can inhibit the adoption of data-driven services due to its sensitive nature. This also has implications for the technology to be used for service provisioning: "Our customers are [...] constantly afraid that data will be stolen, that know-how will be lost. [...] I'd rather not put [data] into any clouds." (lean consultant, automotive company I).

Short-Term Focus. Organizations show a misperception of the potential long-term gains of data-driven services. We find that there is a strong short-term focus on generating revenues and profits driven by, for instance, investor demands. This is especially true for companies listed on the stock exchange that have to report quarterly earnings. Depending on the kind of service, the implementation period may consume a considerable amount of time and resources. Our evidence suggests that organizations oftentimes do not show the patience needed for data-driven services to pay-off in the long-term. When it comes to making investments decisions, organizations aim for short payback periods and shy away from funding new IT infrastructure that is sometimes required to make the services work. Few customers recognize the introduction of data-driven services as an opportunity to leave legacy systems behind and move to a more scalable platform. This may, in part, be fueled by the aforementioned data privacy and security concerns in cloud environments. Summarizing, we find that customers prefer generating a certain short-term profit rather than focusing on uncertain long-term gains aided by the introduction of data-driven services.

4.2 Overcoming the Resistance

While understanding the reasons that influence organizational change resistance is a critical first step, we now turn to solution approaches. We outline five practices that organizations have successfully applied to drive adoption of their data-driven services in the market. Those approaches are starting small, transparent revenue model, great user experience, value-based selling approach, and building trust. We further point out how the practices address the four factors of change resistance as is shown in Fig. 1.

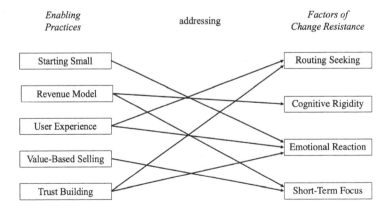

Fig. 1. Enabling practices addressing factors of change resistance

Starting Small. The scope of functionalities of a data-driven service can inhibit or accelerate its adoption process. We find that in order for customers to get familiar with the service, a limited scope is advisable. This allows that trust into the performance and added value of the service can grow. "I think that it is important in general that companies that are interested [in data-driven services] just need to start to get a feel for it [...]. For example, start on a small scale and then, when you see that the solution adds value, you can grow from there" is how a lean manager of automotive company I describes their experience when working with clients. Starting on a small scale further limits the financial exposure for the customer and eases the decision to try the service.

As trust builds up, additional and more complex features can be offered to the client. We find evidence that offering multiple service packages with varying scope allows customers to self-select and chose the solution that meets their needs. A growing level of trust and access to customer data – within the scope of the contract – also comes at an advantage for the service provider: by closely monitoring service transactions, the provider can derive insights, learn about customer pain points and offer better tailored solutions in the future.

Radical changes in the organization elevate the stress level of employees – a factor promoting change resistance. By offering data-driven service with basic functionality, changes remain at a minimum and so does the *emotional reaction to imposed change*. Incremental changes to service scope keep the level of stress low and therefore contribute to less resistance.

Transparent Revenue Model. Entering into a contract for data-driven services may be perceived as risky to customers since payment models deviate from what they are used to. While historically purchase decisions mostly involved one-time payments to the supplier, service contracts require the customer to commit to multiple payments over a period of time. This paradigm shift requires service providers to offer transparent revenue models to ease the transition for their customers. A digital strategy manager from manufacturing company II recalls: "First, we will – and that is based on the feedback that we have received – to begin with, we need simple models, because even these simple models will have to trigger a shift in the mindset of customers."

Data-driven services offer a multitude of revenue models. For simplification purposes, we define a revenue model as a combination of a revenue mechanism and pricing scheme. On the one hand, we find that customers perceive a subscription-based mechanism as the most simple to understand since it incurs a fixed and plannable fee to be paid on a regular basis. On the other hand, pay-per-use models increase transparency of the level of service usage. Those two mechanisms may even be combined as described by a strategic sales manager in automotive company I: "[...] we had a fixed price per month and – depending on the usage – an additional fee per use. This was relatively easy to explain [to the customer], since each usage incurred additional costs on our side". When pricing a data-driven service, customers value a tier model for various levels of demand. This incentivizes the customer to increase service usage and may lead to increased profits on the provider-side due to low costs incurred in scaling a data-driven service.

By defining a transparent revenue model, we ease the customer's transition from old habits *(cognitive rigidity)* of one-time charges to contemporary service payment models. Furthermore, by shifting the center of attention to the time perspective of a service contract, we lift the *short-term focus* in favor of long-term gains that the service may provide.

User Experience. The perceived simplicity of using a data-driven service plays as a key role in the adoption process: "If you want to offer a service, this service should also be easy to understand, if possible, also easy to use" (lean consultant – automotive company I). One instrument for ensuring an easy and intuitive use is the user interface that can simplify the daily work of the end user of a data-driven service. If not enough attention is paid to this requirements, the adoption process may be negatively impacted as a strategic sales manager (automotive company I) recalls: "some of the users just said: ok, I lack various things that I would need for my daily easy-of-use". However, we need to keep in mind that decision-makers about purchasing a data-driven service are not necessarily its end users. We find that the requirements of multiple stakeholders need to be taken into account early during the design process of the service. Methods such as design thinking aid in better understanding the needs of the customer and the development of prototypes.

Further advantages, which result from an intuitive user experience, facilitated by an appealing design, are strengthening the users' emotional binding to the service, the acceleration of the employees' training and a reduction of stress due to seamless integration of the service into daily routines *(routine seeking, emotional reaction)*. Thus, to make the user interface as suitable as possible for the task at hand, it is recommended to conduct extensive testing with the (end) users of the service to identify design errors early, to understand how they use the service and finally to avoid disappointed customers.

Value-Based Selling. Customers enter into a service contract with a provider if they believe that there is a net benefit. For the contract to close, the provider must clearly outline the added value by the service and understand if this also perceived by the customer. Since the value perception varies from customer to customer, it is of utmost importance to showcase what the problem is being solved and the benefit derived.

Clearly outlining the unique selling proposition of the service and highlighting its superiority towards competitors' offering allows providers to put a stake in the ground. Given the intangibility of a service, customers value the definition of key performance indicators (KPIs) that show the added value to their business. This allows tracking the value added over time and also addresses the undesirable *short-term focus* of customers. Developing a better understanding of the value of their service also benefits service providers in such that they learn about the net benefit created to use this for future customers negotiations.

The use of show cases and reference clients that faced a similar issue as the potential client has proven to go a long way. Customers develop a better understanding of how the data-driven service creates a benefit in a given situation. Furthermore, providers may be able to highlight the efficiency gains driven by the service and therefore aid business case calculations on the customer side.

Trust Building. Building trust establishes a foundation upon which provider and customer operate. We find that there are two points in time when trust must be established: First, when the customer signs up for the service (trust that the provider can deliver on the promises made instead of just making a quick sale). Second, when the customer needs to consider whether or not to extend the service contract. In the latter, the question is if the parties have built enough trust into the ongoing value added by the service or even to deepen the relationship. The importance of trust building is underlined by the strategic sales manager of the automotive company I: "Relationship is a decisive issue. Services have a lot to do with trust."

A trustful relationship that develops over time addresses an aforementioned fear a majority of customers bring forward: the threat of their data getting stolen or misused. We find that it is of utmost importance to emphasize the integrity of the data and to show transparency in the way that the customer's data is handled and secured. Ensuring that data is not misused by the service provider or even exposed to a third parties is critical for building a long-lasting business relationship. Putting the customer at ease by building a trustful relationship aids in the process of overcoming change resistance by ensuring that the customer maintains emotional control in every situation *(routine seeking)* and is not exposed to a high level of stress *(emotional reaction)*.

5   Conclusion, Limitations and Future Research

Our study explores if innovation-induced customer change resistance also applies to an organizational context. We conducted an explorative qualitative study to (1) better understand how organizational change resistance relates customer change resistance and (2) to derive practices to address those reasons for change resistance. We find that cognitive, and behavioral factors – *routine seeking, cognitive rigidity, emotional reaction and short-term focus* – drive change resistance impacting the adoption rate of data-driven services. To address these resistance factors, we derive five practices that provide guidance to service providers when launching data-driven services. In particular, those are a focus on value-based selling, a transparent revenue model, building a trustful relationship, ensuring great user experience, and offering basic functionality first before scaling up.

The contribution of our paper is two-fold. First, we adapt a change resistance construct from an individual to an organizational level by showing how the factors of routine seeking, cognitive rigidity, emotional reaction, and short-term focus can be applied to explain slow service adoption. Second, we offer five practices that guide practitioners to overcome resistance to change in the context of data-driven B2B services.

5.1 Discussion

Understanding how change resistance influences the adoption of data-driven services on an organizational level opens up new avenues for service providers to address them. In this work, we provide five practices that service providers can use to get data-driven services off the ground. However, understanding the underlying mechanisms allows organizations to define additional practices that address change resistance factors. The results of our study further show that the service quality and technical sophistication of the service is not necessarily an indicator of the success that it will have in the market. Moreover, service providers need to develop an understanding of their customers and the changes it requires on their end to implement the service. Putting customers at ease by addressing their concerns with concrete steps lays the foundation in building a trustful long-lasting relationship.

Data-driven services augment the existing offering portfolio and amend customer value propositions. Overcoming customers' resistance to change and aiding the adoption process of data-driven services is a prerequisite for service providers to establish new sustainable revenue streams and cash in on an improved value proposition.

5.2 Managerial Implications

The results of our study enable companies to develop a better understanding of their customers and to take a first step in establishing a sustainable business model for data-driven services. Our overview of hesitations allows organizations to identify their customers' pain points and to develop practices to address them adequately in their short-term and long-term strategies.

While understanding the customer is an important factor, our study also points out that service provider need to be prepared for selling data-driven services. This includes, among others, educating employees in principles of user-centric design and teaching value-based selling approaches. With our study, we provide initial guidance to companies that intend to introduce data-driven services to the market. However, managers must re-visit the practices applied to overcome change resistance on a regular basis to ensure that they still align with the requirements of the market and customer expectations.

5.3 Limitations and Future Research

We see three major limitations to our study that are directly linked to our call for additional research. First, we conducted all of our interviews with providers of data-driven services. For further research, we encourage exploring the customer perspective, which may yield additional insights. Second, our expert interviews had a strong focus on the automotive and manufacturing industry, which largely build upon value creation through physical core products in well established markets. Broadening this scope would allow verifying the results of our study. Third, all service providers interviewed offered data-driven services that were in an early lifecycle stage. We therefore encourage additional research once data-driven services have reached a more mature stage in their lifecycle to get a better understanding of potentially changing customer perceptions and expectations.

References

1. Talke, K., Heidenreich, S.: How to overcome pro-change bias: incorporating passive and active innovation resistance in innovation decision models. J. Prod. Innov. Manag. **31**(5), 894–907 (2014)
2. Ram, S., Sheth, J.N.: Consumer resistance to innovations: the marketing problem and its solutions. J. Consum. Mark. **6**(2), 5–14 (1989)
3. Andrew, J.P., Sirkin, H.L.: Innovating for cash. Harv. Bus. Rev. **81**(9), 76–83 (2003)
4. Osgood, C.E., Tannenbaum, P.H.: The principle of congruity in the prediction of attitude change. Psychol. Rev. **62**(1), 42 (1955)
5. Baines, T., Lightfoot, H., Benedettini, O., Kay, J.: The servitization of manufacturing. J. Manuf. Technol. Manag. **20**(5), 547–567 (2009)
6. Neely, A.: Exploring the financial consequences of the servitization of manufacturing. Oper. Manag. Res. **1**(2), 103–118 (2008)
7. Jovanovic, M., Engwall, M., Jerbrant, A.: Matching service offerings and product operations: a key to servitization success. Res.-Technol. Manag. **59**(3), 29–36 (2016)
8. Rabetino, R., Kohtamäki, M., Gebauer, H.: Strategy map of servitization. Int. J. Prod. Econ. **192**, 144–156 (2017)
9. Vargo, S.L., Lusch, R.F.: Service-dominant logic: continuing the evolution. J. Acad. Mark. Sci. **36**(1), 1–10 (2008)
10. Schüritz, R., Wixom, B., Farrell, K., Satzger, G.: Value co-creation in data-driven services: towards a deeper understanding of the joint sphere. In: Proceedings of 40th International Conference on Information Systems (forthcoming)
11. Rogers, E.M.: New product adoption and diffusion. J. Consum. Res. **2**(4), 290–301 (1976)
12. Szmigin, I., Foxall, G.: Three forms of innovation resistance: the case of retail payment methods. Technovation **18**(6–7), 459–468 (1998)
13. Oreg, S.: Resistance to change: developing an individual differences measure. J. Appl. Psychol. **88**(4), 680–693 (2003)
14. Stryja, C., Satzger, G., Dorner, V.: A decision support system design to overcome resistance towards sustainable innovations. In: Proceedings of the 25th European Conference on Information Systems, pp. 2885–2895 (2017)
15. Kleijnen, M., Lee, N., Wetzels, M.: An exploration of consumer resistance to innovation and its antecedents. J. Econ. Psychol. **30**(3), 344–357 (2009)
16. Heidenreich, S., Spieth, P.: Why innovations fail-the case of passive and active innovation resistance. Int. J. Innov. Manag. **17**(05), 1–42 (2013)
17. Laumer, S.: Why do people reject technologies - a literature-based discussion of the phenomena "resistance to change" in information systems and managerial psychology. In: Proceedings of 19th European Conference on Information Systems (2011)
18. Heidenreich, S., Handrich, M.: What about passive innovation resistance? Investigating adoption-related behavior from a resistance perspective. J. Prod. Innov. Manag. **32**(6), 878–903 (2015)
19. Laumer, S., Maier, C., Eckhardt, A., Weitzel, T.: Work routines as an object of resistance during information systems implementations: theoretical foundation and empirical evidence. Eur. J. Inf. Syst. **25**(4), 317–343 (2016)

20. Tversky, A., Kahneman, D.: Judgment under uncertainty: heuristics and biases. Science **185**(4157), 1124–1131 (1974)

21. Wixom, B.H., Ross, J.W.: How to monetize your data. MIT Sloan Manag. Rev. **58**(3), 10–13 (2017)

22. Davenport, T.H.: Analytics 3.0. Harv. Bus. Rev. **91**(12), 64 (2013)

23. Schüritz, R., Seebacher, S., Dorner, R.: Capturing value from data: revenue models for data-driven services. In: Proceedings of the 50th Hawaii International Conference on System Sciences, pp. 5348–5357 (2017)

24. Hartmann, P.M., Zaki, M., Feldmann, N., Neely, A.: Big data for big business? A taxonomy of data-driven business models used by start-up firms. A taxonomy of data-driven business models used by start-up firms (2014)

25. Enders, T., Schüritz, R., Frey, W.: Capturing value from data: exploring factors influencing revenue model design for data-driven services. In: Proceedings of 14th International Conference on Wirtschaftsinformatik (2019)

26. Kuckartz, U.: Qualitative Text Analysis: A Guide to Methods, Practice and using Software. Sage, London (2014)

27. Feldmann, N., Fromm, H., Satzger, G., Schüritz, R.: Using employees' collective intelligence for service innovation: theory and instruments. In: Maglio, P.P., Kieliszewski, C.A., Spohrer, J.C., Lyons, K., Patrício, L., Sawatani, Y. (eds.) Handbook of Service Science, Volume II. SSRISE, pp. 249–284. Springer, Cham (2019)

28. Palinkas, L.A., Horwitz, S.M., Green, C.A., Wisdom, J.P., Duan, N., Hoagwood, K.: Purposeful sampling for qualitative data collection and analysis in mixed method implementation research. Adm. Policy Ment. Health Res. **42**(5), 533–544 (2015)

29. Glaser, B.G., Strauss, A.L.: The Discovery of Grounded Theory: Strategies for Qualitative Research. Aldine Publishing Company, Chicago (1967)

30. Corbin, J.M., Strauss, A.: Grounded theory research: procedures, canons, and evaluative criteria. Qual. Sociol. **13**(1), 3–21 (1990)

31. Saldaña, J.: The Coding Manual for Qualitative Researchers. SAGE, London (2009)

Service Design for Business Process Reengineering

Bianca Banica[1(✉)] and Lia Patricio[2]

[1] Faculty of Automation and Computer Science, Politehnica University of Bucharest,
Bucharest, Romania
bianca.banica@stud.acs.upb.ro

[2] INESC TEC, Faculty of Engineering, University of Porto, Porto, Portugal
lpatric@fe.up.pt

Abstract. In the technology enabled, competitive service environment, organizations try to innovate their service while redesigning their processes to increase efficiency. The present study is aimed at developing a design method that brings together, complementarily, constructs and approaches from two fields: Service Design, which offers a human-centered, holistic focus on creating novel services and Business Process Reengineering, mainly organizational, process redesign and process efficiency focused. The Service Design for Business Process Reengineering (SD4BPR) method was developed following a Design Science Research methodology and it was applied in a business environment for the improvement of the Pre-Sale processes of a software development company dedicated to the health area. The development of the method and its process of work are presented and discussed in order to show how SD4PBR can support the design of technology-enabled services while taking into consideration organizational issues and desired business efficiency.

Keywords: Service design · Business Process Reengineering · Design thinking · Design science research · Pre-Sale

1 Introduction

1.1 Research Context

Businesses and organisations from the present competitive, globalised era are forced to become more innovative through adopting new methods, tools, technologies and change approaches. Reorganizing resources and redesigning processes are considered important approaches to change in business environments [1], with great potential in reducing costs and improving customer satisfaction [2]. Business Process reengineering (BPR) is an approach used for redesigning or replacing inefficient processes in order to achieve improvements on performance indicators such as cost, speed or quality. BPR can be applied to the whole organisation, to parts of the organisation or to a single unit [1]. Most often, the

© Springer Nature Switzerland AG 2020
H. Nóvoa et al. (Eds.): IESS 2020, LNBIP 377, pp. 231–244, 2020.
https://doi.org/10.1007/978-3-030-38724-2_17

re-design of the core processes happens using information technology in order to enable improvements [3]. Service Design(SD) is a multidisciplinary approach to creating new services, which integrates elements from marketing, operations, organizational structure, technology, human resources [4]. SD has started by relying on interaction design paradigms, focused on promoting service innovation through the design of touchpoints [5], the orchestration of service interfaces aimed at enhancing customer experiences [6]. To this end, SD involves understanding customer needs and goals (context and social practices) and translating them into requirements and evidence of service system interaction [7].

As stated in [7], there are few SD research initiatives that address organizational dimensions. They mainly look into what enables or inhibits organizational change: how SD can act as a driver for organizational change at different levels (from service interface to organizational values and norms) [8], how SD can have a transformational role in the infusion of an innovative culture in organizations(through SD knowledge and tools for decision making) [9]. Complementary, [10] investigated how SD can be embedded within organizational practices, in order to lead to sustainable design and human-centered services developed in-house. However, how SD methods can be applied in organizations for reengineering of processes (analysis and redesign) is still underexplored. The application of SD as an approach within organizations is formulated among the future research possibilities by [7], mentioning the need of investigating how "human-centered design methods could be extended and adapted for service system development and delivery". The findings of the same study imply that SD capabilities should be extended towards actual practical implementation proposals.

Previous research pointed out that the design and implementation of an organization is enabled and, at the same time, constrained by its information system, so "the effective transition of strategy into infrastructure requires extensive design activities on both sides, organizational and technological" [11]. While technology offers great potential and opportunities, it should be adequately matched with the business objectives, requirements and limitations in order to support process and service innovation and to enable seamless customer experiences [12]. In order to support the design of technology-enabled services in an efficient way, multidisciplinary contributions should be integrated [12], bringing together several constructs (methods, models, techniques, tools) from different fields.

Therefore, integrating SD and BPR can make an important contribution to better connect the creation and implementation of novel technology enabled services. However, these two approaches are usually performed by different teams, and have different focus, either on the process efficiency (BPR) or on the customer experience (SD). As such, further research is needed to explore how these two approaches can be applied in a complementary way. This study makes a step forward this challenge by developing a BPR method, derived from existing SD methods and BPR models, focused on analysing and restructuring business processes in a human-centred way, tackling the specific improvement points that have great impact on the broad processes.

2   Theoretical Framework

2.1   Business Process Reengineering

Initially, the term "reengineering" was related to the Information Technology (IT) field, but has later evolved, gaining the meaning of broader change process. Therefore, as explained by [13], BPR can be used for the analysis of an organization's business processes, for identifying alterations that can be made in order to attain strategic goals and improve performance and for reducing the costs of activities through the redesign of their workflows. BPR is seen as efficient from a managerial point of view, helping organizations to cope with technological and marketing changes while in a competitive, continuously evolving market [14]. Recent studies [15] have shown how BPR has been used by several organizations of different sizes and from different fields.

However, most of the BPR analysis techniques are focused on the process, but fail to tackle other dimensions (such as People, Communication or Structure). Most of the BPR modelling techniques have a very good level of accuracy, but they are dedicated to more technical modelling and so, they fail to offer strategic implications inclusion. On the contrary, the ones based on conceptual modelling are easy to understand, offer a total view of the process and tackle also the strategic implications. Although the level of accuracy cannot be as high as in the case of the other methods, the conceptual models compensate by offering more flexibility to the modelling process.

2.2   Service Design

Service Design integrates contributions from related fields such as service marketing, operations, interaction design, through a holistic approach. SD follows a human-centered vision, starting from understanding the customer and its context, and proceeding to the orchestration of different elements, with the purpose of finding new opportunities of value-creation, the process being an iterative one [16]. The focus is not only on *what* is being designed, but also on *how* it is designed: through understanding, visualizing, designing for user experiences [17] and finally experimenting through prototyping. Therefore, SD has been divided into stages that follow on the Design Thinking phases: *exploration, ideation, reflection* (prototyping and testing), *implementation*. SD adopts a service perspective, focusing on service as an overall solution for value co-creation. Furthermore, SD adopts a service system approach, taking into account the different components of the service system (people, technology, processes, interactions) [18], but also the interactions between them. On this basis, new research questions have arisen such as how should SD methods continue evolving in order to support organisation change, including the transition from concept towards implementation [4]. Furthermore, the same study refers to the development of service networks and systems as one of the priorities in terms of service design, defining several research areas that emerge from it, such as building service systems which are adaptive and flexible enough to respond to dynamic environments or creating modularized service system architectures. The same authors argue that there is a need for better integrating multidisciplinary contributions for designing complex systems.

The examination of BPR and SD literature shows that, while SD is human centered, putting customers at the core of both analysis and design and trying to enhance the customer experience, BPR revolves around the organization and aims at reducing costs, obtaining more efficient and effective processes. SD focuses on the service system as a conglomerate of entities such as processes, people, technology and their interactions, while BPR concentrates on the process as object of work. The integration of the 2 approaches can be materialized through a method that builds on previous research effort as follows. Firstly, SD methods offer flexible, visual and user-friendly models of analysis and modelling across multiple actors from multiple networks (Service Design for Value Networks-SD4VN method [19]) and across different service levels: concept, system, encounter (Management and Interaction Design for Service-MINDS method [12], MultiLevel Service Design-MSD [18]) that can tackle organizational problems and process redesign. However, they have been rarely applied in business reengineering contexts. Secondly, the BPR field offers several future research opportunities that can be addressed through integration with SD: offering more human-centered focus and a holistic approach [15], the derivation of a method from similar methods rather than focusing only on "best practices" and "innovation principles" [2].

3   Methodology - Design Science Research (DSR)

DSR is a research approach that offers a pragmatic world-view, focusing on the analysis and understanding of organizational phenomena in context and on advancing the research field by developing and evaluating artifacts which solve business/organizational problems and advance the existing knowledge base [20]. As stated in [21], the DSR approach is a bridge between design, relevance and rigor, supporting real-world problem solving through development of valid knowledge. For the purpose of the present paper, a new Design Artifact will be developed, thus extending the DSR theoretical foundations and contributing to the DSR knowledge base. The new method developed through DSR makes use of SD and BPR models and techniques and was exercised in a business environment. The DSR stages [22] were followed as presented below.

For the **Problem identification**, literature review on SD and BPR provided the theoretical framework and the support for identifying the research gaps. For the **Definition of objectives**, following on the insights from literature review, the research challenge established was developing a new method that integrates SD and BPR. The method would advance the field of Service Design by making the transition towards implementation of technology enabled services, while building on future research opportunities from the BPR field: offering a holistic approach, using redesign catalysts and tools derived from similar methods. As far as the **Design and development** phase is concerned, the new method was built following on previous research effort: integrating existing SD methods and BPR analysis and modelling techniques. As **Demonstration**, the new method was applied in a business environment, aiming at improving the business process of Pre-Sale from a software development company. The application of the method includes a Qualitative approach [23] used to understand the experience of the different actors for the exploratory study. Regarding **Evaluation**, the method's process of work was cross-checked with the support of a BPR framework, while its relevance in business context was assessed through feedback from the business environment.

4   Service Design for Business Process Reengineering Method (SD4BPR)

SD4BPR is a method developed through DSR methodology, aimed at enabling the analysis and redesign of business processes with a SD approach. Being a service design method, it is based on the design thinking stages, iterating between exploration, ideation, reflection and implementation. Also, it focuses on a service system perspective, offering a holistic view over the process while being oriented towards creating customer experience value. SD4BPR's approach is contained in 3 stages that integrate the SD stages and, implicitly, the design thinking stages: Exploration is represented by Mapping the AS-IS process, Ideation and Prototyping are represented by Modelling the TO-BE process, while Implementation is tackled in Implementation Possibilities (Table 1).

Since business processes are fully built constructs that have become problematic over time or are in need of adaptation to new ways of work or new technology developments, their understanding requires deconstruction. Therefore, the Mapping the AS-IS stage involves an exploratory study based on interviews and documentation review. Problem analysis and Root-cause analysis BPR techniques have been chosen in order to guide the interviews and to extract insights from the documentation, therefore supporting the first step of the process deconstruction. This will result in an identification of the main components of the process, structured with the help of a Mind Map and of a series of improvements suggested by the participants. Switching towards a SD system perspective, the process is split into goals, activities, actor interactions and artifacts in order to understand the correlations between the different components. Furthermore, each activity is deconstructed to the action level. From a organisational point of view, the interactions between processes and, implicitly, between actors from different networks are as important as the processes themselves and can hold valuable information about sensitive points. In order to obtain a full characterization of the process, both the positive and the problematic points are defined based on the previous deconstruction. In this way, the analysis stage of SD4BPR treats the process as human-centered system, while investigating the specific improvements points of the processes being redesigned and what is their impact on the broad process.

In order to ensure consistency between the strategic business level, the system application level and the technology level, SD4BPR's process involves a multi-level design phase, building on the MSD method [12]. Furthermore, to make the step forward towards implementation proposals (key aspect BPR) for achieving better designed technology-enabled processes/services (one of the main research directions in SD), SD4BPR follows up on MINDS method [12], using Interaction Design models. Finally, since BPR asks for solid implementation proposals based on limitations and technology possibilities, SD4BPR includes a stage focused on exploring available technologies and matching them with both the requirements and the constraints of the business environment.

Table 1. Process of the SD4BPR method

SD4BPR phases	Process	Models/techniques
Mapping the AS-IS process	Exploratory study (interviews, content analysis)	Qualitative approach [23]
	User experience understanding (process break-down)	Mind Map Customer Experience Modelling Service Experience Blueprint
	Identification of problems, setting objectives	Problem Analysis Root-cause Analysis
Modelling the TO-BE process 1. *Service Concept* 2. *Service System* 3. *Service Encounter*	Idea generation (on gathered insights)	Requirements per user profile
	Idea validation/ transformation (feedback meetings)	Service System Architecture
	Graphical representation of (non-)functional requirements	Wireframes Functional and Non-Functional requirements
Exploring Implementation Possibilities	Implementation analysis (technologies on the market/already used by the company)	Cross-check of tools and service requirements

4.1   Mapping the AS-IS Process: Qualitative Exploratory Study

Sample Definition. For the empirical study, 8 semi-structured interviews were conducted with 4 Pre-Sale members involved directly in the process, 4 Pivots from adjacent teams, involved indirectly in the process. Also, a workshop and 3 feedback sessions with 2 Pre-Sale members and the team leader were held. The sample was defined following the guidelines of Qualitative approaches [23], taking into consideration the significant actors, the relevance of these actors in the process and their availability. Since the Pre-Sale team has a daily, close communication and collaboration with the Sales, Development and Implementation teams, it was considered relevant for this study to interview pivots who have direct contact with a Pre-Sale member and, consequently, indirect impact on the final proposal.

Data Analysis. All the semi-structured individual interviews were recorded and then were transcribed and analysed using the NVivo 11 software which is dedicated for the qualitative analysis of data. Firstly, the content of the interviews was coded: divided in relevant categories with associated labels that summarize the category. A mind map of the nodes created using coding was generated in order to better visualize the node structure and test its coherence against data. Several cross-queries have been run between categories to identify their relationships and dependencies. A series of internal documents and in-use artifacts were also analysed as part of the qualitative study (ex: templates, process and workflow documentation, KPI monitoring dashboards, e-mails). They represented resources of information for the preparation of the interviews guidelines, for the better understanding of the AS-IS Pre-Sale process and also indicators for the identification of improvement points. Field notes were collected during every visit at the company in order to complement the data analysis with information from spontaneous remarks or observations.

4.2 Mapping the AS-IS: Business Process Break-Down

In order to have a complete overview of the process, a conceptual model is developed, **Customer Experience Diagram**, the result of combining Actor Network Map [24] with Multiactor activities and interactions [25] (Fig. 1). SD4VN Method [19] was chosen as support since it offers many to many interaction analysis, mapping the actors across different networks. This mapping model represents with more accuracy the context of a business process or the organizational environment itself. It supports the understanding of which are the main activities of the process, who is involved in executing them and through which means they are carried out. In terms of interactions, the diagram is centred on the Pre-Sale agent, mentioning with who he/she directly interacts in order to carry out the activities specified. For each activity, an ideal goal has been established and associated in the diagram in order to better perceive the desired outcome of the different parts of the process (making use of Root-cause analysis, Outcome analysis).

The next tool used by the SD4BPR is an adapted **Service Experience Blueprint** (Fig. 2). Building on the Value Network System Architecture model from the SD4VN method, SD4BPR uses an adapted version blueprint for better understanding of the AS-IS process through illustrating the frontstage and backstage actions of the process and their orchestration on different lanes, for different actors. By decomposing the process to its very specifics, the problematic points that delay the process or that make it unnecessarily difficult can be identified and signalled regardless of their nature (actors, artifacts, interactions), in a Root-Cause Analysis approach. SD4BPR concludes the exploratory study through a clear summary of the positive aspects, respectively of the problems identified in the process. Based on these, the objectives of the process improvement are defined and motivated with arguments, becoming guidance of the future solution. This follows both BPR and SD approaches, setting the baseline for the design phase.

4.3   Modelling the TO-BE Process

Building on the MSD [18], the SD4BPR modelling of the To-Be process is conducted on 3 service levels, making use of conceptual models and graphical representations while keeping the consistency between the strategic and the operational level of the business process.

For **Designing the Service Concept**, several **user profiles** are defined. These are associated with **service requirements** based on the results of Mapping the AS-IS process, keeping the human-centered approach while generating the essential functionalities. Therefore, the service concept can be the concept of the business process itself or of a support system.

For **Designing the Service System**, the **Service System Architecture(SSA)** is used, turning the requirements into activities and showing if these activities are supported by people or technology. SD4BPR builds on the SD4VN [19] version of the SSA, showing the interrelated actions of multiple actors, along with how these are supported. This can be the process itself or a process support system (technology enabled service), but it needs to be contextualized by showing the workflows of the people who execute the tasks and also the workflows of the people who support the executions of the tasks.

For **Designing the Service Encounter**, the focus is on the aesthetics and interactions of service interfaces, therefore, the encounter is considered between the different parts of the business process or between the users and the technology-enabled system that supports the business process. Since the SD4BPR is focused on better designing technology-enabled services/processes, the service encounter is represented graphically through **Wireframes**, offering a visual perspective of the user-technology interface interaction (visually signalling the proposed functionalities). Wireframes have been used in the MINDS Method as well, for a better integration between service design and user interaction fields, with the same aim of innovating technology enabled services [12], therefore are considered relevant for supporting the integration between SD models and BPR models.

These models were chosen for their flexibility in adapting to different analysis and modelling contexts and for their suitability to describing business processes. Also, their integration was meant to be a derivation process in two opposite directions. Firstly, from broad to very specific, deconstructing the process layer by layer through conceptual models that follow on the level of detail: Mind Map (categories) → Customer Modelling Experience (correlated components) → Service Experience Blueprint (workflow of actions, interactions and tools). Secondly, from specific to broad, reconstruction of the process through conceptual models that can enrich the design with a new perspective: Use cases (functional actions

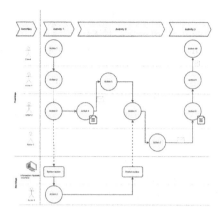

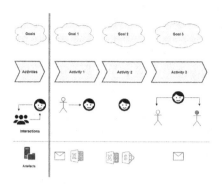

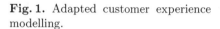

Fig. 1. Adapted customer experience modelling.

Fig. 2. Adapted service experience blueprint.

available for each user role) → Service System Architecture (orchestration of actions from service system perspective) → Wireframes (visual representation of overall system). Therefore, SD4BPR offers a well structured yet enough generalizable analysis and design package of SD and BPR models and techniques that transition smoothly into deconstruction and/or (re)construction of processes or services.

4.4 Exploring Possibilities of Implementation

SD4BPR is meant to evolve Service Design methods towards focusing more on the implementation part of the design process.Therefore, the purpose of this phase is to propose well defined possibilities of implementation of the designed solution, taking into consideration technology opportunities, business constraints and limitations. These are considered redesign catalysts, used by the SD4BPR in order to advance the BPR methods, since they support a more complete exploration of process alternatives, as stated in the future research possibilities from [2]. Therefore, a research effort on existing technologies as well as an analysis of the business constraints is conducted, generating two alternatives: one focused on the existing market technologies and their latest developments and one focused on the implementation of the system/process through tools/technologies already used by the company. The possibilities of implementation are presented at a conceptual level and are matched with the requirements and objectives generated previously.

5   Application Results

This section will present the results of the DSR demonstration phase: the method was applied in a business environment, namely for the improvement of the Pre-

Sale process at Glintt, and several artifacts were generated. Because of confidentiality reasons, the results of this cannot be explicitly shown, but they will be briefly mentioned in order to support the purpose of this paper, as empirical work. The first stage, *Mapping the AS-IS process*, was focused on the results of the interview coding and the modelling of the Pre-Sale process in its current state. The outputs of this stage are a brief process description, based on the tree categories, an activities workflow, a **Customer Experience Diagram**. This helped contextualize the Pre-Sale components of different natures (people, existing physical/digital tools, time-consuming activities, desired goals, unilateral/bilateral connections) for better understanding of the core elements and the intermediate outcomes. Each activity was broken-down into touchpoints in a **Service Experience Blueprint** (MINDS Method, MultiLevel Service Design). The difference between Frontstage actions and Backstage actions was illustrated (manually performed versus system support actions). The artifacts used were associated to each touchpoint or to the transition they serve. To conclude the Exploration stage, the core problems were identified for the Pre-Sale process and the solution objectives associated were defined.

5.1   Modelling the TO-BE Process

Workshop. A workshop was conducted, with the purpose of co-designing the format and content of a Pre-Sale file, the participants being 3 members of the Pre-Sale team. The final output of the workshop was redefining the concept behind the file as a sub-process of the Pre-Sale process which could benefit from several tools, used on-request, or on a continuous basis.

Support System. Based on the results of the AS-IS phase, 4 user profiles or personas have been identified as relevant for the Pre-Sale services: Pre-Sale admin, Pre-Sale member, Sales Agent, Pivot. For each of them, dedicated functionalities have been formulated into use-cases. These have been translated into frontstage and backstage actions and have been orchestrated in a **Service System Architecture**, showing how the actions are supported by the different service interfaces. Figure 3 illustrates a possible continuous logical flow, but the actions can be executed concurrently as well, depending on the situation. Taking into consideration user feedback, a series of functional and non-functional requirements have been identified as base for a Pre-Sale support service system. The visual representation of the service solution that would support the Pre-Sale process was made through **Wireframes**. The Wireframes (Figs. 4 and 5) were developed with the help of Balsamiq software (free trial) and are considered low fidelity prototypes of the system being designed. This option was chosen in order to be able to develop and easily modify the wireframes according to user feedback, focusing on structure, functionality, content and not on graphics or aesthetics.

Possibilities of Implementation. For identifying the most suitable possibilities of implementation based on market technologies, a review of existing tools/technologies has been made. The design catalysts taken into consideration were the level of development of the technologies, their features and their capabilities of matching the requirements generated in the previous phase. The tools/technologies have been divided into the three categories suggested by the requirements: Document Management, Business Intelligence and Project management.

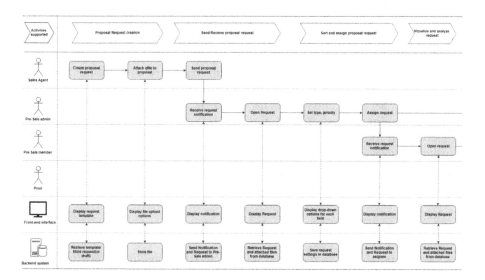

Fig. 3. Extract from the service system architecture of the Pre-Sale support system.

Fig. 4. Wireframe ex: general view Pre-Sale admin.

Fig. 5. Wireframe ex: general view pivot.

6   Method Evaluation

Following the DSR methodology [11], the SD4BPR method was evaluated in terms of process of work and in terms of relevance/efficiency for the business environment in which it was exercised. The results of the application are going to be implemented (according to feedback and organisational constraints), validating the artifacts created through applying the SD4BPR in a business environment, which is a good indicator of the potential of applicability of the method. During the application, 3 feedback meetings and one workshop took place. In the meetings, the participants were asked for their opinion on the intermediate results, while in the end of the workshop, they were asked for their opinion on the process of work and final result. The feedback was positive, the materials and intermediate results presented were considered relevant, proving a good understanding of the process. The final result was considered useful in terms of information systematizing and solution proposal for future implementation. The method was also validated using the BPR framework developed by [2], tackling all its dimensions (Aim, Actors, Tool, Input, Technique, Output). The methodological framework is aimed at enabling researchers to develop or evaluate/validate methods focused on business process improvement and it states the need of more methods that build on existing research effort (avoid focusing just on "best practices" or "innovation principles", but also on similar methods).

7   Conclusion and Future Research

This study presents the development of the SD4BPR method which uses constructs from SD and BPR fields, building on existing research effort or suggested research opportunities. Therefore, it offers a multidisciplinary approach, both process-focused and human-centred, generating customer experience value while aiming for better process efficiency and effectiveness. Following a DSR methodology, the method represents the generated artifact, meant to help solving an organizational problem. By applying it in empirical cases, the method itself enables the generation of artifacts that support organizational processes, improving them. SD4BR uses a qualitative approach, offering a well-structured analysis and design process of work, aimed at business process improvement. The analysis component brings together an exploratory study and mapping models in order to identify the positive/problematic points of the process, while the design component covers modelling on 3 service levels (concept, system, encounter).

The method was applied successfully in a business environment, supporting the improvement of Pre-Sale processes inside Glintt. It facilitated the mapping of the current process for analysis purposes and the modelling of solutions based on the results of the analysis. Two artifacts were generated: the model of an informational support system and the requirements of a Pre-Sale subprocess. These artifacts were perceived as relevant and useful. Their actual implementation will mean better time performance for the Pre-Sale process and more dynamic use of tools and more efficient collaboration for the people involved in the process (BPR desired outcome). Also, the artifacts represent a source of analytical information and can serve as guidance for future implementation, generating customer value (SD desired outcome).

Both the development of the method and its application reveal potential for further integration between SD and BPR. Further research can include how can the method be applied in different organizational contexts from different industries, how can the method be evolved from implementation possibilities to implementation modalities, how should the transition from the AS-IS to the TO-BE process occur in a business environment and how can the method be used for designing this operational process.

References

1. Bhaskar, H.L.: A critical analysis of information technology and business process reengineering. Int. J. Prod. Qual. Manag. **19**(1), 98–115 (2016)
2. Vanwersch, R.J.B., et al.: A critical evaluation and framework of business process improvement methods. Bus. Inf. Syst. Eng. **58**(1), 43–53 (2016)
3. Tuomisaari, H., et al.: Business model transformation as an explanation of dramatic demand-supply change events. In: Proceedings of 2012 IEEE International Conference on Service Operations and Logistics, and Informatics, pp. 345–349. IEEE (2012)
4. Ostrom, A.L., Parasuraman, A., Bowen, D.E., Patrício, L., Voss, C.A.: Service research priorities in a rapidly changing context. J. Serv. Res. **18**(2), 127–159 (2015)
5. Clatworthy, S.: Service innovation through touch-points: development of an innovation toolkit for the first stages of new service development (2011)
6. Sangiorgi, D.: Transformative services and transformation design (2011)
7. Eun, Y., Sangiorgi, D.: Service design as an approach to new service development: reflections and futures studies. In: ServDes 2014, Fourth Service Design and Innovation Conference "Service Futures", pp. 194–204 (2014)
8. Junginger, S., Sangiorgi, D., et al.: Service design and organisational change. Bridging the gap between rigour and relevance. In: International Association of Societies of Design Research, pp. 4339–4348. KOR (2009)
9. Pinheiro, T., Alt, L., Mello, J.: Service design creates breakthrough cultural change in the Brazilian financial industry. Touchpoint: J. Serv. Des. **3**(3), 18–23 (2012)
10. Bailey, S.G.: Embedding service design: the long and the short of it. In: ServDes 2012, Conference Proceedings Co-Creating Services, the 3rd Service Design and Service Innovation Conference, Espoo, Finland, 8–10 February, pp. 31–41. Linköping University Electronic Press, Linköpings universitet (2012)
11. Hevner, A., Chatterjee, S.: Design Research in Information Systems: Theory and Practice, vol. 22. Springer, Heidelberg (2010). https://doi.org/10.1007/978-1-4419-5653-8
12. Grenha Teixeira, J., Patrício, L., Huang, K.H., Fisk, R.P., Nóbrega, L., Constantine, L.: The minds method: integrating management and interaction design perspectives for service design. J. Serv. Res. **20**(3), 240–258 (2017)
13. Fasna, M.F.F., Gunatilake, S.: A process for successfully implementing BPR projects. Int. J. Prod. Perform. Manag. **68**(6), 1102–1119 (2019)

14. Omidi, A., Khoshtinat, B.: Factors affecting the implementation of business process reengineering: taking into account the moderating role of organizational culture (case study: Iran air). Proced. Econ. Finance **36**, 425–432 (2016)
15. Grant, D.: Business analysis techniques in business reengineering. Bus. Proc. Manag. J. **22**(1), 75–88 (2016)
16. Patrício, L., Gustafsson, A., Fisk, R.: Upframing service design and innovation for research impact (2018)
17. Blomkvist, J., Holmlid, S., Segelström, F.: Service design research: yesterday, today and tomorrow (2010)
18. Patrício, L., Fisk, R.P., Falcao e Cunha, J., Constantine, L.: Multilevel service design: from customer value constellation to service experience blueprinting. J. Serv. Res. **14**(2), 180–200 (2011)
19. Patrício, L., de Pinho, N.F., Teixeira, J.G., Fisk, R.P.: Service design for value networks: enabling value cocreation interactions in healthcare. Serv. Sci. **10**(1), 76–97 (2018)
20. Hevner, A., March, S.T., Park, J., Ram, S.: Design science research in information systems. MIS Q. **28**(1), 75–105 (2004)
21. van Aken, J., Chandrasekaran, A., Halman, J.: Conducting and publishing design science research: inaugural essay of the design science department of the journal of operations management. J. Oper. Manag. **47**, 1–8 (2016)
22. Peffers, K., Tuunanen, T., Rothenberger, M.A., Chatterjee, S.: A design science research methodology for information systems research. J. Manag. Inf. Syst. **24**(3), 45–77 (2007)
23. Charmaz, K.: Constructing Grounded Theory: A Practical Guide Through Qualitative Analysis. Sage, Thousand Oaks (2006)
24. Morelli, N., Tollestrup, C.: New representation techniques for designing in a systemic perspective. In: Nordes, no. 2 (2007)
25. Teixeira, J., Patrício, L., Nunes, N.J., Nóbrega, L., Fisk, R.P., Constantine, L.: Customer experience modeling: from customer experience to service design. J. Serv. Manag. **23**(3), 362–376 (2012)

The SDCS Method: A New Service Design Method for Companies Undergoing a Servitization Process

Laís Lima[1](✉) and Jorge Grenha Teixeira[2](✉)

[1] Faculty of Engineering, University of Porto, Rua Dr. Roberto Frias,
4200-465 Porto, Portugal
laissaidsouza@gmail.com
[2] INESC TEC and Faculty of Engineering, University of Porto,
Rua Dr. Roberto Frias, 4200-465 Porto, Portugal
jteixeira@fe.up.pt

Abstract. To cope with the fierce business competition and the increasing challenges brought with it, manufacturing companies have been demonstrating a growing interest in extending their service business. It is in this context that companies seek servitization strategies, i.e., developing the capabilities to add services to their traditional product offerings, to increase value to the customers and to differentiate themselves from the competition. However, companies pursuing a servitization strategy often lack methods and tools to design new services adapted to their context. Thus, this article seeks to cover this gap through the development of a new service design method, the (S)ervice (D)esign method for (C)ompanies undergoing a (S)ervitization process: SDCS. The development of this method followed Design Science Research (DSR) methodology. This article also presents the application of the SDCS method in a company undergoing a servitization process.

Keywords: Service design · Servitization · Design science research

1 Introduction

The demand for products often becomes stagnated, turning them less attractive [1]. That said, and in order for companies to still compete in the market, it becomes crucial to implement service-led growth strategies [2]: the concept of servitization emerges. Servitization can be defined as adding services to products [3] and is often seen as a way to increase the value to customers and differentiate value propositions from competitors [1].

The reason for companies to extend their service business usually proceeds along three lines: financial, marketing and strategic opportunities [4]. According to Anderson and Narus [5], companies can enhance their profits by using services more effectively to meet customers' requirements. Thus, they can increase their customer portfolio, gain more business and consequently reduce the costs of providing the services. Neely [6] mentions that larger firms are more likely to engage in servitization efforts than smaller

© Springer Nature Switzerland AG 2020
H. Nóvoa et al. (Eds.): IESS 2020, LNBIP 377, pp. 245–258, 2020.
https://doi.org/10.1007/978-3-030-38724-2_18

firms. In addition, there tends to be more servitized firms in highly developed economies than in industrializing economies.

To assist companies with their servitization strategies, the methods and tools from service design can be applied. Service design has been characterized as the systematic application of design thinking and other design methodologies and principles to the services field [7]. However, despite all the advances in the service research field, there is still little guidance for companies to develop and implement servitization strategies, and service design is still very focused in the early stages of the service development, with no specific method for the implementation stage [8]. Thus, there are many challenges regarding the implementation of servitization efforts.

Supporting servitization in product-centric companies requires reconstructing their organizational structures, including cultural change, internal communication, retaining service expertise and promoting inter-department collaboration [9]. West and Gaiardelli [10] also state that companies undergoing servitization processes may face financial challenges, as they are usually required to spend money in new infrastructure, physical assets and knowledge during the first stages of this process.

Taking these challenges into consideration, companies that intend to expand their service business, need to have appropriate methods and tools to support their servitization efforts. Thus, the present work aims to fill this gap by proposing a new (S)ervice (D)esign method for (C)ompanies undergoing a (S)ervitization process: the SDCS method. The next section presents a brief introduction to servitization, followed by a further review on service design and its methods and tools. The subsequent section presents the DSR methodology, followed by the SDCS method and its stages. Finally, this article concludes by showing how the SDCS was applied and supported a company undergoing a servitization effort.

2 Servitization

Over the last few decades, the global economy has witnessed a significant increase in the importance of services, while the importance of products has declined [11]. Coined by Vandermerwe and Rada [12], the term servitization is now widely recognized as the process of creating value by adding services to products [3]. Manole and Bier [13] state that the term servitization means more than increasing a company's service business orientation and adding services to its portfolio, but also encompasses the transformation from product-centric to customer-centric approach.

For Ruiz-Alba et al. [14] servitization strategies are only successful when companies engage in the co-creation of design and delivery of services, or in other words, if service propositions are in fact adjusted to customers' needs. However, this may require a lot of effort, since in order to do so, a company must reconfigure its resource base, organizational capabilities and structure; redefine its mission and update its routines, norms and values [15]. Nevertheless, when well applied, servitization can provide several benefits to manufacturing companies [16]. Besides, service differentiation that is, when companies differentiate their services from competitors, stimulates innovation as it reduces customers' perceived purchase risks and therefore helps to attract more customers who are willing to try the innovative value propositions [17, 18]. Thus, by providing differentiated services, organizations create additional value to customers and undergird their competitive advantage [19].

3 Service Design Process, Methods and Tools

Service design is a framework that aims to design services that are useful, usable and desirable from the user perspective and efficient, effective and different from the provider perspective [20]. With the expansion of services, authors have developed over the years several service design methods. However, although each method has its own characteristics, they commonly recognize the first step as being an exploration and understanding of the customers' needs; and the last one as delivering the customers a solution [21]. Thus, they follow the service design process composed by four stages: exploration, ideation, reflection and implementation [22], which is meant to be highly iterative and non-linear [23].

The first stage, exploration, consists of a deep study of the customer experience. Meyer and Schwager [24], define customer experience as the internal and subjective response that customers have to any direct or indirect contact with a company. A good customer experience though is not measured by the number of offered features but by embedding them in a value proposition [25]. Therefore, some of the methods and tools that can be used in this stage are research data, personas [26] and customer journey mapping [27].

After developing a deep understanding of the customer experience, it is time for the project team to generate and develop as many ideas as possible in the ideation stage [28]. However, in service design, ideas are just a starting point of a bigger process as new ideas help on the development of new service concepts. In the development of the service concept organizations should be careful to ensure that decisions are taken consistently and focused on delivering the right service to target customers [29]. Several methods and tools have been developed for this stage, namely customer value constellation, service system architecture and service experience blueprints [30].

After ideas are generated and the service concepts of the most promising ones designed, it comes to prototyping the service experience, in the reflection stage. Nevertheless, it is worth noting that, unlike products that can be prototyped in a way which people can hold the object and get the real feeling, service prototypes need to be experiences of the interaction with the multiple points [31]. In this context, pilot tests can be performed to help confirming whether the proposed method or instruments are inappropriate or too complicated [32]. In addition, some tools can also facilitate and be very useful in the prototyping of services, namely: videos, mock-ups and storyboards [25].

In the last stage of the service design process, implementation, the service is finally offered to the customers. It is important to note that the implementation of a service design project can be very complex since it may involve many different fields and expertise, such as, change management for the elaboration of new procedures and processes; software department, for the development of apps and software related activities or even engineering and product development in case of being necessary to produce physical objects [22]. Yet, the available literature focuses mainly on the initial stages of the service design process: exploration and ideation. That said, there is little literature on how the services can be prototyped [33] and successfully implemented [9], with no specific tool or method indicated for the implementation stage [8]. Besides, and according to the guidelines proposed by [34] there should be a principle of continuous improvement in the service development process which is not addressed by the traditional service design process. Thus, Fig. 1 summarizes the methods and tools proposed for each service design stage.

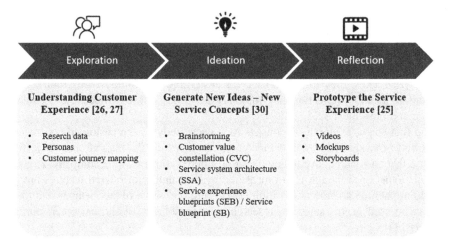

Fig. 1. Service design methods and tools for each stage of the service design process.

4 Methodology

With its roots in information systems discipline, Design Science Research (DSR) has emerged in recent years as a way to ensure rigorous and relevance research [35]. Originally positioned as a technology-oriented methodology, DSR is a methodology that aims to understand organizational phenomena by developing and rigorously evaluating artifacts that can solve organizational problems [36]. Such artifacts may be constructs, models, methods or instantiations. While constructs can be defined as the basic language for concepts to characterize a phenomenon, the models are used to describe tasks, situations or artifacts. Methods are ways of performing goal-directed activities and, lastly, instantiations regard the physical implementations to perform certain tasks [37]. Thus, since this study is focused on developing a method to solve an organizational problem - companies still lack the necessary knowledge, methods and tools to support their servitization efforts - DSR was considered the most appropriate methodology.

There are several processes in the literature that can be adopted for sustaining a DSR research, such as the ones proposed by Hevner [38] and Peffers et al. [39]. For the SDCS development, however, it was chosen the process proposed by Hevner [38], since it emphasizes the iterative process of DSR and allows a better distinction between aspects related to relevance and rigor. Hence, the process consisting of three main cycles: the relevance cycle, the design cycle and the rigor cycle [38]. In the relevance cycle the requirements (problems and opportunities) are presented and the artifact is evaluated in a field testing following specified criteria. The rigor cycle on the other hand presents the foundations that is, the scientific methods, theories and engineering methods that add rigor and robustness to the artifact. Lastly, the design cycle, is the main piece of a design science research project as it consists of the continuous improvement of the artifact, with the refining of the artifact according to the artifact evaluation and stakeholders' feedback.

Therefore, this study follows Hevner's [38] DSR process to develop a new service design method capable of supporting companies undergoing servitization efforts. As can be seen in Fig. 2, the relevance cycle used as inputs the requirements from a qualitative study conducted with the organization's employees in order to better understand the environment in which the artifact was developed. Subsequently, and still in the relevance cycle, the developed artifact was empirically tested at a company undergoing a servitization process in order to be evaluated.

On the other hand, the literature review provided the research gaps and theoretical support important for the rigor cycle. Besides, it provided the foundations for keeping the necessary methodological rigor in the data collection performed in the relevance cycle. Still in the rigor cycle, the current article was elaborated, adding new knowledge to the knowledge base. Based on the relevance to the environment and the acquired knowledge base, the developed artifact is constantly refined at the design cycle. The following section presents the artifact developed in the design cycle.

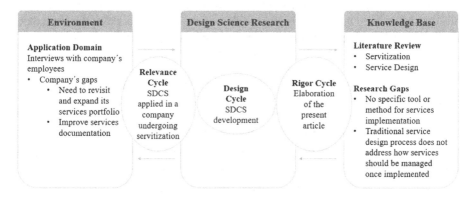

Fig. 2. DSR process.

5   The SDCS Method

Following the DSR process proposed by Hevner [38] and after further understanding the artifact's environment and its requirements, this section details the design of the artifact in the design cycle. For the creation of the artifact, the results of interviews with employees from a company undergoing a servitization effort (environment) and literature review (knowledge base) were used as inputs.

In total 15 interviews were performed, with the data being fully transcribed in the NVivo software and analyzed according to qualitative analysis principles [40]. The interviews highlighted the need for improving service documentation, improving technicians' knowledge and the wish for modular and standard contracts. Below is a summary of the sample characteristics, regarding the performed interviews (Table 1).

Table 1. Interviewed employees sample characteristics

By job function	# of employees
Top management	7
Middle management	4
Low management	4
By company's antiquity	# of employees
More than 10 years	7
Less than 5 years	7
Between 5 and 10 years	1

The SDCS method is composed by 5 main stages: exploration, development, pilot, implementation and service management. While the first 4 are based on the stages of the service design process, the 5th stage is an addition to the traditional service design process based on the principles proposed by Harland et al. [34]. As in the service design process, the SDCS begins by understanding the customer experience in the exploration stage, followed by the design of the service concept in the development stage (ideation stage in the service design process). The pilot stage from SDCS, however, differs from the reflection stage of the service design process in the sense that the service is not prototyped but piloted on a real client. Nevertheless, the objective of visualizing the service prior to its implementation remains the same. Next, SDCS presents the implementation, which is little addressed in the service design process, and adds the service management stage for the services' continuous improvement. Figure 3 illustrates SDCS conceptual framework.

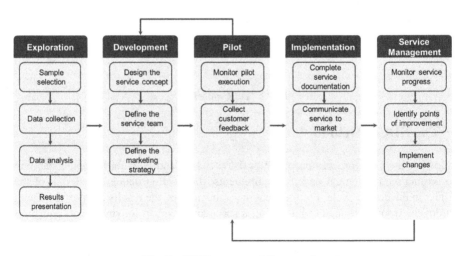

Fig. 3. SDCS conceptual framework.

The first stage, exploration, has the same objective of this stage in the service design process: understand the features that add value to the service value proposition in the perception of the customers. Thus, this stage consists of 4 main sequential steps: sample selection, data collection, data analysis and results presentation. The first step of this stage, sample selection, consists in defining the customers to interview. When selecting the customers to interview, the sample should be as diverse as possible and customers should be interviewed until theoretical saturation is achieved [38]. After the sample being selected, the script for conducting the interviews is elaborated and the meetings with the customers scheduled. Next, the data should be fully transcribed into a software for qualitative analysis such as NVivo and analyzed according to the principles of qualitative analysis [40]. Finally, following service design participatory approach and the need to communicate a service mindset to product-oriented companies, the results presentation consists in a workshop with the most relevant people of the company for the service development, in which the findings are presented, and actions plan defined and prioritized. The action plans may consist on improvements in current services or development of new services. In case of improvement in current services, the following stage should be service management.

In the following stage, development, the goal is to elaborate all the essential service features necessary for piloting it. Thus, it consists of 3 main sequential steps: design the service concept, define the service team and define the marketing strategy. To assist in this stage, SDCS suggests that the team begins by elaborating the customer value constellation model (CVC). This will help the team to design the service concept, defining the new service value proposition, its scope and objectives as well as analyzing if there may be any interdependencies with other existing services. Next, there should also be designed the service system architecture (SSA), for a better understanding of all main activities and organizational departments involved in the service provision. Once the overall departments are identified, it is time to define the service team, that is, those that will be responsible for the service execution. It is important to identify the necessary skills for performing the service and choose the resources based on these skills. Next, and as a requirement from the employees' interviews, the service documentation should be elaborated. The appropriate documentation format, however, may vary according to the service, being for example a very detailed technical manual or service experience blueprints (SEB) [25]. At this moment the service is almost ready to be tested, missing only the definition of the marketing strategy: pricing and target segments. Thus, it is important to define some criteria and segment the customers in order to choose to whom the company intends to offer the service, besides selecting the pilot customers.

In the pilot stage, as in the reflection stage of the traditional service design process, the goal is to promote the necessary changes prior to spending more effort in its implementation. Hence, customers' feedback should be collected in order to promote continuous improvement until the service is ready to enter the market. A pilot test is a way to confirm whether the proposed method or instruments are inappropriate or too complicated [32] and customer's feedback should enable the identification of any adjustment needs and consequent action plans. The service should only move to next stage if no further need of improvements is identified. By that, this stage consists of 2 main sequential steps: monitor pilot execution and collect customer feedback. Thus, while the service is being executed at the chosen customers, the team should be monitoring the pilot execution in order to identify if the service is being executed as defined or if any additional steps are necessary and the service documentation adjusted accordingly.

Regarding the implementation stage, there is still very little literature to support it [41]. Thus, for the elaboration of this stage, the inputs from the interviews were especially important. Therefore, SDCS proposes that the implementation stage be divided into 2 sequential steps: complete service documentation and communicate service to the market. In the first step, all the documentation elaborated in the previous stages should be reviewed and the service level agreements (SLA) defined, so that the standard contracts can be elaborated. With all the service documentation complete, it is time to communicate the new service to the market, being this activity extremely important for greater customers engagement.

The last stage, service management, is based on the aiming for perfection phase proposed by Harland et al. [34] and consists of 3 main steps: monitor service progress, identify points of improvement and implement changes. Thus, in the first step, monitor service progress, the team should define the most important KPI's in order to measure the service's performance. Once the service is monitored, and according to the results, it may be time to promote some changes. Therefore, the second step of SDCS proposes workshops, where the results obtained are presented to the most appropriate people in the organization and actions plans are defined based on the need of improvements identified. Lastly, in the third step, the identified points of improvement are implemented, the resources are trainned in accordance with the changes and the service's documentation is updated.

6 Evaluating the Artifact: Applying the SDCS Method in a Company Undergoing a Servitization Process

Once the SDCS was designed, and following the principles of DSR, the method was applied in a company undergoing a servitization process. This application is crucial to evaluate the method's usability and relevance, so that it can be applied by other companies undergoing servitization.

Thus, for the first stage of SDCS, exploration, 13 customers of different nationalities were interviewed, in which 5 customers were interviewed in person, while the other interviews were held by telephone. The purpose of these interviews was to identify the services that add more value for the customers, and therefore should be included in the company's portfolio. It is worth mentioning that the data collected was fully transcribed and inserted in the NVivo12 Software for data analysis. As stated by Basit [42], the analysis of qualitative data can be arduous, since it is not a fundamentally mechanical or technical activity. That said, using a software for coding the data is crucial to allow the researcher to really connect to the data and be able to understand the phenomenon in an easier way and generate meaningful conclusions. Hence, as a result of the interviews, customers found both predictive maintenance and obsolescence management equally important services to be developed. Below is a summary of the sample characteristics (Table 2).

Table 2. Interviewed customers sample characteristics.

By job function	# of employees
Top management	5
Middle management	4
Low management	4
By industry segment	# of employees
Durable manufacturing	4
Retail & Wholesale	3
Nondurable manufacturing	3
Food & Beverage	2
Pharma & Healthcare	1

Following SDCS, a workshop was then conducted with the company's top management in order to present these results and define the next steps. Hence, the team chose to develop the predictive maintenance service as they believe it would add more competitiveness to the company's service portfolio.

Next, to begin the discussion of the service concept, a workshop was scheduled with some key people for the service development, in which the CVC model (Fig. 4) was presented. It is noteworthy that the colored boxes represent the company's concept of the predictive maintenance service while the grey boxes are extended service offerings, which will not be offered by the company at the moment.

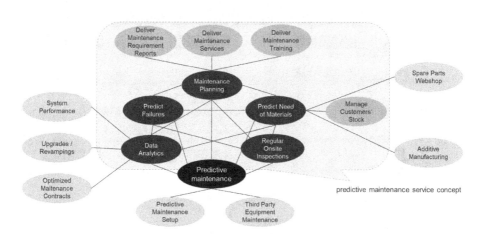

Fig. 4. Service concept positioning in the customer value constellation for predictive maintenance.

Following SDCS, the team proceeded with the definition of the service value proposition, scope, objectives and analyzed service interdependencies. With regards to the service scope, the team proposed some different maintenance plans, as presented below:

- PM Bronze plan – customer is responsible for scheduling maintenance with the company's technicians, or doing maintenance internally, and for spare parts management.
- PM Silver plan – The company is responsible for contacting the customer and for scheduling maintenance. Customer is responsible for spare parts management.
- PM Gold plan – The company is responsible for contacting the customer, for spare parts management and for scheduling maintenance. In this plan, the company must guarantee availability for intervention within 24 h.
- PM Platinum plan – The company is responsible for everything mentioned in the gold plan plus the customer has a spare parts package established in the contract and does not need to purchase them individually.

It is noteworthy, however, that although the proposed plans follow a progressively logic, the different features (alert detection, spare parts management and intervention time frame) can be combined in several different ways, according to each customer's preference.

Next, the team developed the service system architecture (SSA), illustrated in Fig. 5, in which on the left of the model are represented the main people involved in the service provision, both those that interface with the customer (service interfaces), and those that work internally in the company (backstage support). The top row indicates the main activities involved in performing predictive maintenance while the other boxes indicate who participates in each activity.

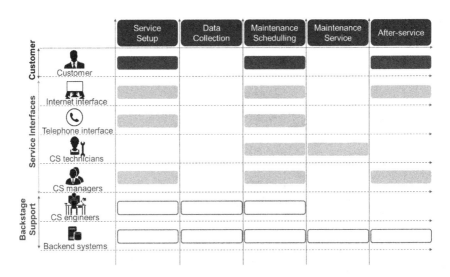

Fig. 5. Service system architecture for predictive maintenance.

Still in the workshop, and in the next step of the development stage of the SDCS method, the team defined the main skills necessary to perform the service as well as the specific people that will be involved in the service provision, followed by the elaboration of the service documentation. The team chose the service experience blueprint (SEB) as the most appropriate documentation. This is because the predictive maintenance service is mostly automated and when human intervention is needed, technicians must follow the technical manuals of other types of maintenance: preventive and corrective. Thus, no additional documentation is required.

Ideally there should be one SEB for each activity depicted in the SSA. However, for the purposes of exemplification, only one activity will be detailed: maintenance scheduling, which concerns the scheduling of a maintenance service with the company once a potential failure is detected. Thus, the SEB of the most advanced plan, Platinum Plan, is presented below (Fig. 6).

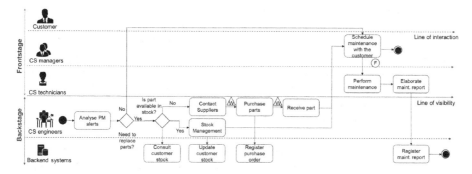

Fig. 6. SEB of the platinum plan.

It is worth mentioning that the activities demonstrated are specific to the service in question and according to the context of the company being studied. However, it is possible to visualize that in the most advanced plan, illustrated in Fig. 5, there is almost no interaction with the customer, which is expected from a highly industrial environment. For classic service contexts, on the other hand, the value co-creation through active customer participation is prioritized.

Once the different SEB were elaborated, the execution team should receive the appropriate training. At the present moment, this training is being scheduled. However, once training is provided, the team will need to define the pricing and target segments in which the company wishes to focus as well as the pilot customers. The pilot phase is important to identify if the SEB reflects the reality or if needs any adjustments. Furthermore, from the pilot customers feedback, opportunities for improvement can be identified and actions plan prioritized. It is worth emphasising that these improvements will need to be completed for the service to move to the next stage: implementation. Next, all the service documentation will have to be revised, including the pricing strategies previously defined and the standard contracts developed in accordance with the different service plans identified. Finally, the Marketing and Sales team will develop the sales and social media contents so that the service can be communicated to the market. At this moment, the service will enter into a continuous improvement phase, in the service management stage.

7 Conclusion

This research aims at developing a new service design method for companies undergoing a servitization process, providing these companies the tools and knowledge for the design and implementation of a new service.

Following a DSR approach, to ensure the relevance and rigor of the new method, a literature review was performed in the most relevant subjects for the method development as well as a thorough study, following a qualitative approach, with a company undergoing a servitization process.

Hence, the literature review revealed that the existing literature on service design is focused on the early phases of service development and very vague regarding its implementation, with no specific method or tool for this purpose. Besides, the service design process addresses only new services development, without specifying how they should be managed once implemented. The conducted interviews highlighted the employees' concern with improving service documentation, the poor service strategy and establishing standard contracts. Thus, all these aspects were considered in the design of the proposed SDCS. Next, the method was applied in the design of a new service for a company undergoing a servitization process.

However, there are also some other limitations to this study which are worth mentioning. The method is being applied in only one company, in a B2B and highly industrial context, therefore other applications should ensue to further evaluate and improve the method. Also, once the SDCS method has not yet been fully applied, results on the later stages are scarce. Further research can report all the phases of the method application. However, this research offers a first approach to propose and define a new service design method for companies undergoing a servitization process. In addition, with the method application and detail of each performed activity, the project adds a practical contribution, by not only establishing the new method, but also exemplifying in practice how companies in a servitization process can apply it. Furthermore, the method also adds a further theoretical contribution by presenting phases dedicated exclusively to the implementation and management of new services, subjects little or not addressed by the traditional service design processes. It is also worth mentioning the relevance of the SDCS research contribution as one of the few methods, following Teixeira et al. [43] and Patrício et al. [44], that explores the creation of a service design artifact by adopting the DSR methodology.

References

1. Martín-Peña, M.L., Bigdeli, A.Z.: Servitization: academic research and business practice. Universia Bus. Rev. **49**, 18–31 (2016)
2. Kowalkowski, C., et al.: What service transition? Rethinking established assumptions about manufacturers' service-led growth strategies. Ind. Mark. Manag. **45**, 59–69 (2015)
3. Baines, T.S., et al.: The servitization of manufacturing: a review of literature and reflection on future challenges. J. Manuf. Technol. Manag. **20**(5), 547–567 (2009)
4. Gebauer, H., Friedli, T., Fleisch, E.: Success factors for achieving high service revenues in manufacturing companies. Benchmarking: Int. J. **13**(3), 374–386 (2006)

5. Anderson, J.C., Narus, J.A.: Capturing the value of supplementary services. Harv. Bus. Rev. **73**(1), 75–83 (1995)
6. Neely, A.: Exploring the financial consequences of the servitization of manufacturing. Oper. Manag. Res. **1**(2), 103–118 (2008)
7. Holmlid, S., Evenson, S.: Bringing service design to service sciences, management and engineering. In: Hefley, B., Murphy, W. (eds.) Service Science, Management and Engineering Education for the 21st Century. SSRI, pp. 341–345. Springer, Boston (2008). https://doi.org/10.1007/978-0-387-76578-5_50
8. Overkamp, T., Holmlid, S.: Views on implementation and how they could be used in service design. In: Service Design Geographies, Proceedings of the ServDes, 2016 Conference, Copenhagen, Denmark, 24–26 May. Linköping University Electronic Press (2016)
9. Zhang, W., Banerji, S.: Challenges of servitization: a systematic literature review. Ind. Mark. Manag. **65**, 217–227 (2017)
10. West, S., Gaiardelli, P.: Driving the servitization transformation through change management: lessons learnt from industrial cases. In: Proceedings of the Spring Servitization Conference, Aston University, Birmingham (2016)
11. Berry, L.L., et al.: Creating new markets through service innovation. MIT Sloan Manag. Rev. **47**(2), 56 (2006)
12. Vandermerwe, S., Rada, J.: Servitization of business: adding value by adding services. Eur. Manag. J. **6**(4), 314–324 (1988)
13. Manole, M., Bier, K.: From a product-centric to a customer-centric brand identity through servitization: the case of FOSS, a Danish food safety, and quality solution provider (2018)
14. Ruiz-Alba, J.L., et al.: Servitization strategies from customers' perspective: the moderating role of co-creation. J. Bus. Ind. Mark. **34**(3), 628–642 (2019)
15. Kowalkowski, C., et al.: Servitization and deservitization: overview, concepts, and definitions. Ind. Mark. Manag. **60**, 4–10 (2017)
16. Tan, K.H., et al.: Riding the wave of belt and road initiative in servitization: lessons from China. Int. J. Prod. Econ. **211**, 15–21 (2019)
17. Gebauer, H., Gustafsson, A., Witell, L.: Competitive advantage through service differentiation by manufacturing companies. J. Bus. Res. **64**(12), 1270–1280 (2011)
18. Zhang, J., Wang, J., Yang, H.: Service differentiation under mixed transit oligopoly: a graphical analysis. In: Transport and Society-Proceeding of the 22nd International Conference of Hong Kong Society for Transportation Studies, HKSTS 2017, Leeds (2018)
19. Opresnik, D., Taisch, M.: The value of big data in servitization. Int. J. Prod. Econ. **165**, 174–184 (2015)
20. Mager, B., Sung, T.J.D.: Special issue editorial: designing for services. Int. J. Des. **5**(2), 1–3 (2011)
21. Foglieni, F., Villari, B., Maffei, S.: How to (re)design services: from ideation to evaluation. In: Foglieni, F., Villari, B., Maffei, S. (eds.) Designing Better Services. BRIEF-SAPPLSCIENCES, pp. 27–45. Springer, Cham (2018). https://doi.org/10.1007/978-3-319-63179-0_3
22. Stickdorn, M., et al.: This Is Service Design Doing: Applying Service Design Thinking in the Real World. O'Reilly Media Inc., Newton (2018)
23. Menor, L.J., Tatikonda, M.V., Sampson, S.E.: New service development: areas for exploitation and exploration. J. Oper. Manag. **20**(2), 135–157 (2002)
24. Meyer, C., Schwager, A.: Understanding customer experience. Harv. Bus. Rev. **85**(2), 116 (2007)
25. Patrício, L., Fisk, R.P.: Creating new services. In: Serving Customers Globally, pp. 185–207 (2013)

26. Miaskiewicz, T., Kozar, K.A.: Personas and user-centered design: how can personas benefit product design processes? Des. Stud. **32**(5), 417–430 (2011)
27. Rosenbaum, M.S., Otalora, M.L., Ramírez, G.C.: How to create a realistic customer journey map. Bus. Horiz. **60**(1), 143–150 (2017)
28. Brown, T.: Design thinking. Harv. Bus. Rev. **86**(6), 84 (2008)
29. Goldstein, S.M., et al.: The service concept: the missing link in service design research? J. Oper. Manag. **20**(2), 121–134 (2002)
30. Patrício, L., et al.: Multilevel service design: from customer value constellation to service experience blueprinting. J. Serv. Res. **14**(2), 180–200 (2011)
31. Polaine, A., Løvlie, L., Reason, B.: Service Design: From Insight to Inspiration. Rosenfeld Media, New York (2013)
32. Van Teijlingen, E.R., Hundley, V.: The importance of pilot studies (2001)
33. Blomkvist, J., Holmlid, S.: Service prototyping according to service design practitioners. In: Conference Proceedings, ServDes 2010, Exchanging Knowledge, Linköping, Sweden, 1–3 December 2010. Linköping University Electronic Press (2010)
34. Harland, T., et al.: DIN SPEC 77007:2018-08, Leitfaden Lean Services - Professionalisierung des Dienstleistungsgeschäfts (2018)
35. Weber, S.: Design science research: paradigm or approach? In: AMCIS (2010)
36. Greenfield, A.: Against the Smart City: The City is Here for You to Use, Part I, pp. 1–153. Do Projects, New York City (2013)
37. March, S.T., Smith, G.F.: Design and natural science research on information technology. Decis. Support Syst. **15**(4), 251–266 (1995)
38. Hevner, A.R.: A three cycle view of design science research. Scand. J. Inf. Syst. **19**(2), 4 (2007)
39. Peffers, K., Tuunanen, T., Rothenberger, M.A., Chatterjee, S.: A design science research methodology for information systems research. J. Manag. Inf. Syst. **24**(3), 45–77 (2007)
40. Charmaz, K., Belgrave, L.: Qualitative interviewing and grounded theory analysis. SAGE Handb. Interview Res.: Complex. Craft **2**, 347–365 (2012)
41. Overkamp, T., Holmlid, S.: Views on implementation and how they could be used in service design (2016)
42. Basit, T.: Manual or electronic? The role of coding in qualitative data analysis. Educ. Res. **45**(2), 143–154 (2003)
43. Teixeira, J.G., Patrício, L., Tuunanen, T.: Bringing design science research to service design. In: Satzger, G., Patrício, L., Zaki, M., Kühl, N., Hottum, P. (eds.) IESS 2018. LNBIP, vol. 331, pp. 373–384. Springer, Cham (2018). https://doi.org/10.1007/978-3-030-00713-3_28
44. Patrício, L., de Pinho, N.F., Teixeira, J.G., Fisk, R.P.: Service design for value networks: enabling value cocreation interactions in healthcare. Serv. Sci. **10**(1), 76–97 (2018)

Understanding Service Design and Design Thinking Differences Between Research and Practice: An Empirical Study

Ana Torres[1]([✉]) and Cátia Miranda[2]

[1] LIAAD - INESC TEC and Polithecnic Institut of Porto, Porto, Portugal
ana.torres@inesctec.pt
[2] Faculty of Engineering, University of Porto, R. Dr. Roberto Frias,
4200-465 Porto, Portugal
catia.4.miranda@hotmail.com

Abstract. Service Design (SD) and Design Thinking (DT) evolved in the last decade and have become popular in the research field of service science. However, the application of SD and DT research outcomes into practice is still scarce. To help understanding the differences between research and practice, we conducted 20 semi-structured interviews with professionals and trainees from four organizations that are involved in service innovation projects. The results reveal several similarities and complementarities, (dis)advantages, requests and obstacles, which hinder companies from implementing and using structured SD and DT approaches. The findings present some challenges for both researchers and practitioners on actions they could take to overcome barriers and foster the SD and DT practice within organizations.

Keywords: Service Design · Design Thinking · Service innovation · Qualitative analysis

1 Introduction

Services industries have expanded fast in the past decades and the rise of service-based business models become crucial to offer services as a company to stay competitive grows [1]. Moreover, in today's digital economy there is a huge demand in managing product-service offerings using structured and digitalized methods and IT tools in companies [2]. This trend, often referred to *service innovation*, can be capitalized on by using Service Science (SS) methodologies. More recently, many producing companies have shown an increasing interest in service innovation due to changing business models, value co-creation, and profit margins and, therefore, have evolved a demand for SS methodologies over the recent years [3].

As such, service design and service innovation activities should create the value constellations [4], beneficial for both business and customer. The knowledge about the benefits of using structured methods and procedures in service innovation, in general can be assumed. However, although many SS methodologies and models are

© Springer Nature Switzerland AG 2020
H. Nóvoa et al. (Eds.): IESS 2020, LNBIP 377, pp. 259–272, 2020.
https://doi.org/10.1007/978-3-030-38724-2_19

extensively discussed in relevant literature, only few structured service procedure methodologies are actually implemented and used in practice.

Service Science (SS) evolved in the last years and has become a popular and interdisciplinary field of research that focuses on the structured development of services. SS is defined as the application of scientific, management, engineering and design on tasks that a person, organization or system perform with another person, organization or system, and thus growing the service innovation research [5].

Service design (SD) and design thinking (DT), being part of the SS discipline [6], both represent an interdisciplinary/holistic approach that aims to systematically design service value propositions using frameworks, methods and tools [4, 5, 7].

SD and DT approaches with a strong focus on user centricity, customer integration and multidisciplinary collaboration have recently become highly relevant in SS, driven by the growing emphasis on customer orientation and service systems [7, 8]. Consequently, these methods have been highly suitable for the customer-centered development of smart service innovation.

Being a recent research stream of SS there is a few knowledge about how organizations and businesses understand, implement and apply SD and DT approaches, (e.g. framework, principles, object, processes and tools). Beyond the growing body of literature on service science innovation, empirical exploratory research focusing on SD and DT approaches remains scarce. Even though there is a mass of contributions discussing the concept and the process of DT and SD, there is lack of clarification regarding each approach. In addition, there is little research about how these two approaches relate with each other and how they are being applied in practice, thereby constituting the main challenge of this study. Therefore, we determine the following research questions (RQ) for this contribution:

1. What are the main principles and approaches of SD and DT and how are they understood in practice?
2. What are the differences, shared ground and complementarities between SD and DT?
3. Which are the (dis)advantages of SD and DT in practice?

To help understanding the gap between research and practice of SD and DT approaches, we conducted 20 semi-structured interviews with 7 experts from several organizations, and 13 trainees involved in structured development of services-and-product-innovation. To address the research questions, this paper is organized as follows: Sect. 2 presents the theoretical underpinnings of this study, reviewing the state-of-the-art with respect to SD and DT research. Methodology and sample characteristics of the empirical study are detailed described in Sect. 3. The results of the interviews are presented in Sect. 4. Section 5 discusses the main findings of this study, as well as the main faced challenges. Finally, recommendations for practitioners and some topics for future research are suggested.

2   Service Design and Design Thinking

In the past decades the Service Science research has increased and attracted the attention of the academic community [5–7]. More recently, service innovation is on the agenda of a lot of different research streams [9] e.g. business literature, service management & marketing [10], IS literature, and more recently on smart services literature [11, 12], as well as in practice.

Service innovation refers to the creation of new and/or enhanced service offerings, service processes, and service business models [13]. Innovations can be driven by a detailed understanding of people's needs and their preferences. Hence, with the change of the overall business landscape, human-centered, creative, iterative and practical approaches are required to produce innovative ideas, developing integrated solutions of product-services so that companies gain competitive advantage [14, 15].

This perspective is embedded in the *Service-dominant* (S-D) logic which is supported by the value-in-use and co-creation of value rather than the value-in-exchange and embedded-value, concepts of the goods-dominant (G-D) logic [16]. Although, the service is designed aiming to provide experiences to the customer, these experiences cannot be predesigned [17]. Hence, a service must be designed in a holistic and adaptable way, allowing customers to co-create experiences regarding their preferences [18]. As such, services are increasingly created in service value networks with multiple agents that cooperate and exchange resources [7, 19]. This view on *service design* offers new possibilities for innovation based on resource sharing, but it also signifies that the network of multiple agents has to be taken into account in service engineering [20].

Service Design (SD) is understood as an interdisciplinary, creative and holistic approach which is becoming commonly used to improve and create services. This approach considers the customer or user as the starting point for launching a new service or improve an existing one [21]. Therefore, the main focus of SD is to provide a holistic and well-planned customer experience, always taking into consideration the customer problems and needs [22], and the processes involving the service, the tangible evidences and the technology solutions supporting the experience/system/service [18].

Design Thinking (DT) is a human-centered approach to innovation based on design tools to integrate the needs of people, the potential technologies and the requirements in order for businesses to have success [14]. It is useful for any type of organization, as it allows to work with open and complex problems [23, 24].

SD and DT are a new field of design investigation where the vocabulary and paradigm are still developing. Both are *human-centered* and can be applied regarding problem framing, information gathering and interpretation, solution ideation and evaluation in the development of an existing service or designing a new service solution. Equally, SD and DT are involved in the called *Human-Centered Design* which captures insights and produces innovative solutions that reflect the needs of the consumers [14]. Therefore, in this perspective SD and DT presents similarities and complementarities.

There is a sort of different models in literature, as well on practice (e.g. on companies' websites) regarding the SD and DT process and, these process models vary according to the number of steps or the precision identified in each phase. In general, these models and methodologies aim to assist companies in executing the service innovation development process in a structured manner and reduce the difficulty of the project by ex-ante defining the different phases.

A description of the state-of-the-art in research of SD and DT regarding both concept, principles, object, models, process, methods and tools is summarized, respectively on Tables 2 and 3, on Sect. 4.

3 Research Method

To address the research questions, an exploratory study using qualitative research was conducted, as it is suitable to acquire in-depth understanding about a subject [25–27]. In accordance to the authors, we conducted semi-structured interviews with practitioners to obtain qualitative information regarding the knowledge and usage of SD and DT approaches on service innovation projects. Semi-structured interviews have been chosen as a suitable methodology to address the stated research questions, since they promise exploring, understanding and learning about experiences and behaviours with openness [28], providing good practical data from experts of leading organizations in service innovation field. To achieve our goal, some principles and practices of Grounded Theory were used to acquire in-depth understanding about the emergent concepts during the interviews when compared with the ones in theory [29]. In a first stage, an interview guideline has been prepared to ensure the coverage of topic-related information. Secondly, we evaluate the first iteraction of the guideline within a pre-test with other researchers to prevent ambiguity, complexity of terms and to obtain first insights about time spent and other constraints, as scientific questionnaires are extensive in nature. The questions and structure of the interview guideline is divided into three sections: (1) the interviews started with an introduction about the goal and scope of the research and general questions regarding the practical background of the interviewee and which innovation project(s) the participants were involved. Besides, the interesting insides about the interviewee, it introduces them about the basic questions to the situation and reduces barriers in terms of the relationship between interviewer and interviewee. In addition, a contextualisation was also made and some documentation with definitions shown, in case of hypothetical doubts about the subject. Afterward, the main part of the interviews (2) focus on participant's point of view and insights regarding the usage of SD and DT approaches involved for application on service innovation projects in practice, namely: the followed principles, steps, processes, methods and tools, as well as its perceived advantages and disadvantages. Closing with (3) final comments and opinions of the interviewees about the expected results. Interviews took on average 28–30 min.

Sampling. The interviewees are primarily professionals/teaching assistants and former students/trainees of Porto Design Factory (PDF) working directly in services innovation projects, where students from different areas and nationalities cooperate in order to develop innovative projects with business organizations. The interviewed former students are from the ME310 – Product and Service Innovation Post-Graduation which is focused on teaching students the innovation methods and processes for designers, engineers, and future project managers, and the SQUAD program focused on digital design and experience design. Additionally, experts from leading enterprises on service innovation were included, to obtain a richer understanding regarding the SD and DT practice, as well, to allow a comparison of the results between practitioners with different levels of expertise (e.g. novices vs experts). To get a variety of insights enterprises experts have been chosen, since they were proficient and are spread in different service areas (ICT, consultancy, health-care, customer service). The sample design is split into professionals and trainees to understand the different points of view of practitioners with different levels of expertise regarding service innovation approaches. As showed in Table 1, a list of all organizations and their interviewed position is presented. In total, 20 in-depth interviews were conducted: 13 trainees and 7 professionals were interviewed. ME310 students have already finished their degree and have background in design and engineering. SQUAD students have finishing their degree and have background in design, marketing and social sciences. Regarding the professionals, the teaching assistants have already attended a course at PDF and the remaining interviewees have background in engineering, design and economics and work at companies that use DT and SD to develop their projects.

Table 1. Sample characteristics.

Interviewee and position		Number
Trainees	ME310 – product and service post graduation	9
	SQUAD program	4
Professionals	Teaching assistant at PDF	2
	Arts and design teacher	1
	Healthcare industry lead	1
	Service line manager	1
	UX specialist	1
	Former ME310 student and teacher	1
Total		20

Data Analysis. To evaluate the transcribed interviews, we follow the several steps of the qualitative content analysis, according with Charmaz [28]. The first step of the process was the transcription of the interviews in several segments. Afterwards, the process called initial coding started, that aims coding and reducing the statements into a set of relevant and meaningful categories and subcategories, helping in the analytic process. This process was iterative as it requires the constant analyses, as the research evolves, and data was coded several times in different categories and concepts. Next step, called axial coding, is the process of relating the codes (categories and subcategories) to each other, creating a hierarchy. For example, when an interviewee described a principle within the Service Design approach, that information is coded in the following way: "Service Design; Principles; Human-Centered", where Principles and Human-Centered are subcategories of the category Service Design. All steps are performed using the software NVivo12 (NVivo Transcription). To address the research questions, the SD and DT categories are divided into seven subcategories: definition, principles, object of study, process, tools, advantages and disadvantages. The results of our analysis will be following explained.

4 Results

4.1 Design Thinking Research Versus Practice

To address the RQ, a description of the state-of-the-art for DT in research, regarding both concept, object, models, process and tools, and the results of the interviews for the DT practise, on the five subcategories are summarized in Table 2.

Table 2. Design thinking comparison between research and practice

Categories	DT research	DT practice	Common
Concept	Conceptualized as a human-centered approach as it applies the tools of a designer and embraces human skills to stimulate transformation and development [14]	Considered a human-centered mindset which uses visual tools to promote the abductive thinking in order to solve problems	Human-centered
Principles	Empathy; collaboration; testing; human-centered [32]	No criticism; testing; build on top; build to think, don't think to build; collaborative; iterative; real; human-centered	Testing; collaboration; human-centered
Objects	Product; service; experiences [23]	Service; product service system; products	Product; service
Process	3 I Model: inspiration, ideation and implementation [33]; HCD model: hearing, creating and delivering [33]; Double diamond: divergent and convergent phases [30]; The design thinking model by the Hasso-Plattner-Institute: empathize; define; ideate; prototype; test [31]	Double diamond empathize; define; ideate; prototype; test	Double diamond empathize; define; ideate; prototype; test
Tools	Interviews; observation; conversations; personas; role objectives; explore customer's pain points; body storming; mind mapping [31]; sketching; value chain analysis; brainstorming; concept development assumption; customer co-creation; learning launch [34]; stakeholders map; customer journey map; service blueprint, business model innovation; rapid prototyping tools [35]; "How might we?" questions [32]	Prototypes; interviews; ideation techniques mainly brainstorming; benchmarking; observation; personas; empathy map; business model canvas; user journey map	Interviews; prototyping; user journey map

When compared DT between research and usage by practitioners, as results show, the main differences essentially arise in the definition of the concept, since in literature DT is considered an approach and in interviews is referred as a mindset. Regarding the principles, in practice the respondents see DT more broadly, however they also focus on essential points present in literature. Concerning the objects, process and tools responses have many similarities with the literature. Regarding the process, participants mainly use the "Double Diamond" [30] and the steps of The Design Thinking Model by the Hasso-Plattner-Institute [31]. The most common used tools are "Interviews; Prototyping and User Journey Map". Globally, in practice DT is not referred as a methodology, since people do not see DT as something to follow rules, but rather a way of freely thinking without censorship, being iterative in nature and promoting abductive thinking.

4.2 Service Design Research Versus Practice

Table 3 summarizes the most important findings of SD comparison between research and practice, regarding the five subcategories: concept, principles, object, process, methods and tools. The results of the interviews concerning SD in practice are only from professionals, since interviewees of PDF were not aware of SD practice.

Basically, the main conceptual difference is that respondents define SD as a creative process, since it allows access to physical evidences and thus enable the solutions to provide the desired form. Regarding the SD principles, interviewees strongly agree that SD has iterative stages of receiving feedback, designing the service, prototype and test. They also refer SD as a cyclic process, allowing to do quick changes and refinements, which in turn enables to create other development opportunities. Besides, it permits to improve the quality of the solution since there are continuous improvements (e.g. *"Stages can be repeated according to the context and needs of the project."* Professional, Healthcare Industry Lead, about SD principles).

Furthermore, the respondents add some principles, referring SD an open-mindset and co-creative, which are not denoted in research.

Concerning the process, interviews practice is similar with the research reviewed. However, results show a new one: respondents add the "Design Sprints" used in the ideation phase, when projects have a short time frame.

Concerning the methods and tools interviews results have many similarities with research. SD in practice uses several common methods and tools, although the large number of methods and tools described in research. SD being a Human-Centered process mainly uses observation, focus groups, prototypes and service Blueprint. Unexpectedly, experts recognize "Design Thinking" as a creative process within SD tools. This finding is not in line with literature, which we discuss further on conclusions.

Table 3. Service design comparison between research and practice

Sub-categories	SD research	SD practice	Common
Concept	It is a creative, iterative and human-centered approach, once it seeks for understanding the users and stakeholders and their context [1]	Human-centered process used to develop innovative services through maximizing results by the integration of everything that is part of the services such as people and processes	Human-centered
Principles	Human-centered; collaborative; iterative; sequential; real; holistic [22]	Open mindset; human-centered; co-creative; collaborative; real; iterative	Human-centered; collaborative; real; iterative
Objects	Service; service systems [1]	Service	Service
Process	Service design process; exploration, ideation, reflection; implementation [18, 23] double diamond: discover, define, develop, deliver [36]; TISDD service design framework: research, ideation, prototyping and implementation [22]; four design activities of a design process: analysing, generating, developing and prototyping [37]	Exploration; ideation; prototype and test; implementation design sprints	Exploration; ideation; prototype; implementation
Methods and Tools	Affinity diagram; blueprint; brainstorming, character profiles; conjoint analyses; contextual interview; customer journey map; cultural probes; documentaries; empathy tools/probes; ethnographic user research; focus group; immersion (workshop); observations; prototyping; questionnaires/surveys; role play; scenarios; service prototype; shadowing; stakeholders map; storyboarding; task analysis grid [38]	Benchmarking; user journey map; eye tracking; focus groups; ideation techniques; interviews; personas; prototypes; service blueprint; service system architecture; service system navigation; service value constellation; design thinking as a creative process	Observation; prototypes; focus groups; service blueprint

4.3   Advantages and Disadvantages of DT and SD in Practice

Final RQ specifically address the advantages and disadvantages of the usage of DT and SD (respectively) in practice, explicitly expressed by interviewees. Results are given in Tables 4 and 5, which are explained below.

Which refers DT practice, the interviews revealed nine advantages and six disadvantages. As demonstrated in Table 4 - panel A, regarding the advantages of DT, the majority of respondents point out that DT as an iterative (70%) and explorative (55%) approach, allows to explore several times the problem and the solution in order to know if the solution fits the resolution of the problem. Additionally, almost half of the interviewees state that DT allows testing product-service solutions with the user (45%). Tests are considered fundamental, and it is critical to spend time with the end-user, in order to get some understanding of their problems and needs. Most of respondents (40%) also find interesting working in multidisciplinary teams since everyone is essential at a point of the project. Further stated advantages (with 35% each) are open-minded approach with focus on innovative ideas. Open-minded refers to a mind-set to freely generate innovative ideas, since it does not censor anything and all ideas that arise are analysed. Respondents also refer that DT verbs are more open when innovating which allows to extend hypotheses. Lastly, DT allows to have different perspectives (25%) in the sense that it is an innovative approach that enables to see problems in a different perspective, being iterative, permitting, in turn the improvement of the solution. Therefore, these results strengthen the reasons to foster DT use by practitioners.

Table 4. Results of DT (dis)advantages in practice/number of respondents

A - Advantages	Professionals (n = 7)	Trainees (n = 13)	Total
Iterative approach	5	9	14 (70%)
Explorative approach	3	8	11 (55%)
Testing with the end user	4	5	9 (45%)
Working in multidisciplinary teams	2	6	8 (40%)
Understanding the user	2	5	7 (35%)
Improves the creation of ideas	3	4	7 (35%)
Open hypothesis	2	5	7 (35%)
Do not censor anything	1	5	6 (30%)
Allows to have a different perspective	0	5	5 (25%)
B - Disadvantages			
Time consumer	2	8	10 (50%)
Seems like a vicious cycle	0	7	7 (35%)
Sometimes seems inefficient	1	5	6 (30%)
Get lost in ideas	0	5	5 (25%)
Lack analytical rigor	1	3	4 (20%)
Subjective approach	0	3	3 (15%)

From the results of the stated disadvantages, presented in Table 4 - panel B, particularly one DT disadvantage stands out, which is time consumer (50%). Half of the respondents state that it is spent a lot of time on the ideation phase that could sometimes be shortened. However, they suggest that this can be solved through practice and with the improvement of the project management. Furthermore, respondents (only trainees) refer that sometimes DT looks like a vicious cycle (35%, 7 respondents), since it is necessary to do a lot of tests, which do not necessarily achieve a positive result and sometimes seems inefficient, being difficult not to get lost with ideas (25%, 5 respondents), as they are always discovering new ones and dropping others. These disadvantages were stated only by trainees, which may suggest the influence of less experience of novices on these difficulties. The last two disadvantages are related to statements concerning the lack of analytical rigor on DT approach (20%), as it works with insights and with what people say, but sometimes the number of elements of these focus groups is not enough to validate the results. Hence, respondents find DT subjective (15%), as it depends a lot on the personal hint of each one gives to and, it depends on how people understand for example the observation they are making.

Therefore, these results highlights shortcomings within the DT usage, specifically rejection.

Concerning SD, the conducted interviews revealed 5 advantages and 3 disadvantages that were mentioned by the questioned experts with respect to the practical applicability of SD within their companies. As mentioned before, currently no trainees interviewed indicated that they are aware of SD approach (n = 13; 0%). As such, results of the expressed (dis)advantages of SD, presented on Table 5, are only from professionals.

Table 5. Results of SD (dis)advantages in practice/number of respondents

A - Advantages	Professionals (n = 7)
Achieve results	3 (43%)
Close relation with the user	3 (43%)
Allows to improve constantly	3 (43%)
Iterative approach	2 (29%)
Allows creativity	1 (14%)
B - Disadvantages	
Expensive	3 (43%)
Time consuming	2 (29%)
Constant restarts can lead to frustration	1 (14%)

As shows Table 5 – panel A, the interviewees are split equally (43%, 3 respondents each) over the most cited advantages of SD, and therefore, either find SD an approach/process which achieves good results since every phase is co-created with the several stakeholders involved, which allows to better understand their needs. Naturally, this advantage relates to the next one, since participants think that is valuable to have a close contact with the user. Also, interviewees find positive that SD is an iterative process (29%), since it is possible to test several times before implementing the solution. With these tests, solutions are constantly being improved, so they could meet the needs of the end user. The last advantage relates to a single statement (14%) which refers SD allowing creativity, since to develop an innovative solution, it is important to have also innovative ideas. Therefore, during the process it is possible to become more involved with the problem and starting to generate original ideas.

Therefore, these results set additional light into reasons that lead to SD practice within organizations.

Regarding the results of SD disadvantages, presented in Table 5 - panel B, interviews refer that sometimes the process can be expensive (43%) and consequently, the owners of a project may not be willing to pay. Besides, experts point out that SD takes time (29%) and therefore, the innovative solution risks to become obsolete (e.g. *"When the design ends, although it has planned current and future needs, it may happen that the political or economic moment no longer is the same."* Professional, Healthcare Industry Lead). The explanations thereby were when the service design is planned and structured too much in advance, companies express fear of being obsolete and loosing flexibility and adaptability in terms of service features and its execution. Lastly, one participant (14%) refers that the SD processes may lead to some frustration, as it is necessary to deal with constant restart, in consequence of potential failures.

Resulting the insights given, the stated disadvantages could hinder the use of SD by experts, within their organizations.

5   Conclusion

The results of our study reveal that there are no relevant differences between SD and DT in research and practice, except for students/trainees participating in service innovation projects.

Indeed, experts reveal a strong knowledge about DT and SD, despite trainees do not know about SD approach or acknowledge the benefits of its processes, methods and tools. Beneficial for both research and practice this *"knowledge gap"* with respect to SD approach among novices, can be tackled with further information provision and training.

Beneficial for both research and practice the most dominant differences refer SD as an open mind-set and co-creative process, adding "Design Sprints" in the ideation stage of short time frame projects. Unexpectedly, experts also recognize Design Thinking as a creative process within SD methods and tools. Despite this finding is not consensual with literature, the novelty of the result could trigger a future research challenge.

Additionally, our findings demonstrate that DT presents several similarities with SD perceptions. As such, these overlaps also suggest the integration of the DT approach on SD processes, methods and tools.

In terms of the advantages of SD and DT for businesses, its iterative nature, collaboration with multidisciplinary teams and stakeholders, allows to achieve better results driving customized service innovation and preventing future service encounters failures. Forcibly, the ongoing iterations have several costs (e.g. time, financial, opportunity) that may risk the market entry of service innovations and business competitiveness (e.g. first to the market).

Still, some obstacles identified from the disadvantages and SD "knowledge gap" have to be overcome:

From a practical perspective, regular education and trainings, especially in the field of SD and the development of service innovation offerings, need to be implemented in order to build a sustainable knowledge base about advantages, benefits and best-practices of SD and DT. Moreover, experts on SD and executives need to spread these outcomes in order to benchmarking, strengthen on the corporate benefits obtained by SD and DT practices. Also, top level management is encouraged to develop corporate research in partnership with external I&D centres, with multidisciplinary teams integrating the final user/customer (e.g. to provide a proof-of-concept, prototype and test highly customer-experience product-service; living labs cooperation), as well as, investing on highly and especially educated employees on SD.

From a research point a view, additional effort is needed on the integration of DT approach on SD practice, since the results demonstrate some overlaps between both approaches and are still mostly perceived as indistinctive. Therefore, research need to identify and clarify the boundaries and complementarities of both SD and DT, in order to obtain synergies and remove redundancy.

Since the identified disadvantages could hinder the use of SD and DT practice, existing project management methods should build the base of the procedure method to improve and optimize time of iteration stage and reduce complexity. For example, for smaller projects there is no need for a systematic SD, since the costs might exceed the benefits. The costs themselves, related to time needed to affect to systematic design, in the case of project complexity, and the associated expenses (e.g. budget restrictions) have been referred to as an obstacle for SD practice.

However, a major limitation of this study need to be considered: The findings of SD practice are based on a small number of interviews with experts from companies. Further research needs to verify the results within a larger scope of experts. Hence, expanding the interview towards a larger set of interviewees coming from a more diverse background that is more dissimilar seems to be a fruitful direction for successive work.

Another interesting research setup would be to take practitioners, e.g. business people in charge of developing services in some innovative companies, identify those with training and research knowledge in SD/DT and those without and, then compare what they accomplish and how they do it. Also, comparing the definition of SD and DT that could be found in the job description (if exist), could be beneficial to have a complete picture concerning SD and DT: (i) literature, (ii) competences required by the business, (iii) what people (differentiating per role) think and know about the concepts. Moreover, examining how SD and DT have really contributed for value creation, for both service firms and customers, could be an interesting future research challenge.

Acknowledgments. We are grateful to interviewees from Porto Design Factory, IBM, Microsoft and BIT Sonae for their collaboration in the data gathering of this study.

References

1. Patrício, L., Gustafsson, A., Fisk, R.: Upframing service design and innovation for research impact. J. Serv. Res. **21**, 3–16 (2018)
2. Gembarski, P.C., Lachmayer, R.: Mass customization und product-service-systems: vergleich der unternehmenstypen und der entwicklungsumgebungen. Smart Service Engineering, pp. 214–232. Springer, Wiesbaden (2017). https://doi.org/10.1007/978-3-658-16262-7_10
3. Pezzotta, G., Cavalieri, S., Romero, D.: Engineering value co-creation in PSS: processes, methods, and tools. In: Rozens, S., Cohen, Y. (eds.) Handbook of Research on Strategic Alliances and Value Co-creation in the Service Industry, pp. 22–36. IGI Global, Hilliard (2017)
4. Alter, S.: Metamodel for service design and service innovation: integrating service activities, service systems, and value constellations. In: ICIS 2011 Proceedings, Shanghai (2011)
5. Maglio, P.P., Spohrer, J.: Fundamentals of service science. J. Acad. Mark. Sci. **36**(1), 18–20 (2008)
6. Chesbrough, H., Spohrer, J.: A research manifesto for services science. Commun. ACM **49**(7), 35–40 (2006)
7. Spohrer, J., Maglio, P.P., Bailey, J., Gruhl, D.: Steps toward a science of service systems. Computer **40**(1), 71–77 (2007)
8. Schallmo, D.R.A.: Vorgehensmodell des design thinking. Design Thinking erfolgreich anwenden, pp. 41–60. Springer, Wiesbaden (2017). https://doi.org/10.1007/978-3-658-12523-3_4
9. Maglio, P.P., Kwan, S.K., Spohrer, J.: Commentary - toward a research agenda for human-centered service system innovation. Serv. Sci. **7**(1), 1–10 (2015)
10. Kleinschmidt, S., Burkhard, B., Hess, M., Peters, C., Leimeister, J.M.: Towards design principles for aligning human-centered service systems and corresponding business models. In: International Conference on Information Systems (ICIS), Dublin (2016)
11. Maglio, P.P., Lim, C.: Innovation and big data in smart service systems. J. Innov. Manag. **4**(1), 11–21 (2016)
12. Wünderlich, N., et al.: Futurizing smart service: implications for service researchers and managers. J. Serv. Mark. **29**(6/7), 442–447 (2015)
13. Ostrom, A.L., et al.: Moving forward and making a difference: research priorities for the science of service. J. Serv. Res. **13**(1), 4–36 (2010)
14. Brown, T.: Design thinking. Harv. Bus. Rev. **86**(6), 252 (2008)
15. Evenson, S., Dubberly, H.: Designing for service: creating an experience advantage. In: Salvendy, G., Kalntrod, W. (eds.) Introduction to Service Engineering, pp. 403–413. Wiley, Hoboken (2010). https://doi.org/10.1002/9780470569627.ch19
16. Vargo, S.L., Lusch, R.F.: Inversions of service-dominant logic. Mark. Theory **14**(3), 1–10 (2014). https://doi.org/10.1177/1470593114534339
17. Patrício, L., Fisk, R.P.: Synthesizing service design and service science for service innovation. J. Serv. Des. **3**(2), 14–16 (2011)
18. Patrício, L., Fisk, R.: Creating new services. In: Serving Customers: Global Services Marketing Perspectives, Chap. 10, pp. 185–207 (2013)
19. Vargo, S.L., Lusch, R.F.: Institutions and axioms: an extension and update of service dominant logic. J. Acad. Mark. Sci. **44**(1), 1–19 (2016)
20. Maglio, P.P., Vargo, S.L., Caswell, N., Spohrer, J.: The service system is the basic abstraction of service science. IseB **7**, 395–406 (2009)

21. Holmlid, S., Evenson, S.: Bringing service design to service sciences management and engineering, pp. 341–345 (2008). https://doi.org/10.1007/978-0-387-76578-5_50
22. Stickdorn, M., Hormess, M., Lawrence, A.: This is Service Design Doing. O'Reilly Media Inc., Sebastopol (2017)
23. Stickdorn, M., Schneider, J.: This is Service Design Thinking. BIS Publisher, Amsterdam (2011)
24. Dorst, K.: The nature of design thinking. In: Design Thinking Research Symposium Proceedings, pp. 131–139 (2010)
25. Jebb, A.T., Parrigon, S., Woo, S.E.: Exploratory data analysis as a foundation of inductive research. Hum. Resour. Manag. Rev. **27**(2), 265–276 (2017)
26. Neuman, W.: Social Research Methods: Qualitative and Quantitative Approaches. Pearson Education Limited, Harlow (2014)
27. Boyce, C., Neale, P.: Conducting in-depth interviews: a guide for designing and conducting in-depth interviews for evaluation input. Pathfind. Int. Tool Ser. Monit. Eval. **4**(2), 207–215 (2006). https://doi.org/10.1080/14616730210154225
28. Charmaz, K.: Constructing Grounded Theory: A Practical Guide through Qualitative Analysis. SAGE Publications Limited, London (2006)
29. Gioia, D., Corley, K., Hamilton, A.: Seeking qualitative rigor in inductive research: notes in the Gioia methodology. Organ. Res. Methods **16**(1), 15–31 (2012)
30. Tschimmel, K.: Design thinking as an effective toolkit for innovation. In: Proceedings of the XXIII Conference: Action for Innovation: Innovating from Experience, p. 20, Barcelona (2012)
31. Plattner, H.: An introduction to design thinking process guide. In: Institute of Design at Stanford, pp. 1–15 (2015). https://doi.org/10.1007/978-1-4302-6182-7_1
32. Brown, T.: Change by Design. HarperCollins e-Books, New York (2009)
33. Brown, T., Wyatt, J.: Design thinking for social innovation. Dev. Outreach **12**(1), 29–43 (2010)
34. Liedtka, J., Ogilvie, T.: Designing for Growth. Columbia University Press, New York (2012)
35. Chasanidou, D., Gasparini, A.A., Lee, E.: Design thinking methods and tools for innovation. In: Marcus, A. (ed.) DUXU 2015. LNCS, vol. 9186, pp. 12–23. Springer, Cham (2015). https://doi.org/10.1007/978-3-319-20886-2_2
36. Foglieni, F., Villari, B., Maffei, S.: How to (re)design services: from ideation to evaluation. Springer Briefs in Applied Sciences and Technology, pp. 27–45 (2018). https://doi.org/10.1007/978-3-319-63179-0_3
37. Meroni, A., Sangiorgi, D.: Design for Services. Gower Publishing Limited, Surrey (2011)
38. Alves, R., Jardim Nunes, N.: Towards a taxonomy of service design methods and tools. In: Falcão e Cunha, J., Snene, M., Nóvoa, H. (eds.) IESS 2013. LNBIP, vol. 143, pp. 215–229. Springer, Heidelberg (2013). https://doi.org/10.1007/978-3-642-36356-6_16

Upgrading the Data2Action Framework: Results Deriving from Its Application in the Printing Industry

Oliver Stoll[1(✉)], Shaun West[1], Mario Rapaccini[2], Cosimo Barbieri[2], Andrea Bonfanti[3], and Andrea Gombac[3]

[1] School of Engineering and Architecture, HSLU Lucerne University of Applied Sciences and Arts, Lucerne, Switzerland
{oliver.stoll, shaun.west}@hslu.ch
[2] School of Engineering, University of Florence, Florence, Italy
{mario.rapaccini, cosimo.barbieri}@unifi.it
[3] Ricoh Italia, Vimodrone, Italy
{Andrea.Gombac, Andrea.Bonfanti}@ricoh.it

Abstract. This paper describes the application of the Data2Action (D2A) framework for the development of Smart Print Services. The context for the development was printing-as-a-service in Italy. The firm in this study wanted to create new Smart Services to improve customer experience and drive out waste. The analysis in this paper is based on an Action Research methodology to understand the application of the D2A framework. The results describe the lessons learnt from its implementation, show examples of tools used and reflect on the application of the framework and the individual tools. Using reflections from the development process, an improved framework is proposed that further increases its applicability for use in complex product service systems.

Keywords: Smart services · Servitization · Advanced services

1 Introduction

Many firms across different industrial segments use digital technologies to enable servitization of complex product service systems (PSS) [1], for example, Canon Ink, with its eMaintenance® allows Canon to provide advanced proactive services, e.g., remote diagnostics and consumable management. MAN Trucks [2] uses similar approaches to support owners and operators of lorries to improve operational performance. For this to take place effectively, those developing the digital services have to understand the complex ecosystem of people and equipment that the firm is part of, as well as the operational objectives and the strategy of the customer, whether the customer is the facility owner and/or operator or some other actor.

Collecting field data from an installed base of smarter/connected products is the first move. But more important is how these data can be leveraged to help decision making. Field data from connected products that provide raw information about product condition and usage, are a gold mine, especially if properly integrated with other company data (e.g., sales data, contract entitlements) and external data (e.g., social

© Springer Nature Switzerland AG 2020
H. Nóvoa et al. (Eds.): IESS 2020, LNBIP 377, pp. 273–286, 2020.
https://doi.org/10.1007/978-3-030-38724-2_20

media, weather forecast). The goal is to transform the information coming from these 'data lakes' into insights that suggest the best actions to be taken, prevent problems, or respond to any need of the people involved in the value creation process. This creates a very complex problem: the right data must be transformed in the right way, so that the right information, in the right form can be delivered to the right person (i.e., the decision-making unit, DMU) [3] at the right time. As the digitization/digital era has flourished, managers have shown growing interest in developing data-driven decision support systems and business intelligence applications, that show analytics and dashboards. These latter are human-machine interfaces (HMIs) [4] that are specifically designed to provide overviews of the key performance indicators (KPIs) and are essential to understanding how business operations are run. Today, the development of data-driven models for descriptive, diagnostic or predictive purposes, which is part of the more articulated practices of Business Performance Management (BPM), can be facilitated by specific methods such as CRISP-DM of IBM [5]. To embrace enlarged views of value co-creation (e.g., business ecosystems, value chain, value constellation) and the paradigms of user-centered design (e.g., design thinking), as well as to focus on smart (digitally-enabled) services as the unit of value creation, the authors have developed the Data2Action (D2A) framework [3]. D2A provides a structure to support understanding of how value is created as key competences, knowledge, data, resources and technologies (e.g., industrial equipment) from different actors in the business ecosystem are integrated. Then D2A helps define how Smart Services can prevent problems/pains and/or produce benefits/gains. Core to this process is the decisional 'cockpit' (i.e., the dashboard) that helps people in shaping the smart service, deciding the actions to be taken and the workflows to be activated.

The paper aims to answer the following research questions (RQs): "How can structured frameworks - such as D2A –support the development of data-driven Smart Services? What are the benefits of adopting these frameworks, against unstructured approaches, in particular in large/complex organizations and processes?" In order to answer the RQ, an action research methodology was applied.

2 Background

The next section briefly reviews the extant theories underpinning the development of data-driven smart services and explains the need for adopting structured approaches such as the D2A framework.

2.1 Smart Services and Data-Driven Decisions

The term 'Smart Services' can be defined combining the perspective of service dominant logic (SDL) [6] with all the enhancements and innovation that come from digital technologies such as smart connected products, IoT, cloud computing, or predictive analytics [7]. A service is smarter the more it creates value, combining and recombining the available (but scarce) resources in sustainable ways. Technologies are crucial for a smart service business or process to be sustainable [8, 9]. As the installed base of manufacturing companies includes more connected products that create huge amounts of field data [10], the focus usually shifts from technologies to how the manufacturer can create value with data. Numerous strategies and possible transformations have been proposed, ranging from selling data [11] to proving smart product-service systems or digital platforms [12], such as in the case of GE Digital. It is agreed that data cannot be used on their own for decision making [7, 13, 14], but must be elaborated and transformed into information and knowledge [15, 16], according to well-shaped processes. The beneficiaries of these processes are people that combine the knowledge they already possess with the newly generated knowledge to develop insights, make conjectures and hypotheses around given facts. Hence, decision making is based on both sense-making (i.e., creating value in the customer's eyes through human judgement) and knowledge-creating processes (i.e., generating insights from huge amounts of field data, as far as they are collected and made available from fleet, people, assets, equipment, etc.). The problem is that value is a subjective, context-dependent construct, that can be largely influenced by the phenomena taking place in non-controllable external environments. This cannot be avoided, as value is generated as value in use, as far as a user interacts and experiences something (e.g., a product, a service process). Therefore, value is created once decisions are acted out [17, 18]. In other words, data-driven decisions are effective if the corresponding actions (e.g., workflows, processes, tasks) can be viewed as smart services that create value for some beneficiaries. The adoption of structured frameworks can be thus helpful to develop these approaches in complex organizations [19, 20], limiting the implications of having data of poor quality [21], or no data at all in a given domain. Structured approaches, in fact, facilitate the linkages between the endpoints of any smart service initiative: what data and for what purpose. In the next subsection, the D2A framework is briefly explained.

2.2   Data2Action Framework

The Data2Action framework [3] evolved from three projects related to value creation from data in complex systems. It provides a structure to support understanding of the ecosystem, the actors and the equipment within it, as well as the interactions between the assets and actors necessary to support the innovation of Smart Services. The framework consists of four steps and continuous refinement: understand; ideate; prototype; and test (see Fig. 1).

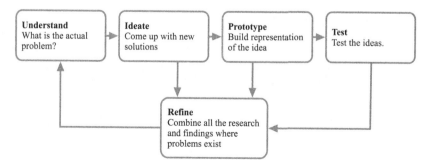

Fig. 1. Overview of the D2A framework

3 Methodology

In order to answer the RQ, it is necessary to analyze which improvements the Data2Action framework needs in order to support the decision-making process needed for the development of Smart Services based on the data owned by the company. These processes are becoming increasingly complex, so it is essential to address management research with a qualitative approach [22], which gives the possibility of gaining an understanding of managerial actions and processes in real-life organizational settings [23]. In order to conduct qualitative research, case studies can be applicable especially when the research interest is concerned with the dynamics of a phenomenon [24]. Nonetheless, in order to understand the effect of the Data2Action framework, it has to be applied with the aim of supporting the Smart Services development and improving the performance of the business. For this reason, this paper can be characterized as an action research case study [25].

The action research methodology was applied with the collaboration of Ricoh Italia, the Italian affiliate of Ricoh, a multinational manufacturer of printers. In particular, we collaborated with the service department, which wanted to understand how to develop new Smart Services using the data collected from the individual printers, focusing on how the order fulfillment system could be used to improve customer experience and at the same time drive out waste (i.e., cost or inventory) from their service system.

The collaboration involved a multidisciplinary (service design experts and data scientists) and international (HSLU and University of Florence) team of researchers, and a team of different specializations (IT, service, operation) and different levels of responsibility (director, technician, project manager). Once the whole team was set up, the Data2Action framework was applied starting from the understanding phase and proceeded for a whole year. A continuous conversation was maintained in order to put the contributions of the research into action and relate the theoretical and empirical domains. Every week, the two teams had a meeting of 1–2 h where feedback was provided from both ends in terms of achievements reached and difficulties and problems emerging, for these, deeper discussions were needed to analyze them in detail. Notes and the follow-ups of the meetings were shared in order to validate the research results and develop the framework further.

4 Results and Initial Findings

This section first follows the development of the dashboards based on the problem descriptions, then leads on to the reflections and the lessons learnt from the development process. The term dashboard has been used to describe a visual interface that provides information to the consumer of the information.

4.1 Understanding the Problem and Development of Decision-Supporting Dashboards

The development process started with an analysis of the business ecosystem and the service supply chain (Fig. 2). Here, a persistent loss of printer toner cartridges was identified as a problem by following the D2A framework. This was referred to as the "Bermuda Triangle": a place between the reception point of the toner, the stock and the printer at the customer's site.

The "Bermuda Triangle" was investigated closely, revealing different ways for the toner to disappear. For example, the customer stores the toner for later use and simply forgets where it is stored. Alternatively, another person who needs the toner does not know that there is a spare toner stored at the customer site and orders a new one. In this example, the customer does a poor job regarding their management of toners. The customer pays a monthly flat rate (with an agreed minimum number of copies) plus a fee per copy (when the minimum of copies has been reached). Consequently, the customer is not concerned about this problem as long as there are no cost implications for them – they worry only about loss of print services on a printer due to lack of toner.

Another example of wasting toner is that when the toner reaches a certain threshold, for example 20%, the printer either automatically orders a new toner or the customer does. Since the customer has spare toners on-site, the customer can replace the toner immediately. By doing so, the customer wastes 20% of toner in the cartridge, which is paid for by the supplier.

The supplier has no direct influence on the "Bermuda Triangle" and trying to change the customer behavior is a delicate and difficult task. Therefore, the focus was laid on the supply chain, which could be optimized in a way that reduces the risk of toners getting lost in the "Bermuda Triangle", without cutting the availability of the printing service provided.

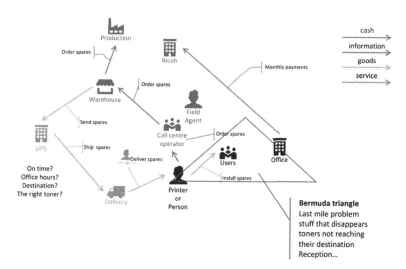

Fig. 2. Analysis of the business ecosystem from a macro perspective

The service delivery was analyzed with the purpose of understanding how the customer or the printer orders toner cartridges. The results should then provide a better understanding of the customer's toner ordering possibilities and Ricoh's toner-request validation process, including the validation support systems (Fig. 3).

The customer has the option of actively ordering the toners by phone, e-mail, eService or by using the automated ordering service. The service either automates the process of ordering or facilitates the ordering process. In the automated service the printer orders the toners autonomously for the customer. In the facilitation version of the service, the machine provides the relevant information the customer needs to provide during the validation process. All toner requests are validated by the "Validation Agent" and by the TVT validation tool from Microsoft.

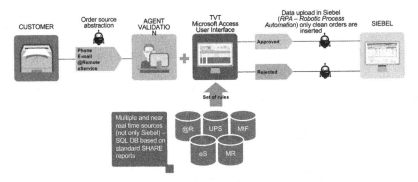

Fig. 3. Toner-request validation process including the validation support systems

The "Validation Agent" for office customers is the Service Operator Call-center (SOC) (Fig. 4), with the SOC being the actor that validates customer requests and the user of the validation tool. The SOC as an actor was considered the target for potential digital solutions. The objective of the solutions is to improve the decision-making of the SOC and the decisions must reduce toner waste and/or improve customer experience.

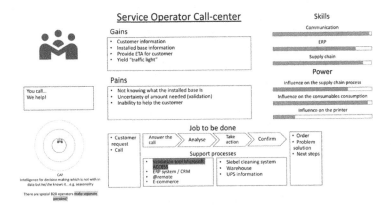

Fig. 4. Most relevant actor for the customer request validation analysis

The final dashboard used Microsoft BI, the solution developed shows the delta of the "Reception Point in Time" (RPiT) and the "Delivery Point in Time" (DPiT) for each serial number of a printer. A negative value represents a delayed delivery and vice versa; the DPiT accuracy was high, however the accuracy of the predicted values was not sufficient for the implementation of an on-time delivery micro-service. However, based on the first coded prototype it was possible to define what data of which quality needs to be collected in the future.

4.2 Lessons from the Analysis of the Development Process

The theoretical frameworks that supported the application of the D2A framework were the three dimensions of SD logic [6], which are relevant for the understanding of the system. The lessons learnt from the understanding phase are shown in Table 1. The key lesson was that a common understanding of the problem could be visually described, allowing experts and managers alike to understand it. The time taken to describe the problem was well spent and formed the basis for the following steps in the D2A framework.

The results of the ideation phase are shown in Table 2. The ideas for solutions included pain relievers and gain creators for the actor, and the description of how the actor solves the problem. The co-ideation provided much better results through the combination of knowledge from the business and technical sides. The level of detail of the ideas grew with the increasing number of iterations. Soon, many ideas were created, but not all could be prototyped.

The ideas were rated on a qualitative basis, using a 2×2 decision matrix, this assessment tool was developed directly out of the reflection process. The tool uses two dimensions for the qualitative value assessment of ideas: feasibility and value. The feasibility valuation was mainly driven by the data-savvy members of the team. The business-people drove the value impact estimation of the ideas within the project.

The Source Target Link Matrix (Fig. 5) is a tool developed and implemented for the D2A framework during the project. The problem was that the desired information from the conceptual dashboards could consist of combinations of different target-data. The connections between the source-data, the target-data and the information desired in the dashboards can be complicated. These connections need to be managed during the development process. The Source Target Link matrix aims to transform the conceptual solutions into a technical perspective. It should help to structure and manage the source-data, target-data, analysis methods, and dashboards. The management of source-data includes a list of data that is needed but does not yet exist. Moreover, the matrix allows the combination of different sources and target data, allowing the data to be reused.

The testing phase is shown in Table 4. This phase of the D2A framework was not conducted in sequence to the prototyping phase, rather it was performed in parallel. For reasons of simplicity, the results of this stage are presented separately. Testing the prototypes with the individuals for whom they were developed corroborates their value. People outside the team, who were not the prototype's target audience, struggled to recognize the value of the application. To communicate the value, the context in which the prototype was developed needs to be known to the audience and here an additional tool could be used to support the communication of value (Table 3).

Table 1. Learnings from the understanding phase

Iteration	What worked...	What did not work...
...iteration 1	The theoretical frameworks such as the three dimensions SD logic and Service Design Thinking	Initial usage of the tools by Ricoh (e.g., job-to-be-done ecosystem mapping, personas and journey mappings)
...adjustments after iteration 1	Pre-define tools with content to initialize discussion	
...iteration 2	The discussions about the pre-fitted content provided better results	Cross-disciplinary understanding
...adjustments after iteration 2	None	
... iteration 3	Improved cross-disciplinary problem understanding	People from outside the project group could not understand the complex ecosystem visualization
...adjustments after iteration 3	Dividing the complex ecosystem into smaller ecosystems that could be aggregated into the complex system	
... iteration 4	Reduction of complexity of the ecosystems	Lack of understanding of data science and advanced analytics by business-people
...adjustments after iteration 4	Presentation by Smart Operations on data science and data analysis methods	

Table 2. Lessons from the ideation phase

Iteration	What worked...	What did not work...
...iteration 1	Pre-define the tools with content to initialize discussion	Low quality of ideas
...adjustments after iteration 1	Co-ideate	
...iteration 2	New ideas and problem understanding through co-ideation	Generic ideas for solutions
...adjustments after iteration 2	Refine the solutions by visualizing them and adding more details	
... iteration 3	The refinement of the solutions revealed more root causes of the problems	Too many ideas to choose from for the prototyping phase; what to do first?
...adjustments after iteration 3	Rating the ideas based on a rating matrix	
... iteration 4	Structured overview of what ideas to do first	Measurable value estimation of the ideas

Table 3. Lessons from the prototyping phase

Iteration	What worked...	What did not work...
...iteration 1	Prototyping conceptual dashboards with pen and paper	Poor feasibility regarding coding the dashboards was assessed, also because of the wrong data available or missing data
... adjustments after iteration 1	Co-development with Smart Operations and HSLU of conceptual prototypes	
...iteration 2	Co-development increased the quality and variety of the conceptual prototypes	Handover of the conceptual prototypes to the data analysts and coders
... adjustments after iteration 2	Development of the Source Target Link matrix	
... iteration 3	The handover with the Source Target Link matrix worked	Some of the data was not usable for certain types of analysis which was a surprise for the OEM
... adjustments after iteration 3	The D2A needs a section for preliminary data assessment	
... iteration 4	Working prototypes for further refinement	

Table 4. Lessons from the testing phase

Iteration	What worked...	What did not work...
...iteration 1	Testing by presenting the conceptual prototypes	People outside of the project team did not understand the value delivered by the prototype
... adjustments after iteration 1	The presentation of the results needs to offer more context to show its value	
...iteration 2	The actor for whom the solutions were targeted provided valuable feedback to improve or redesign both the conceptual and the working prototypes	The developers lost focus of the value and analysis direction due to new exciting things that were not yet valuable to the actors
... adjustments after iteration 2	The testing sprints between prototyping were closer	

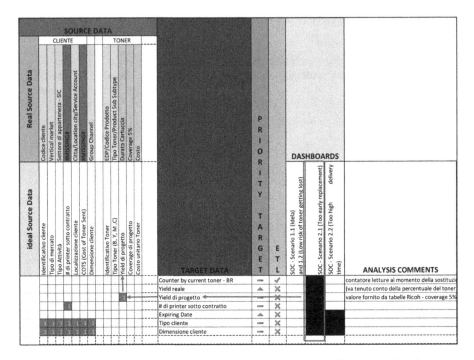

Fig. 5. Source Target Link matrix developed to support coding of dashboards

5 Discussion

This section describes how the D2A framework supported the problem description and how this allowed the solutions to be built and handed over to the data scientists to implement the code in the OEM's service delivery system. The approach is compared to the literature where appropriate and follows the four steps of the D2A framework described in Fig. 1. Figure 6 shows the updated D2A framework, each of the phases will be discussed below with reflections on the changes made.

The understanding phase has been updated to improve its capability to define the problem space. The job-to-be-done tool aimed to capture how the customer creates value for the firm's offering. The tool provided the necessary knowledge for both parties about the value creation process within the OEM's customer. The application of the job-to-be-done tool helped to close the knowledge gap between all of those involved. According to the literature [15] the knowledge gap hinders the development of digital solutions. Ecosystem mapping helped to identify the relevant actors and equipment and their relations to each other; this then led to the discovery of the "Bermuda Triangle". The visual way of illustrating the problem created a common understanding between all parties. Ecosystem understanding is a fundamental aspect of the development of smart services [8] and PSS [26]. Journey mapping helped identify value creation and when the pains and gains emerge along the journey. The journey mapping also revealed other actors that co-create value [6] with the use of equipment.

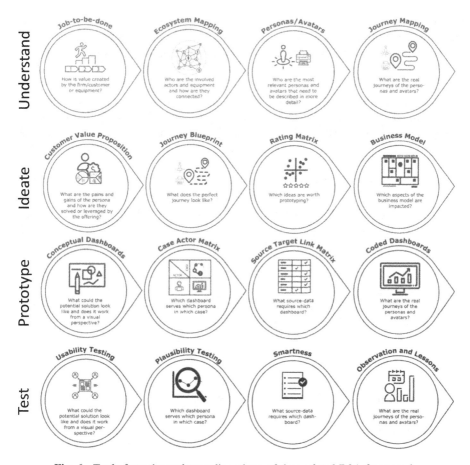

Fig. 6. Tools from the understanding phase of the updated D2A framework

Personas, avatars and journey maps deepened the understanding of the relevant actors and equipment, including what they do and how they do it, and supported building solutions.

The ideation phase was refined, based on the results of this study, by the addition of a rating matrix and the business model canvas. During the development of individual value propositions, the group learned that there are many value propositions. The customer value proposition tool, in combination with journey mapping, provided the most significant impact in the visualization of value co-creation. The visualizations helped to explain the context in which value was created. The rating matrix tool was added to support idea selection; it has two dimensions: value and feasibility. The two dimensions provided justification and structure for the selection of ideas and allowed open conversions between the data scientists (feasibility axis) and the business leaders (value axis).

The prototyping phase was most effective using hand-drawn conceptual dashboards as this supported co-creating of solutions. The coded dashboards take longer to develop, resulting in longer iteration steps. The Case Actor matrix supports the visualization of the interconnected prototypes with the different actors and situations.

Transferring conceptual dashboards to coded dashboards came with the challenge of data complexity and this challenge was overcome through the Source Target Link matrix. This new tool offered a structured link between the data providers and the information consumers, as well as a tool to identify missing (or incomplete data). The matrix provided a technical perspective on the conceptual dashboards that the coders and data scientists preferred. The literature [27] considers the handover from the conceptual phase to coding to be challenging, and here the D2A framework supports this transition.

The testing phase applied two different layers of testing. First, testing the usability of both the conceptual and the coded dashboards and second, testing the plausibility of the data and the information derived from it. The testing has to run in parallel with the prototype phase as this ensures co-creation. The early testing and the redesigning of dashboards helped to reduce waste [28]. However, given that the results of the testing phase were not well documented by either party, more work should be undertaken here to confirm this hypothesis.

To close the discussion, this paper has identified three changes to the D2A framework that were required for its successful application in print-services. These modifications are based on theory and practical lessons from this use case in print-services. The application of the framework has created value for customers of the service (i.e., customer experiences) and for the supplier (i.e., reduced costs).

6   Conclusions and Recommendations

Business impacts were found from the application of the modified D2A framework. The impacts were found in the improved development process and collaboration of the partners as well as in customer experience and the identification and reduction of waste.

The D2A framework supported the common understanding of the problem and highlighted poor assumptions made in the past; it supported the ideation phase in a positive way, and the idea selection/ranking in a pragmatic way through the addition of two additional tools. It helped with the prototyping phase, particularly with the addition of the Source-Target Link matrix; and, it provided a rudimental testing approach.

It is recommended that the updated D2A framework is applied to new use cases to determine its applicability, and that the testing phase is further improved.

Acknowledgements. The authors would like to thank Ricoh Italy, SmartOperations, the University of Florence, the Lucerne University of Applied Sciences and Arts and the Swiss Alliance of Data Intensive Services.

References

1. Mont, O.K.: Clarifying the concept of product-service system. J. Clean. Prod. **10**(3), 237–245 (2002)
2. Evans, D.: What servitization did for MAN Truck and Bus UK (2016)
3. Stoll, O., West, S.S., Rapaccini, M, Mueller-Csernetzky, P.: Understanding wicked problems to build smart solutions. In: FTAL Conference - Industrial Applied Data Science, Lugano (2018)
4. Firican, G.: Best practices for powerful dashboards. Bus. Intell. J. **22**(2), 33–39 (2017)
5. Overgoor, G., Chica, M., Rand, W., Weishampel, A.: Letting the computers take over: using AI to solve marketing problems. Calif. Manag. Rev. **61**(4), 156–185 (2019)
6. Vargo, S.L., Lusch, R.F.: Service-dominant logic: continuing the evolution. J. Acad. Mark. Sci. **36**(1), 1 (2008)
7. Ardolino, M., Rapaccini, M., Saccani, N., Gaiardelli, P., Crespi, G., Ruggeri, C.: The role of digital technologies for the service transformation of industrial companies. Int. J. Prod. Res. **56**(6), 2116–2132 (2017)
8. West, S., Gaiardelli, P., Rapaccini, M.: Exploring technology-driven service innovation in manufacturing firms through the lens of Service Dominant logic. IFAC-PapersOnLine (2018)
9. Beverungen, D., Müller, O., Matzner, M., Mendling, J., vom Brocke, J.: Conceptualizing smart service systems. Electron. Mark. **29**(1), 7–18 (2019)
10. Porter, M.E., Heppelmann, J.: Managing the Internet of Things: how smart, connected products are changing the competitive landscape. Harv. Bus. Rev. **92**(11), 64–88 (2014)
11. Opresnik, D., Taisch, M.: The value of big data in servitization. Int. J. Prod. Econ. **165**, 174–184 (2015)
12. Liu, Z., Ming, X., Song, W., Qiu, S., Qu, Y.: A perspective on value co-creation-oriented framework for smart product-service system. Proc. CIRP **73**, 155–160 (2018)
13. Rowley, J.: The wisdom hierarchy: representations of the DIKW hierarchy. J. Inf. Sci. **33**(2), 163–180 (2007)
14. Liew, A.: DIKIW: data, information, knowledge, intelligence, wisdom and their interrelationships. Bus. Manag. Dyn. **2**(49), 49 (2013)
15. Choo, C.W.: Sensemaking, knowledge creation, and decision making organizational knowing as emergent strategy. In: Strategic Management of Intellectual Capital and Organizational Knowledge (2001)
16. Choo, C.W.: The knowing organization: how organizations use information to construct meaning, create knowledge, and make decisions (2007)
17. Keeney, R.L.: Value-focused thinking: identifying decision opportunities and creating alternatives. Eur. J. Oper. Res. **92**(3), 537–549 (1996)
18. Bumblauskas, D., Nold, H., Bumblauskas, P., Igou, A.: Big data analytics: transforming data to action. Bus. Process Manag. J. **23**(3), 703–720 (2017)
19. Lim, C.-H., Kim, K.H., Kim, M.J., Heo, J.Y., Kim, K.J., Maglio, P.P.: From data to value: a nine-factor framework for data-based value creation in information-intensive services. Int. J. Inf. Manag. **39**, 121–135 (2018)
20. Frost, R.B., Cheng, M., Lyons, K.: A multilayer framework for service system analysis. In: Maglio, P.P., et al. (eds.) Handbook of Service Science, Volume II, Service Science: Research and Innovations in the Service Economy. Springer, Cham (2019). https://doi.org/10.1007/978-3-319-98512-1_13

21. Song, Z., Sun, Y., Wan, J., Liang, P.: Data quality management for service-oriented manufacturing cyber-physical systems. Comput. Electr. Eng. (2017). https://doi.org/10.1016/j.compeleceng.2016.08.010
22. Guercini, S.: New qualitative research methodologies in management. Manag. Decis. **52**(4), 662–674 (2014). Emerald Group Publishing Ltd.
23. Rynes, S., Gephart, R.P.: Qualitative research. Acad. Manag. J. **47**(4), 454–462 (2004)
24. Eisenhardt, K.M., Graebner, M.E.: Theory building from cases: opportunities and challenges. Acad. Manag. J. **50**(1), 25–32 (2007)
25. Denscombe, M.: The Good Research Guide. Open University Press McGraw-Hill Education (2010)
26. Baines, T.S., Lightfoot, H.W., Evans, S., Neely, A., et al.: State-of-the-art in product-service systems. Proc. Inst. Mech. Eng. J. Eng. Manuf. **221**(10), 1543–1552 (2007)
27. Blosch, M., Osmond, N., Norton, D.: Enterprise Architects Combine Design Thinking. Lean Startup and Agile to Drive Digital Innovation, Gartner (2016)
28. Gustafson, J.W., Jones C.H., Pape-Haugaard, L.: Designing a dashboard to visualize patient information. In: The 16th Scandinavian Conference on Health Informatics, Aalborg, Denmark (2018)

Service Ecosystems

Tiers-Lieu for Services: An Exploratory Approach to Societal Progression

Jolita Ralyté[(✉)] and Michel Léonard[(✉)]

Institute of Information Service Science, CUI, University of Geneva,
Battelle - bâtiment A, 7 route de Drize, 1227 Carouge, Switzerland
{Jolita.Ralyte,Michel.Leonard}@unige.ch

Abstract. The progress of Society depends largely on the success of its digital transformation, i.e. on the development of digital infrastructure taking form of information services. Moreover, the challenge consists in building interdisciplinary or even transdisciplinary information services and service systems. Given the variability of activities to consider and the precision needed to build this digital infrastructure, it is essential that all relevant actors contribute to the digital construction. Therefore, a contributory and exploratory approach enabling innovation and co-creation is necessary to develop information services. As a potential solution, we propose a novel approach called Tiers-Lieu for Services (TLS), that we present in this paper. We also report on an exploratory TLS conducted in the context of our continuing education program.

Keywords: Tiers-Lieu for Services · Digital transformation · Digital construction · Contributory approach · Exploratory approach

1 Introduction

The success of digital transformation, or should we say digital construction, is an important factor in the progress of today's Society. The focus of digital construction being on the development of digital infrastructure, we claim that such infrastructure should take the form of information services and service systems [1]. Public and private organizations aim to use new technologies to deliver new digital services and products, innovate their activities, and be more efficient and attractive. The concept of information service emerged as a support for such initiatives. In the context of digital construction, the challenge goes even further, it consists is building interdisciplinary or even transdisciplinary information services [2] enabling new collaborations, new activities and new value propositions, and therefore the progression of Society. Given the variability of activities to consider and the precision needed to build this digital infrastructure, it is essential that all relevant actors contribute to the digital construction, as the collective intelligence is a key for its success. Conventional software engineering approaches do not fit to this context as they intend to develop a solution for a particular problem, satisfy a particular goal. Societal progression is not about

© Springer Nature Switzerland AG 2020
H. Nóvoa et al. (Eds.): IESS 2020, LNBIP 377, pp. 289–303, 2020.
https://doi.org/10.1007/978-3-030-38724-2_21

solving problems, it is about the advancement and innovation. It is defined in terms of intentions rather than precise goals. Therefore, we claim that a contributory and exploratory approach enabling innovation and co-creation is necessary to develop information services in the context of societal progression.

Digital transformation, construction or co-construction, no matter how we call it, is not just about adopting new information technologies and digitalizing human activities. It embraces much broader strategic ambitions and involves fundamental changes in the activities, structure and culture of the organization [3], and even Society. Digital transformation is not only technology driven. New requirements also emerge from societal matters like ecology, energy, democracy, health, gender equality, education, transportation, etc. (see the SDG program of United Nations [4]). However, they should not be a headache of only public institutions. All organizations, public and private, and even individual citizens, are responsible for the success in reaching societal intentions. Therefore, they are also potential contributors to digital value co-creation. The success depends on how well all these actors combine their forces and capabilities, while the motivation lies in the fact that the value of the developed services concerns all of them.

In this context, we formulate the research question as follows: *What would be an efficient approach to foster the contribution of organizations and citizens to the development of information services, and in particular transdisciplinary information services, and thus to contribute to the progression of Society?*

To answer this question, we have elaborated a contributory and exploratory approach for information service co-creation named Tiers-Lieu for Services (TLS). The approach defines a contribution framework: concepts, roles and regulations, and guides the exploration leading to the design and implementation of information services. TLS promotes conceptual modeling as the underpinning technique for co-designing information services [1]. It is driven by creating, discussing, refining conceptual models that allow to reach an informational consensus between contributors and to co-create information assets as information services.

Up to now, we have conducted two exploratory TLS, both in the context of a continuous education program at our university, which allow us to assess the applicability of the approach. In this paper we illustrate one of them.

The rest of the paper is organized as follows: in Sect. 2 we discuss the foundations of our work, while in Sect. 3 we present our approach for driving a TLS. In Sect. 4 we report on one exploratory TLS, and in Sect. 5 we discuss the findings and limits of our study, and draw the research agenda. Section 6 concludes the paper.

2   Foundation

The foundation of the approach Tiers-Lieu for Services lies on two concepts: information service and Tiers-Lieu. We briefly present them in this section.

2.1 Information Service

The concept of information service is a fruit of cross-pollination between the fields of information systems and service science. In service science, the notion of service is defined as "the application of competences for the benefit of others" [5], and service systems are "value co-creation configurations of people, technology, value propositions connecting internal and external service systems, and shared information (e.g. language, laws, measures, and methods)" [6]. These definitions emphasize the value of the service and its co-creation, but does not give a central place to the domain of information, which is essential for any digitalized service. The purpose of this article is to show how the concept of information service allows to consider this essential aspect of the service and to elaborate its value proposition.

The concept of information service was first defined as a refinement of the information system component with the aim to address the challenges of information systems evolution, agility and interoperability [5]. Indeed, in the context of information systems evolution, an information service is defined as "a component of an information system representing a well-defined business unit that offers capabilities to realize business activities and owns resources (data, rules, roles) to realize these capabilities" [7]. And conversely, an information system is seen as built of a collection of interoperable information services [1]. From conceptual point of view, an information service includes four interrelated information spaces: static, dynamic, rule and role. The static space represents the data structure of the service (the set of classes and their relationships), the dynamic space defines the capabilities of the service (the set of actions), the rule space defines the set of rules that govern service activities (business rules, integrity constraints, pre- and post-conditions of actions), and the role space defines the organizational roles that have rights and responsibilities on the service actions.

In the context of digital co-construction to societal progression, we refine the definition of information service as follows: An information service is an autonomous, consistent and interoperable piece of digital infrastructure enabling an intra-, inter-, or even trans-organizational activity and providing value to all involved parties. Moreover, the value of the service exceeds the sum of the values received by each individual party involved.

2.2 Tiers-Lieu

The origin of the "Tiers-Lieu" (TL) concept lies in social science and was mainly developed in France [8,9] as "a social configuration" for co-creation. Indeed, Burret [10] defines a Tiers-Lieu as: "a social configuration where the encounter between individual entities intentionally engages in the conception of common representations". A "common representation" is considered here in a broad sense; it can be a conceptualization of a digital service or system, a new business activity, a bill of law, a public transportation plan, etc. A "social configuration" means that the goal of the TL is of interest to people from different horizons, different activity and social domains, representing public as well as private organizations or just themselves as responsible citizens.

It is important to mention that the notion of Tiers-Lieu is not a literal translation of the "Third Place" introduced in sociology by Oldenburg [12] to identify a social environment which is not the usual one of home ("first place") neither the one of workplace ("second place"). A third place defines any public physical or virtual place that enables a community life and facilitates broader and more creative interactions. It is a place where people relax in public, meet friends and make new contacts. The ambition of a TL is much broader, it aims to offer an environment where people with different backgrounds and having various skills and abilities can meet with the aim to co-create, not only to discuss and relax [9,10]. A TL has a shared intention, and provides a conceptual and regulatory framework for the heterogeneous contributions towards reaching this intention. In [11] a TL is defined as an exploratory environment for service-centered innovation. Therefore, we insist on using the French appellation for this concept, as the literal translation is not appropriate because of its connotation with the Oldenburg's definition.

As main characteristics of TL we highlight the following:

- *TL as a community builder.* TL attracts conscientious and responsible people (citizens, business (wo)men, government agents, politicians) that adhere to the same intention – to make progress in the Society, to advance private, business and societal well-being.
- *TL as a means for coopetition* where business competitors collaborate and combine their forces for innovating, as this is the only way to progress quickly and accountably.
- *TL as a way to co-create.* By providing a dedicated context and conditions TL naturally fosters peoples' creativity and cognitive interaction, and drives the co-creation of common information assets. The co-created value is shared by all contributors.
- *TL as a multi-organizational and multi-disciplinary space* that gathers people from different professions, different organizations, having different responsibilities and skills, that aim to share their expertise and to co-create. They can participate as simple citizens or as representatives of their organizations.
- *TL as a cross-pollination space.* TL is not just about getting together and siting around a table. It explores the collective intelligence of all participants based on a well-defined co-creation protocol. A conceptual frame for cross-pollination space proposed in [13] builds the foundation for the TL.

3   Defining Tiers-Lieu for Services

3.1   The Goal of Tiers-Lieu for Services

The goal of a Tiers-Lieu for Services (TLS) is to build information services in a contributory way involving actors from various disciplines and domains aiming to innovate and co-create value as common information assets, and so, to contribute to the digital progress of Society. Information services built in the context of a TLS are not only novel, they are necessarily accountable and indispensable. As TLS is providing a multi-disciplinary and multi-organizational context, it enables the co-creation of inter-disciplinary and inter-organizational services, and even transdisciplinary and trans-organizational ones [2]. Indeed, the concept of transdisciplinarity is very relevant for TLS. According to Nicolescu [14], transdisciplinarity "concerns that which is at once between the disciplines, across the different disciplines, and beyond each individual discipline". Therefore, transdisciplinary/trans-organizational services do not belong to any particular discipline or organization. Nevertheless, they contribute to the progress of each of the disciplines and organizations concerned, and go beyond their individual concerns – contribute to the societal progression.

TSL allows to overcome situations that would lead to an impasse the development of any organization doing it alone. By acting together, organizations and individual citizens have much better chance to succeed. TLS transforms individual problems into common challenges by building common information assets – information services. To enable such co-creation, TLS provides a regulatory framework for the activities to be done as a part of it, that is for building accountable and inclusive information services. A TLS can be launched by any action fitting into the context of digital progress of Society. The nature of the action should be transdisciplinary and trans-organizational.

3.2   The Approach

Figure 1 depicts the metamodel of the TLS approach, and defines its main concepts. A TLS is launched by an *intention* to overcome *situations* that may lead to dead ends the progression of Society. The initiative is based on the observation of these situations, and identifying societal *issues* that spark the intention to facing them, exploring and building information commons in the form of *information services*. For governing its activities, TLS provides a r*egulatory frame* as a set of *TLS-rules*. In addition to the initiator of the TLS, that can be any responsible citizen or organization, a consortium of organizations, an association, or a group of citizens, TLS requires several contributors. Contributors can play one or more roles in a TLS, and they can play them as independent *citizens* or as *representatives* of their *organizations*. We identify eight contributor roles: initiator, builder, facilitator, observer, historian, developer, regulator, and concierge.

- *Initiator*: this role designates persons who initiate the TLS by identifying situations and challenges they wish to explore and defining TLS intention. They invite the contributors and initiate the exploration. Based on the discussions with contributors, they are the ones who take final decisions during the TLS.
- *Builders* are the contributors that actively participate in the process of co-construction of information services and other common information assets. They are the main explorers and designers.

- *Facilitators* take care of the TLS process: they guide discussions towards the concretization of intentions and help in using various exploration, conceptualization and modeling techniques. They are neutral towards the realization of the TLS intention and do not take decisions themselves.
- *Observers* represent a passive role with regards to the TLS process and achievements. They are invited to attend discussions, but they are not supposed to participate in them. Observers (e.g. students, researchers, teachers) may aim to gain experience or domain knowledge. They can also simply share the general interest of TLS's intention without willing to participate in explorations.
- *Historians* play a supporting role. They seek to ensure that the intention of the TLS and information services under co-construction become more and more explicit and refined. They record continuous service improvements and highlight key elements of discussions and contributions. They provide essential information to the TLS activities, and make sure that the agreed schedule is respected.
- *Developers* are responsible for the development of information services, both at the digital and organizational levels. They must act within the defined framework for service co-construction, and pay attention to their evolution and interoperability with other information services.
- *Regulators* are responsible for the compliance of information services co-created during the TLS with laws and regulation policies governing the domain explored in the TLS. They provide the regulatory information, which is relevant in the context of the TLS.
- *Concierges* manage the reception and animation of the TLS. They take care of the TLS place, security, catering, etc., and set up animations in different formats. They are also responsible for communicating about the TLS and documenting it: taking pictures, creating social media and online discussions, managing relations with the press.

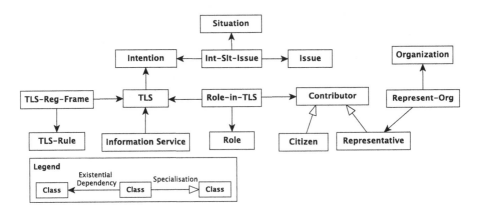

Fig. 1. Metamodel of the TLS approach.

To formalize the guidance for driving a TLS we use Map process modeling formalism [15], which allows to express process models in intentional terms instead of fixed steps and activities. Map provides a representation system based on a non-deterministic ordering of intentions and strategies. In fact, a process is represented as a labeled directed graph where intentions are nodes and strategies are edges between intentions. Since, many strategies can be used for achieving an intention, Map allows to represent flexible and situation-driven process models including multiple techniques to achieve the intentions. The process model of the TLS approach represented in Fig. 2 reveals four main intentions (in addition to the Start and Stop that indicate respectively the beginning and the end of the process execution), and several strategies to reach the intentions. We detail them below.

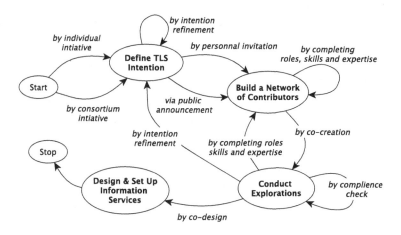

Fig. 2. A process model for driving a TLS.

Define TLS Intention. As mentioned above, the initial definition of the TLS intention can be issued as an individual initiative by a responsible citizen or an organization, or proposed by a consortium of organizations or a group of citizens. In the process model these cases are formalized as two strategies, namely *by individual initiative* and *by consortium initiative*. Both of them consist in analyzing the related situations and revealing societal issues that lead to the definition of the intention. The difference lies in the fact that the consortium has to reach an internal consensus before triggering a TLS action. The intention of the TLS can be refined at any time during the TLS (the strategy *by intention refinement*).

Build a Heterogeneous Network of Contributors. Initiators of the TLS are responsible for constituting the network of contributors. They can collect them *by personal invitation* or *via a public announcement*, seeking to gather representatives from all concernment parties, having the essential skills and expertise in the TLS domain, and representing all necessary roles of the TLS. It is not required to have all the roles listed above, except the builders and at least one concierge. It is essential to have most of the builders since the beginning of the TLS. Other roles can be omitted or completed later during the TLS, at any time when it is considered necessary. This is formalized as a strategy named *by completing roles, skills and expertise*.

Conduct Explorations. This step is crucial for the success of the TLS as it aims at achieving actionable results in the form of information services and related value propositions. The *co-construction* strategy, in fact, encloses various conceptualization techniques that support the exploration process and lead the contributors to the common understanding of the information space and the identification of information services. The choice of the techniques depends on the builders' and facilitators' knowledge and proficiency in using these techniques. We claim that conceptual modeling is the underpinning technique to information service co-construction, and we further argue on this in the following section. The *compliance check* strategy is applied to guarantee the regulatory compliance of the identified services.

Design and Set up Information Services. In this step, contributors decide which of the exploration results should be materialized into information services or even service systems. Conceptual modeling is used again, now as a means to co-design information services, in particular to express their static, dynamic, rule and role spaces. Developers contribute in prototyping, developing and setting up these services.

3.3 The Role of Conceptual Modeling in TLS

The main purpose of a TLS is to explore and co-design information services. In order to take consistent and responsible positions, contributors coming from heterogeneous horizons need to speak a common language allowing them to understand each other and build informational consensus. We claim that such language is conceptual modeling. In fact, this argument is not so original as many researchers claim it. Notably, Olivé [16] insists that "conceptual schema should be the center of the development of information systems". The use of conceptual modeling is also promoted in [17] for developing service-oriented information systems. Similarly, in information service development a common representation is expressed in terms of conceptual models [1,7]. In the context of TLS, the role of conceptual modeling is to ensure a formal and unambiguous representation of a particular domain in view of digitalization.

Several types of models can be used in the information service design for representing the value and the content of information services [2], their context, and how they serve different actors. Each of them expresses a particular view such as: the activity of the organization (including its structure, roles, rules, goals, processes, economic situation, etc.), the digital systems (databases, architecture), the ontological representation of the activity domain, and the most important – the information (data, rules, activities, roles, responsibilities). Therefore, driving a TSL with the help of conceptual modeling allows:

- to facilitate reaching informational consensus of contributors,
- to guarantee responsible and agile digitalization of organizational/social activities,
- to explain the direction pursued by the informational propulsion induced by the co-creation of services,
- to maintain the co-creation action within the framework of the original TLS intention, and
- to facilitate building accountable and inclusive services.

4   Exploring Tiers-Lieu for Services

We have conducted two exploratory TLS, one in November and the other in December of 2018, both in the context of a continuing education program, a Certificate of Advanced Studies (CAS), on Business Development with Information Services. This CAS, is dedicated to the practitioners willing to upgrade their knowledge and skills in the engineering and management of information services and systems, and to become responsible actors of digital transformation in their organizations. The CAS lasts one semester and awards 15 credits ECTS. The participants come from various organizations and have different roles such as business analyst, IS manager, project manager, IS designer, developer, or any other business activity supported by information services. In total, 13 people, participants of this education program, participated in the TLS. Our primary goal was to offer the participants an experience of contributing in a TLS and learning how to conduct one. The second goal was to demonstrate the value of conceptual modeling in a TLS – to assess its contribution to the success of the TLS.

As a preparation, all participants had a short course on TLS. They also had courses and training of various design and creativity techniques such as Business Model Canvas [18], Creativity Triggers [19], Service Model Canvas [20], Value Proposition Canvas [21]. Half of the participants had previously taken another CAS on conceptual modeling of information services and systems, and therefore had acquired corresponding skills (e.g. data, process, goal modeling).

The first TLS was dedicated to exploring the technological push – the potential of implementing a novel digital artefact and identifying information services that could be developed based on this technology. The second TLS, took the opposite way by exploring the societal push to information service development. In this paper we report on the second TLS.

4.1   A TLS for Exploring Societal Push

The aim of this TLS was to explore how societal goals could be reached with the help of information services. It had as subject "Digital service and regulation development for the collective design of Neighborhood Local Plans (NLP) in the Canton of Geneva". The TLS took place in a co-working space in Geneva, and lasted 15 h (10 h on Friday and 5 h on Saturday morning). The TLS was initiated by a representative of the NLP initiative. All participants of the continuing education program were invited as contributors and had the role of builders. One of the teachers played the role of facilitator, especially for conceptual modeling, while the other represented an historian. The initiator also played the role of concierge as he set up all the organization and prepared the required material, as well as he took care of the documentation and media.

The initiator presented the subject and the main issues related to defining NLP. In particular, he stressed on the difficulty of establishing collaboration between all parties involved in elaborating plans and related regulations, the length and rigidity of the process, the missing digital support, and so on. The contributors were invited to answer the following questions:

– How to write a regulation (a normative document) in collaboration? What information services would be needed to support such collaboration?
– Can information services propel the collective conception of normative texts?

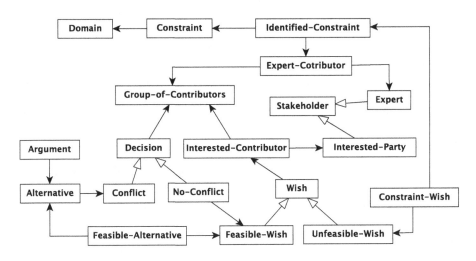

Fig. 3. An excerpt of the conceptual model produced during the TLS

After a short brainstorming session, the participants organized themselves into 3 groups. The rest of the time was naturally split in 3 periods of co-creation with intermediate presentations of the progress in group work, and then the final presentations. The group No2 had the most advanced skills in conceptual modeling. So, they naturally started with elaborating a conceptual domain model (see Fig. 3), which appeared to be a valuable means for identifying information services. They designed a transdisciplinary service system – a platform providing services for managing roles, ideas and initiatives, collecting needs of inhabitants, studding feasibility, co-deciding, etc. The group No1 focused on technical aspects of collaborative work and undertook the elaboration of an architecture

for a platform supporting collaborative editing, change log, document versioning
and multichannel notification. The real progress was observed once the group
elaborated a conceptual model with the help of the facilitator. The group No3,
investigated the roles and responsibility domains in NLP projects. During the
last period of the first day they decided to merge with the group No2. Together,
based on both inputs (the conceptual domain model and the role and responsibility model) they elaborated the responsibility space of each role and created
a service model canvas [20] highlining the value of the information service platform for NLP (see Fig. 4). They also identified that the TLS approach would be
the most appropriate to involve all these roles in the co-creation of neighborhood plans and related regulations. During the final brainstorming, participants
agreed that most of the ideas developed during the TLS could serve to design
generic purpose information services supporting the co-creation of regulations.
Because of the space limit we cannot show all the conceptual artifacts produced
during this TLS.

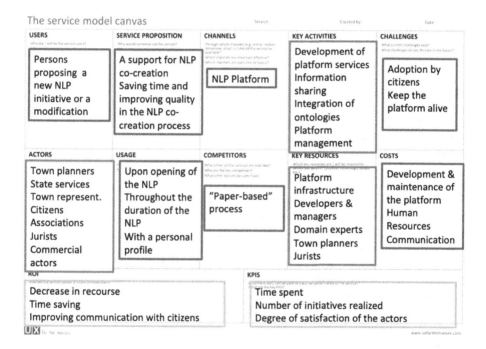

Fig. 4. An excerpt of the conceptual model produced during the TLS

4.2 Analysis of the TLS Results

Based on our observation and the appreciation of the contributors and initiators
of the TLS, we consider the results of this TLS as successful. From academic point

of view, participants learned a new approach to co-constructing information services that they found very effective. Quite obviously, conceptual modelling was an efficient way to facilitate group intelligence, to trigger most of the ideas, to reach informational consensus and to better scope the identified services. The TLS process also demonstrated that conceptual modeling experts were more efficient than those who did not yet have this capability.

A potential threat to validity of our study is the selection of contributors. Indeed, we skipped the step of building the network of contributors as we already had our participants. There is no statistics on the best number of participants in a TLS. The main guideline is to select/invite contributors representing all concerned parties. Our participants already known each other and had collaborated on small group projects in the context of the education program. Besides, they already learned a few design techniques and some of them had conceptual modeling practice. We do not know if similar results could be reached if none of the TLS members had design and conceptual modeling skills. However, the origin and background of the participants were sufficiently different (they work in different organizations and have different responsibilities) to consider that the context of our experiments were close to a real TLS setting.

5   Discussion

While the concept of Tiers-Lieu is gaining in awareness and finds more and more followers, its adaptation to the information service co-creation, that we call Tiers-Lieu for Services, is our original contribution. The TLS approach is not yet fully formalized, and needs further exploration, experimentation and validation. In this paper we present our exploratory work in the domain, our first steps towards a method for conducting a TSL. Our definition of TLS is based on the definition of TL as given in [8, 11] and on the works in the domain of information service engineering [1, 2, 7]. We are convinced that TLS approach allows to put in practice the inter-disciplinary and even transdisciplinary co-creation of information services. We also consider it as an accountable and inclusive approach as it allows to take into consideration opinions and initiatives of contributors from various horizons and having different responsibilities (private and public organizations, governmental institutions, citizens), everyone concerned by digital transformation with the help of information services. Our explorations demonstrate that TLS suits well societal as well as business purposes, and can be driven by a technology push as well as by societal goals. TLS allows to explore larger horizons and go beyond the capabilities of one person, one profession (discipline), one organization. It allows to create transdisciplinary and trans-organizational digital infrastructure. Our explorations also demonstrate that having skills in conceptual modeling has a direct impact on the success of a TLS. Not only conventional data and process modeling techniques are useful, all conceptualization and exploration techniques fostering collective intelligence (e.g. Design Thinking, Creativity Triggers, Business Model Canvas, Service Model Canvas, Value Proposition Canvas, etc.) are also very useful. Altogether, the conceptualization techniques allow to better share the understanding of the domain and to co-create. The two TLS we have conducted are certainly not a formal proof of the effectiveness of the TLS approach. Our intention was rather to explore the potentialities of this approach, with the aim of further formalizing it. Each TLS was based on a real case and was presented by its promoter, which, in our opinion, gives credibility to our study.

Based on these explorations and foundations in service science [6], we aim to develop a method for conducting a TLS. Such method has to be situation-driven. By situation-driven (or situational [22]) we mean taking into consideration that the context of each TLS will be different, e.g. will have a different intention and scope, a different set of contributors (including their number, origin, responsibilities, skills), a different location, etc. Accordingly, it will require a specific composition of method chunks to fit the situation. For example, if the TLS participants do not know each other, the method would recommend applying team building techniques at the beginning of the TLS. In another situation, where participants would have no conceptual modeling skills, the method would insist oh inviting a facilitator with such expertise. A TLS is not a one-off meeting, it usually requires several sessions, each having a specific purpose and a variation of contributors (involving new ones, losing others). Therefore, the method has to be easily adaptable by selecting on the fly the most suitable method chunks. Finally, the contributory principle is fundamental for guiding a TLS. It aims to ensure that every participant has the opportunity to express his/her opinion and to be heard, that the opinion of each participant is valuable and seriously taken into account, and that there is no hierarchy or any other clustering of participants excepted their roles.

6 Conclusion

In this paper we present the Tiers-Lieu for Services (TLS) as a novel approach to guide digital co-construction in an exploratory and responsible way. We argue that the TLS approach helps to co-create transdisciplinary and transorganizational information services and therefore allows to contribute to the societal progression. As first step to the formal definition of our approach, we define the concept of the TLS (the meta-model), the roles and the guidance to drive a TLS (the process model).

With the aim to explore the applicability of the approach, we have organized two TLS, both in the context of a continuing education program. Each of them had a different intention and a different initiator, but the same contributors. The first TLS was driven by a technology push – an introduction of a new device in the health care domain. Different ways to promote and use the device were explored and materialized in terms of information services. The second TLS, presented in this paper, faced a societal challenge – a co-development of regulations, and came up with a proposal of a dedicated service platform. Both TLS demonstrated the value of conceptual modeling as an underpinning technique to succeed. Our current preoccupation is to further formalize the approach in terms of a situational method easily adaptable to different types of TLS.

References

1. Ralyté, J., Khadraoui, A., Léonard, M.: Designing the shift from information systems to information services systems. Bus. Inf. Syst. Eng. **57**(1), 37–49 (2015)
2. Ralyté, J.: Towards a method family supporting information services co-creation in the transdisciplinary context. Int. J. Inf. Syst. Model. Des. **4**(3), 50–75 (2013). https://doi.org/10.4018/jismd.2013070103. IGI-Global
3. Baker, M.: Digital Transformation. CreateSpace (2014)
4. Sustainable Development Goals. https://www.un.org/sustainabledevelopment/
5. Vargo, S., Lusch, R.F.: Service-dominant logic: continuing the evolution. J. Acad. Mark. Sci. **36**(1), 1–10 (2008)
6. Maglio, P.P., Spohrer, J.: Fundamentals of service science. J. Acad. Mark. Sci. **36**(1), 18–20 (2008)
7. Arni-Bloch, N., Ralyté, J., Léonard, M.: Service–driven information systems evolution: handling integrity constraints consistency. In: Persson, A., Stirna, J. (eds.) PoEM 2009. LNBIP, vol. 39, pp. 191–206. Springer, Heidelberg (2009). https://doi.org/10.1007/978-3-642-05352-8_15
8. Burret, A.: Tiers-lieu. Et plus si affinités, FYP EDITIONS (2015)
9. Le manifeste des Tiers Lieux. https://movilab.org/index.php?title=Le_manifeste_des_Tiers_Lieux
10. Burret, A.: Étude de la configuration en Tiers-Lieu - la repolitisation par le service (Tiers-Lieu Configuration Study - Repolitization Through Services). Ph.D. Thesis, Université des Lumières, Lyon, France (2017)
11. Léonard, M., Yurchyshyna, A.: Tiers-Lieu: exploratory environments for service-centred innovations. In: proceedings of of the 5th International Conference on Emerging Network Intelligence - EMERGING 2013, Porto, Portugal (2013)
12. Oldenburg, R.: The Great Good Place: Cafes, Coffee Shops, Bookstores, Bars, Hair Salons, and Other Hangouts at the Heart of a Community. Da Capo Press, Cambridge (1998)
13. Yurchyshyna, A., Opprecht, W.: Towards an ontology-based approach for creating sustainable services. In: Morin, J.-H., Ralyté, J., Snene, M. (eds.) IESS 2010. LNBIP, vol. 53, pp. 136–149. Springer, Heidelberg (2010). https://doi.org/10.1007/978-3-642-14319-9_11
14. Nicolescu, B.: Manifesto of Transdisciplinarity. State University of New York (SUNY) Press, New York (2002)
15. Rolland, C., Prakash, N., Benjamen, A.: A multi-model view of process modelling. Requir. Eng. J. **4**(4), 169–187 (1999)
16. Olivé, A.: Conceptual schema-centric development: a grand challenge for information systems research. In: Pastor, O., Falcão e Cunha, J. (eds.) CAiSE 2005. LNCS, vol. 3520, pp. 1–15. Springer, Heidelberg (2005). https://doi.org/10.1007/11431855_1
17. Thomas, O., vom Brocke, J.: A value-driven approach to the design of service-oriented information systems - making use of conceptual models. Inf. Syst. E-Bus Manage **8**(1), 67–97 (2010)
18. Osterwalder, A., Pigneur, Y.: Business Model Generation: A Handbook for Visionaries, Game Changers and Challengers. John Wiley and Sons Inc., Hoboken (2010)
19. Burnay, C., Horkoff, J., Maiden, N.: Stimulating stakeholders' imagination: new creativity triggers for eliciting novel requirements. In proceedings of the 24th IEEE International Requirements Engineering Conference, RE 2016. IEEE Xplore (2016)

20. Turner, N.: Introducing the service model canvas. http://www.uxforthemasses.com/service-model-canvas/
21. Osterwalder, A., Pigneur, Y., Bernarda, G.: Value Proposition Design: How to Create Products and Services Customers Want. John Wiley and Sons Inc., Hoboken (2014)
22. Henderson-Sellers, B., Ralyté, J., Ågerfalk, P.J., Rossi, M.: Situational Method Engineering. Springer, Heidelberg (2014). https://doi.org/10.1007/978-3-642-41467-1

A Service Ecosystem Ontology Perspective: SDG Implementation Mechanisms in Public Safety

Salem Badawi[1], Sorin N. Ciolofan[1] (ID), Nabil Georges Badr[2] (ID),
and Monica Drăgoicea[1(✉)] (ID)

[1] Faculty of Automation and Computers, University Politehnica of Bucharest,
Splaiul Independenţei 313, 060042 Bucharest, Romania
salem.badawi@acse.pub.ro, sorin.ciolofan@cs.pub.ro, monica.dragoicea@upb.ro
[2] Higher Institute for Public Health, Saint Joseph University of Beirut,
Beirut, Lebanon
nabil@itvaluepartner.com

Abstract. Taking a Service Dominant Logic perspective, this paper proposes steps towards a better understanding on how actors' practices can support the sustainable development targets, and the role of digital technologies in facilitating such practices. A service ecosystems ontology customized for the development of data-intensive digital services in public safety is presented. Semantic reasoning is used to develop new knowledge in the domain of expertise of the involved Actors. The methodology presented here envision the creation of the Information Common Goods, fundamental resources in value co-creation, fuelling the lifecycle management activities that increase the value of the Service Delivery in the ecosystem. A network of actors, using resources (including Information Common Goods), co-create value from services, under the guidelines of institutions inside a service ecosystem. This helps in leveraging technology and Information Common Goods, as a main resource for value creation through integrative actors' practices.

Keywords: Service ecosystems · Actors practices · Ontology · Semantic reasoning · Data quality

1 Introduction

In 2015, United Nations Assembly adopted a list of 17 Sustainable Development Goals (SDG) that are set to drive universal actions toward ending poverty, achieving prosperity, ensuring equality and saving the planet [1,2]. They are defined by one of the most widely accepted definition of sustainable development as *"meeting the needs of the present without compromising the ability of future generations to meet their own need"* [3]. Even though this definition does not provide a clear vision and techniques to achieve sustainability, it points the compass towards the delivery of value for a broad range of Actors in a long-term

© Springer Nature Switzerland AG 2020
H. Nóvoa et al. (Eds.): IESS 2020, LNBIP 377, pp. 304–318, 2020.
https://doi.org/10.1007/978-3-030-38724-2_22

perspective as a base of sustainable development. Various initiatives are put in place today to provide guidance on the coherent implementation of the SDG, from an institutional perspective [4], and to explore the SDG evolution based on various digital technologies [5], as a means of developing rich information-based services in public safety [6,7].

This article argues that, by adopting a systems thinking that integrates the Service Dominant Logic (S-D Logic) perception of service ecosystems [8,9], it is possible to understand practices of different Actors in their quest for survival in the surrounding dynamic environment, tightly connected to the integration and coherence of the SDG implementation. The offered perspective aims to present how these practices may lead to the survival of the environment in the long-term.

Since Vargo and Lusch presented a new vision of S-D Logic in 2004 [10], the perception on service and market has evolved and revolutionized. Their work has shifted the paradigm to service as the primary unit of exchange, and classifies other exchanges as a frontage for this primary one. The role of service beneficiary turned to be vital in service provisioning. Thus, value realization has been stated because of a co-creation process between multiple parties, including the beneficiaries [11]. More importantly, their recent work has shed light on operant resources, i.e. institutions, competence (knowledge and skills), as the core sources of economic growth [12].

Henceforth, we propose an ontology-based approach to explain Actors' practices and the value co-creation process implementation in a service ecosystem to advance the implementation of the Public Safety as a Service vision described in [6,13]. Specific details on using semantic reasoning and ontologies for the development of collaborative knowledge sharing in the perspective of the 2030 Agenda for Sustainable Development [1] and the 17 Sustainable Development Goals (SDGs) defined by the United Nations General Assembly [2] are detailed in our previous work [6]. Within its scope, the paper shows how Information Common Goods as a service is a fruitful position that induces substantial discoveries, as service delivery lifecycle for self-managed sustainability.

The remainder of the paper is organized as follows: Sect. 2 presents general considerations on sustainability and the expansion though S-D Logic. Section 3 proposes a working methodology and specific ontology development aspects. Section 4 addresses a specific application case study in water resources management based on a real scenario related to accidental river water pollution compiled from [14] and [6]. Finally, a short conclusion and further development perspectives are formulated in Sect. 5.

2   Sustainability and S-D Logic

Sustainability has a broad conceptualization: starting from the basic view of sustainability as a source of competitive advantage, by playing a vital role in increasing the service delivery life cycle [15,16], to a wider spectrum that considers the economic, social and environmental factors [17]. In the economic view, sustainability is concerned with value co-creation in network of actors and value in-exchange as the value capture element [18,19]. In networks of actors, value is created through actor practices including interaction, service exchange between actors with mutually beneficial relationships [20]. Service exchange is a practice that integrates resources [21] through decoupling and recoupling of actors available resources with other resources provided via the interaction within the network [22]. The environment and the availability of resources, as well as the levels of perceived benefit influence actors' interaction, across the actors' network [20]. This influence may often lead to value creation through the betterment of the circumstances of other actors in the service ecosystem [20]. Ultimately, sustainability is not only an issue of a single actor survival, but also of the survivability and prosperity of the surrounding environment and of the service ecosystem as whole [23].

Sustainability views that focus on environmental factors are concerned with the conservation of resources for future generations [24], in other words, resulting in a long-term service provided with minimal operand resources and multiple actors involved in the value co-creation processes. This makes S-D Logic a perfect approach for sustainability, by considering service as the base of economic and social exchange [25], where the focus shifts from production as a means to deliver value towards services as a means to co-create value.

Consequently, building further on the review conducted by Geissdoerfer et al. [18] for business models that support sustainability, a *sustainable service ecosystem* can be described as *a system that maintains Actors' relationships in resource integration networks*, and *co-creation of value for a broad range of involved Actors, in a long-term perspective*. This approach is in line with Sustainable Business Model that emphasizes on the importance of proactively engaging with all the actors in the service ecosystem and ensure their long-term well-being [26].

In their first presentation of S-D Logic, Vargo and Lusch [10] emphasised on the important role of co-Learning in improving service provision by understanding better ways to serve their customers, and the role of co-Learning in improving operant resource (knowledge) [27], value proposition, value co-Creation and actors performance in service ecosystem (Table 1).

co-Learning, or "learning with market" [30], has a vital role in service systems sustainability and competitive advantage by helping Actors to adapt to market dynamic needs and service unpredictable outcomes [31]. It develops three learning capabilities [32]: (a) the ability to *develop* and *maintain* Actors' *networks*; (b) the ability to *interpret* and *understand value co-creation processes* that influence Actors' practices in their networks; (c) the ability to *understand* Actors' *practices* (i.e. *integration, normalization,* and *representation*) that support actors' interaction and service exchange.

Even though we emphasize most of the time on the role of actors' collaboration and interaction in actors' network, to maintain sustainability, we should not ignore the role of an individual actor in service system survival. At the dyadic level, the ability to customize service to meet individual beneficiary needs is vital to service system viability [33]. Additionally, it is important to consider

Table 1. Value co-creation processes and techniques (adapted from [28,29]).

Processes	Definition
co-Learning	≻ The ability to learn from actors experience, how to improve the service (what was good vs. could be better)
co-Design	≻ Granting actors the ability to design their own services by participating in the development process and the ability to customize their services
co-Delivery	≻ Actors granted the ability to participate in service delivery by encouraging them to share their experience, and the ability to help others with their experience
co-Production	≻ Granting actors the ability to contribute to the product development cycle, considering the product as a service delivery mechanism
co-Innovation	≻ Actors granted the ability to participate in service evolving by suggesting new ideas and participate to implement them [29]

the beneficiary point of view through service life cycle, starting from the design through the development and provision [33].

3 Working Methodology

Henceforth, we present an approach to integrate the above-mentioned evaluation of value co-creation in service ecosystems through *actors' integrative practices*. The suggested working methodology addresses a general situation related to the evaluation of data quality in data-intensive digital services, in the development of environment-oriented information services as common goods to support public safety services discovery and development, as formulated in [7].

Step 1. *Reasoning on Information Common Goods in a service context.*

To introduce our reasoning on how *Information Common Goods* (ICG) may be defined as a resource in a service context, Fig. 1 describes a Service Delivery mapping diagram introducing the main steps to describe this process. Further, the following three-level concept taxonomy is used in the foundation of a service delivery life-cycle for self-managed sustainability:

- Actor
  Provider: Concept Developer; Designer, etc.
  Customer: User, etc.
- Resource
  Tools: Data collection, Data Manipulation, etc.
  Data: Raw data, Structured data, etc.

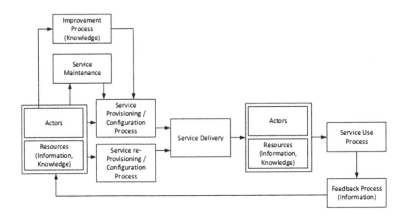

Fig. 1. Information Common Goods: A Service Delivery mapping diagram.

– Process
  Conceptualization: Service design, Service Configuration, Service Provisioning, Service Reprovisioning
  Delivery: Service Delivery
  Consumption: Service Use
  Lifecycle management: Service Feedback, Service Maintenance, Service Improvement

Within the perspective of this paper, the term *Information Common Goods* refers to data collected, formatted for sense-making, and stored in a data warehouse, ready to be used by multiple authorities/providers/decision makers (actors) to provide planning/action/services (artefacts) using a *common good* platform of the *public safety as a service* [13] (service). ICG are therefore, resources of information for knowledge building, useful to actors (Providers, Concept developers, Designers, etc.) to conceptualize a service through Design and Configuration processes, and provision it to be of value to other actors (Customers, User, etc.).

The latter would consume the services, provide potential feedback used as additional information by the providers and increase the value/benefit of the service to the ecosystem. Information in the ICG is then analysed, transformed and updated in a cyclic process of value creation that hones the data/information into valuable knowledge. In addition to the use-end of the service delivery, two additional processes of maintenance and continuous improvement are envisioned. These processes benefit from the ever-increasing information resources to tune the outcome into self-managed sustainability. Therefore, Actors (Providers, Concept developers, Designers, etc.) engage in lifecycle management activities that increase the value of the Service Delivery. Other Actors (Customer, User, etc.) participate in feedback and additional collection of data that could be used for service improvement, thus increasing the value provided in the service ecosystem.

Step 2. *Conceptualizing the service ecosystem ontology.*

To discover and evaluate the actors' integrative practices towards the evolution of the ICG supporting public safety services, a second step to follow is the collection of information about concepts, entities, and relations among them in the domain of discourse. Based on this collective knowledge, a service ecosystem ontology has been proposed to be created. As a starting point, the glossary of the core concepts related to service ecosystems **BoK**, possibly growing towards a full domain taxonomy [34] has been compiled. This glossary proposes to integrate the following core concepts: Actors, Resource, Institutions, Service, Service Ecosystem, Service System, Value, Actors Practices, Technology.

Actor. This term is generally defined as a human (an individual) or a group of humans that represents social and economic organisation which is legally recognized as an enterprise that collaborates to achieve mutual goal [35]. The Actor has a role in service outcome or can have multiples roles in service exchange [25] as apart of the society or the service ecosystem [9, 36]. Actors have characteristics like Adaptability, Flexibility and Resiliency to ensure their viability and sustainability in the service ecosystem. Adaptability refers to Actors ability to unbundle and re-bundle their available resources when it is necessary [20]. Flexibility refers to actors' ability to switch between different resource sources and different roles [37]. Resiliency refer to the ability to consider possible environmental influence or unfortunate events [38].

Resource. This term refers to anything, tangible or intangible, internal or external, operand or operant, which the actor can draw on for increased viability or draw on for support [10, 39]. Resources are generally classified in S-D Logic as operant (those that are capable of acting on other resources to contribute to value creation, often referring to competences (knowledge and skills)) and operand (those that require action taken upon them to be valuable, often referring to natural resources) resources [10].

Institutions and institutional arrangements. An Institution refers to the rules and guidelines that coordinate and constrain service exchange [9]. Institutions define the value of resources at a particular place and time by defining meaning for what is considered a valuable resource, and providing guidelines for accessing, adapting, and integrating of such resources in a specific context [19].

Service. Service is the *"application of competences for the benefit of another entity or the entity itself"* and *"as the fundamental basis of any value creation in an exchange"* [10]. According to the S-D Logic foundations that are rooted in the political-economic work of Adam Smith [40], service is a process of exchange, where exchanges, as the foundations of the economy, consist of specialized knowledge and skills [25, 41].

Service Ecosystem. Service ecosystems are *"relatively self-contained, self-adjusting system of resource integrating actors connected by shared institutional arrangements and mutual value creation through service exchange"* [9]. Self-contained means that each ecosystem consists of sub ecosystems whose goal is to survive in a much bigger ecosystem [22]. Whilst, self-adjusting refers to the ecosystem ability to reconfigure the sources of resources by integrating resources from nested and overlapping subsystems [22]. A service ecosystem is a constellation of service systems that starts from a dyadic relation between two parties and emerges into a vast network of collaborated service systems creating the society. The focus of each entities within the network is to survive by achieving resource density that refer to the effective combination of resources to solve a specific problem at a specific time and place (context) [19,42].

Service System. It is defined as *"a value co-creation configuration of people, technology, value propositions connecting internal and external service systems, and shared information"* [43,44]. In other words, a service system is a configuration of resources, mainly operant ones, shared information and technology, for creating value in a *collaborative process* for the benefit of human actors. Both definitions emphasize on the importance of four core concepts: Actor, Technology, mutual value, and value proposition, in activities that connect entities within the service system as well as the service system with other service systems.

Value. In S-D Logic, at least two actors, including the beneficiary, always create value that is determined by the beneficiary [45]. In another perspective, value is in form of experience derived and determined by the context and environment [46]. This view associates with S-D Logic of value nature, as unique and phenomenological [41,47].

Actors Practices. This concept refers to actors' practices in a service network (ecosystem) that supports and formulates actors' interaction and resource integration [32]. These practices are grouped as representational, normalizing, and integrative.

Representational practices refer to the actors' effort in communicating in the service ecosystem using a standardized language. They help maintaining the relationship between actors in their network, and foster effective, efficient actors collaboration and value co-creation [22]. On the other hand, *normalizing practices* are activities that develop guidelines or parameters for actors' interaction in the service ecosystem, including the formation of institutions, the definition of job performing standards and interaction protocols, and the coordination of effective and efficient actors' interaction [22]. Lastly, *integrative practices* refer to the actors' effort to integrate resources to co-create value. This includes unbundling/re-bundling of operant resources (skills and knowledge) with private and public resources to co-create value in the service ecosystem [22], and the ability to access, adapt, integrate resources, being influenced by actors' networks via establishing exchange relationships and shaping the social context [19].

Technology. Akaka and Vargo developed their definition of technology as *"a combination of practices, process and symbols that fulfill a human purpose"* [39], building on Mokyr's definition of technology as *"useful knowledge"* [48] and Arthur's definition as *"an assemblage of practices and components that are means to fulfill human purposes"* [49]. Following these definitions, technology can be classified as practice, components or both of them. It is clear that technology and human knowledge and skills are inseparable [50]. There is no use of technology as component (device) without human competence, therefore it should be considered as one unit, as a combination of operant and operand resources. Technology resources improve communication and collaboration between actors, the development of technology led to new methods of communication that reach widely with minimal cost and effort, and communication foster collaboration between actors which increase service ecosystem viability [19]. Technology facilitate the access to resources through exchange in actors networks and foster integrative practices (Access/Adapt/Integrate Resources) by helping actors to obtain needed knowledge for creating value via shared information.

Pakkala and Spohrer [38] argue for the ability of some cognitive technologies to act and interact with people and environment. Other perspectives argue for the classification of technology as operant resources that has the ability to act on other resources to create value [39]. Technology resources lead to new Normalization Practices by creating institutions. For example, some practices that were not accepted long time ago - or at least in some cultures - become more and more accepted, such as ride sharing in case of Uber [51] and sharing personal information in case of social media [52]. However, even though a technological artefact may be considered to have agency to act in its environment, it is difficult to argue in favour of technology's ability to realize mutual benefit by having conscious intent to act on its own. But we may argue that, despite the debate about the classification of technology, both views emphasize on the role of technology in facilitating actors practices.

Step 3. *Formalizing the service ecosystem ontology.*

The main focus of this step is the definition of the classes and the classes hierarchy, and the relations between classes, to integrate this analysis of the BoK on service ecosystems. During the development of the suggested service systems ontology (presented in Fig. 2), we have followed the criteria specified in [53], in a continuous evaluation process suggested by [34]. More details on class hierarchy and individuals by type are presented in Fig. 3.

ServiceClient, ServicePrincipal, ServiceProducer, ServiceProvider, and Stakeholders [36] are Actors that participate in service exchange. They develop practices in service delivery that involve specific processes, where various resources are integrated.

The amount of information, knowledge, and other resources that an Actor possesses at any moment in time is measured by density [54]. Actors resource density (**Actor** *hasDensity*(**Resource**)) is increased by the ability to access, adapt and integrate resources (Table 1), which create value constellations [19]. In the process of value co-creation, Actors should have access to network resources, and access to knowledge on how to adapt and integrate them. This collaborative process lead to new Integrative practices that further lead to increase density and create value constellation, which refer to value creation in network of actors (constellation) [55]. co-Innovation refers to changing the way value is created and create innovative solution for actors survival.

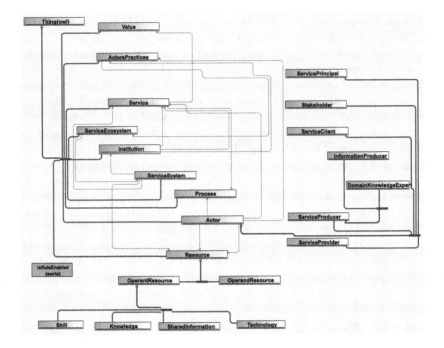

Fig. 2. An ontology to describe the service ecosystems context.

DomainKnowledgeExpert and InformationProducer are two types of ServiceProducer that are able to create collaboratively knowledge resources in the domain of interest, through contributory participation. Therefore, Actors confederated value co-creation initiative may lead to the creation of service sustainably coherent with the ecosystem complexity, i.e. sustainable services [6]. An Institution [9] coordinates the process of exchange through service, while coordinating ActorsPractices integrated in ServiceEcosystem. A ServiceSystem entity is enforced via Institution, it includes various resources, and creates services. Eventually, groups of such entities create specific configurations as service ecosystems [56]. Within the scope of this paper, Value relates to *resilience* and *sustainability*, two concepts that are considered today as complementing each other to drive universal actions for collective progress inside society [1,57].

4   Demonstration Case Study

To exemplify the exploitation of the proposed service ecosystems ontology, we have created a very simple real scenario related to accidental river water pollution compiled from [14] and [6]. This implies using semantic reasoning together with the service ecosystems ontology, to promote collaborative knowledge sharing in the required domain of expertise to solve data quality issues in Cyber Physical Systems for water resources management. Two use cases have been created: (a) definition of specific rules for the deduction of new knowledge (based on

Classes	Properties	Individuals
Actor	*exchangeService*(**Service**) / *hasPractices*(**ActorsPractices**) / *integrateResource*(**Resource**) / *performProcess*(**Process**)	
ServiceClient		citizens; emergency_Team
ServicePrincipal		Ministry_of_Environment
ServiceProducer		
DomainKnowledgeExpert		Sorin_Ciolofan; Aurelian_Draghia; Monica_Dragoicea; Radu_Drobot
InformationProducer		ANAR
ServiceProvider		IT_experts
Stakeholder		
ActorsPractices	*create*(**ServiceEcosystem**)	collaboration_between_universities; contributes_to_social_awarness; environmental_regulation_creation; mitigation_of_pollution
Institution	*coordinate*(**ActorsPractices**) / *coordinate_exchange_of*(**Service**)	CollaborationProtocol; EmergencyProtocols; EnvironmentlRegulationProtocols; pollution_color_coding
Process		DataCleaning
Resource		
OperandResource		Dambovita_River
OperantResource		
Knowledge		hydro_technical_knowledge; service_science_knowledge
SharedInformation		TSQA_ontology
Skill		Mike11
Technology		CPS_Water
Service	*create*(**Value**) / *hasProcess*(**Process**)	PollutionAlerts
ServiceEcosystem	*is_a_configuration_of*(**ServiceSystem**)	
ServiceSystem	*arrangedVia*(**Institution**) / *hasResource*(**Resource**)	
Value		resilience / sustainability

Fig. 3. Case study: service ecosystem creation for accidental water pollution.

the existing knowledge database), rules that are executed in Protégé by invoking the Pellet reasoner; and (b) definition of specific queries, implemented in Protégé, to retrieve existing knowledge based on certain criteria. Specific details on implementation issues are detailed in [6].

a. New knowledge creation. For the sake of our example, we assume that (a) the Knowledge Base (KB) contains the individuals represented in Fig. 3; (b) CPS_Water is a Cyber Physical System for water pollution [6]; (c) ANAR is an abbreviation for National Administration "Romania Waters"; and (d) Ministry of the Environment is a typical example of a Service Principal individual.

Semantically reasoning on existing knowledge implies invoking the Pellet reasoner in Protégé and firing the SWRL-encoded rule **R1**. We assume that the variables **?x** and **?y** from Listing 1 are bounded to existing individuals in the KB.

R1. *If there is a ServicePrincipal that integrates the resource Dambovita River and which creates value resilience and if exists an InformationProducer that creates value resilience and integrates the same resource (Dambovita River) then the InformationProducer should integrate the resource CPS_Water and should exchange service pollution alerts.*

Listing 1: SWRL coded rule R1

```
integrateResource(?x,Dambovita_River),ServicePrincipal(?x),
cocreateValue(?x,resilience),integrateResource(?y,Dambovita_River),
InformationProducer(?y),cocreateValue(?y,resilience)->
integrateResource(?y,CPS_Water),exchangeService(?y,PollutionAlerts)
```

As a result of executing this simple, demonstrative rule, new knowledge is inferred: ANAR should use the CPS_Water based on which it should trigger pollution alerts.

b. Querying the database. For this use case, we assume that exists an InformationProducer that integrates resource CPS_Water. We want to search for an expert that has skills in MIKE11 software. We assume that prior to execution of query Q1, the rule R1 has been executed, and the new fact that ANAR integrates resource CPS_Water was already inferred as the outcome of this rule execution (according to use case a.)

Q1. *Find the domain knowledge expert that has skills in MIKE 11 and can be useful for the Information Producer which integrates CPS_Water.*

Listing 2: SQWRL coded query Q1

```
DomainKnowledgeExpert(?y),integrateResource(?x, CPS_Water),
hasOperantResources(?y, Mike11),InformationProducer(?x)->
sqwrl:select(?y)
```

As a result of executing this simple, demonstrative query we obtain: the expert that has a skill in MIKE11 software is `ing.Aurelian_Draghia`.

5   Short Discussion and Conclusions

Data Quality is essential in creating Information Common Goods that contribute value in complex services in public safety [7]. Nevertheless, it can be argued that Data Quality parameters must be considered when we talk about developing a Big Data model for Information Common Goods definition in public safety [58]. By now, certain foundational models divide data quality into four essential attributes, Accuracy, Consistency, Currency, and Uniqueness [59], and others focus on attributes of Objectivity, Trust, Relevancy, Completeness, and Timeliness.

The approach presented in this paper aims to enable a common, shared understanding of information structure exchanged among domain experts, through service provisioning activities. This understanding allows to agree on a common language and vocabulary in a coherent and consistent manner, and to create knowledge level protocols for communication within the domain of interest, in this case service ecosystem research and development. The suggested working methodology is exemplified on a real scenario in public safety services related to water resources management applying semantic reasoning to embed actors practices, such as that raw data is transformed into information and knowledge, and then managed as Information Common Goods.

Our future work will aim to understand the possible expansion of the taxonomy of terms in the service ecosystems ontology, and to fully fill in the useful properties of the integrated concepts. As well, it is worth to deepen the discussion on the management and maintenance of the Information Common Goods, that ought to address means and concepts of maintaining Data Quality, as detailed along with the brief demonstration presented here. As well, this paper leaves open discussion on the applicability of this study for Open (and public) data initiatives and the integration of the ontology to these open data projects. It is also interesting to discuss about the knowledge development process in the Information Common Goods and how this process is compared with the other approaches in the literature such as the data-information-knowledge-wisdom (DIKW) pyramid. A further validation of the proposed approach based on a research method (such as Design Science research) is envisioned as further work.

References

1. The 2030 Agenda for Sustainable Development. United Nations (2015). https://sustainabledevelopment.un.org/
2. Sustainable Development Knowledge Platform. Goal 11: Make cities inclusive, safe, resilient and sustainable. United Nations. https://sustainabledevelopment.un.org/sdg11
3. Cassen, R.H.: Our common future: report of the world commission on environment and development. Int. Affairs **64**(1), 126 (1987)
4. Institutional and Coordination Mechanisms: Guidance Note on Facilitating Integration and Coherence for SDG Implementation. United Nations Development Programme (2017). https://www.undp.org/
5. Wu, J., Guo, S., Huang, H., Liu, W., Xiang, Y.: Information and communications technologies for sustainable development goals: state-of-the-art, needs and perspectives. IEEE Commun. Surv. Tutorials **20**(3), 2389–2406 (2018)
6. Drăgoicea, M., Léonard, M., Ciolofan, N.S., Militaru, G.: Managing Data, information, and technology in cyber physical systems: public safety as a service and its systems. IEEE Access **7**, 92672–92692 (2019)
7. Drăgoicea, M., Badr, N.G., Manea, L.M.: Emerging information common goods for the development of complex services in public safety. In: Proceedings of the 23rd International Conference on System Theory Control and Computing, Sinaia, Romania. IEEE, 9–11 October 2019 (In Press). http://icstcc2019.cs.upt.ro/
8. Ng, I.C., Vargo, S.L.: Service-Dominant (SD) Logic, service ecosystems and institutions: bridging theory and practice. J. Serv. Manag. **29**(4), 518–520 (2018)
9. Vargo, S.L., Lusch, R.F.: Institutions and axioms: an extension and update of service-dominant logic. J. Acad. Market. Sci. **44**(1), 5–23 (2016)
10. Vargo, S., Lusch, R.: Evolving to a new dominant logic for marketing. J. Market. **68**(1), 1–17 (2004)
11. Vargo, S.L., Lusch, R.F.: Service-dominant logic 2025. Int. J. Res. Market. **34**, 46–67 (2017)
12. Akaka, M.A., Koskela-Huotari, K., Vargo, S.L.: Further advancing service science with service-dominant logic: service ecosystems, institutions, and their implications for innovation. In: Maglio, P., Kieliszewski, C., Spohrer, J., Lyons, K., Patrício, L., Sawatani, Y. (eds.) Handbook of Service Science, Volume II, pp. 641–659. Springer, Heidelberg (2019). https://doi.org/10.1007/978-3-319-98512-1_28
13. Drăgoicea, M., Badr, N.G., Falcão e Cunha, J., Oltean, V.E.: From data to service intelligence: exploring public safety as a service. In: Satzger, G., Patrício, L., Zaki, M., Kühl, N., Hottum, P. (eds.) IESS 2018. LNBIP, vol. 331, pp. 344–357. Springer, Cham (2018). https://doi.org/10.1007/978-3-030-00713-3_26

14. Ciolofan, S.N., Militaru, G., Draghia, A., Drobot, R., Drăgoicea, M.: Optimization of water reservoir operation to minimize the economic losses caused by pollution. IEEE Access **6**, 67562–67580 (2018)
15. Aaker, D.A.: Managing assets and skills: the key to a sustainable competitive advantage. California Manag. Rev. **31**(2), 91–106 (1989)
16. Williams, J.R.: How sustainable is your competitive advantage? California Manag. Rev. **34**(3), 29–51 (1992)
17. Jeurissen, R.: Cannibals with forks: the triple bottom line of 21st century business. J. Bus. Ethics **23**(2), 229–231 (2000)
18. Geissdoerfer, M., Vladimirova, D., Evans, S.: Sustainable business model innovation: a review. J. Cleaner Prod. **190**, 712–721 (2018)
19. Archpru Akaka, M., Vargo, S.L., Lusch, R.F.: An exploration of networks in value cocreation: a service-ecosystems view. In: Special Issue - Toward a Better Understanding of the Role of Value in Markets and Marketing, pp. 13–50. Emerald Group Publishing Limited (2012)
20. Vargo, S.L., Akaka, M.A.: Value cocreation and service systems (re) formation: a service ecosystems view. Serv. Sci. **4**(3), 207–217 (2012)
21. Fujita, S., Vaughan, C., Vargo, S.: Service ecosystem emergence from primitive actors in service dominant logic: an exploratory simulation study. In: Proceedings of the 51st Hawaii International Conference on System Sciences (2018)
22. Lusch, R.F., Vargo, S.L.: Service-Dominant Logic: Premises, Perspectives, Possibilities. Cambridge University Press, Cambridge (2014)
23. Vargo, S.L., Maglio, P.P., Akaka, M.A.: On value and value co-creation: a service systems and service logic perspective. Eur. Manag. J. **26**(3), 145–152 (2008)
24. Rennings, K., Wiggering, H.: Steps towards indicators of sustainable development: linking economic and ecological concepts. Ecol. Econ. **20**(1), 25–36 (1997)
25. Vargo, S.L., Akaka, M.A.: Service-dominant logic as a foundation for service science: clarifications. Serv. Sci. **1**(1), 32–41 (2009)
26. Bocken, N.M., Short, S.W., Rana, P., Evans, S.: A literature and practice review to develop sustainable business model archetypes. J. Cleaner Prod. **65**, 42–56 (2014)
27. Fragidis, G., Tarabanis, K.: Towards an ontological foundation of service dominant logic. In: Snene, M., Ralyté, J., Morin, J.-H. (eds.) IESS 2011. LNBIP, vol. 82, pp. 201–215. Springer, Heidelberg (2011). https://doi.org/10.1007/978-3-642-21547-6_16
28. Caridà, A., Edvardsson, B., Colurcio, M.: Conceptualizing resource integration as an embedded process: matching, resourcing and valuing. Market. Theory **19**(1), 65–84 (2019)
29. Russo-Spena, T., Mele, C.: "Five Co-s" in innovating: a practice-based view. J. Serv. Manag. **23**(4), 527–553 (2012)
30. Storbacka, K., Nenonen, S.: Learning with the market: facilitating market innovation. Ind. Market. Manag. **44**, 73–82 (2015)
31. Ng, I., Badinelli, R., Polese, F., Nauta, P.D., Löbler, H., Halliday, S.: SD logic research directions and opportunities: the perspective of systems, complexity and engineering. Market. Theory **12**(2), 213–217 (2012)
32. Sigala, M.: A market approach to social value co-creation: findings and implications from "Mageires" the social restaurant. Market. Theory **19**(1), 27–45 (2019)
33. Aurich, J.C., Fuchs, C., Wagenknecht, C.: Life-cycle oriented design of technical Product-Service Systems. J. Cleaner Prod. **14**(17), 1480–1494 (2006)
34. Fernandez, M., Gómez-Pérez, A., Juristo, N.: Methontology: from ontological art towards ontological engineering. AAAI Technical report SS-97-0 (1997). http://www.aaai.org/Papers/Symposia/Spring/1997/SS-97-06/SS97-06-005.pdf

35. Katzan, H.: Principles of service systems: an ontological approach. J. Serv. Sci. **2**(2), 35–52 (2009)
36. Lyons, K., Tracy, S.: Characterizing organizations as service systems. Hum. Factors Ergon. Manuf. Serv. Ind. **23**(1), 19–27 (2013)
37. Quero, M.J., Ventura, R.: Value proposition as a framework for value cocreation in crowdfunding ecosystems. Market. Theory **19**(1), 47–63 (2019)
38. Pakkala, D., Spohrer, J.: Digital service: technological agency in service systems. In: Proceedings of the 52nd Hawaii International Conference on System Sciences (2019). http://hdl.handle.net/10125/59628
39. Akaka, M.A., Vargo, S.L.: Technology as an operant resource in service (eco) systems. Inf. Syst. e-Business Manag. **12**(3), 367–384 (2014)
40. Smith, A.: An Inquiry into the Nature and Causes of the Wealth of Nations. W. Strahan and T. Cadell, London (1776)
41. Vargo, S.L., Lusch, R.F.: Service-dominant logic: continuing the evolution. J. Acad. Market. Sci. **36**(1), 1–10 (2008)
42. Vargo, S.L., Lusch, R.F.: It's all B2B... and beyond: toward a systems perspective of the market. Ind. Market. Manag. **40**(2), 181–187 (2011)
43. Maglio, P.P., Spohrer, J.: Fundamentals of service science. J. Acad. Market. Sci. **36**(1), 18–20 (2008)
44. Maglio, P.P., Vargo, S.L., Caswell, N., Spohrer, J.: The service system is the basic abstraction of service science. Inf. Syst. e-business Manag. **7**(4), 395–406 (2009)
45. Akaka, M.A., Vargo, S.L., Schau, H.J.: The context of experience. J. Serv. Manag. **26**(2), 206–223 (2015)
46. Prahalad, C.K., Ramaswamy, V.: Co-creation experiences: the next practice in value creation. J. Interact. Market. **18**(3), 5–14 (2004)
47. Mele, C., Spena, T.R., Peschiera, S.: Value creation and cognitive technologies: opportunities and challenges. J. Creating Value **4**(2), 182–195 (2018)
48. Mokyr, J.: The Gifts of Athena: Historical Origins of the Knowledge Economy. Princeton University Press, Princeton (2002)
49. Arthur, W.B.: The Nature of Technology: What it is and How it Evolves. Simon and Schuster, New York (2009)
50. Vargo, S.L.: Situating humans, technology and materiality in value cocreation. J. Creating Value **4**(2), 202–204 (2018)
51. Piscicelli, L., Ludden, G.D., Cooper, T.: What makes a sustainable business model successful? An empirical comparison of two peer-to-peer goods-sharing platforms. J. Cleaner Prod. **172**, 4580–4591 (2018)
52. Zhang, H., Gordon, S., Buhalis, D., Ding, X.: Experience value cocreation on destination online platforms. J. Travel Res. **57**(8), 1093–1107 (2018)
53. Gruber, T.R.: Toward principles for the design of ontologies used for knowledge sharing? Int. J. Hum.-Comput. Stud. **43**(5–6), 907–928 (1995)
54. Normann, R., Ramirez, R.: From value chain to value constellation: designing interactive strategy. Harvard Bus. Rev. **71**(4), 65–77 (1993)
55. Normann, R.: Reframing Business: When the Map Changes the Landscape. Wiley, Hoboken (2001)
56. Spohrer, J.C., Maglio, P.P.: Toward a science of service systems. In: Maglio, P., Kieliszewski, C., Spohrer, J. (eds.) Handbook of Service Science. SSRI, pp. 157–194. Springer, Heidelberg (2010). https://doi.org/10.1007/978-1-4419-1628-0_9
57. Kharrazi, A.: Resilience. In: Fath, B. (ed.) Encyclopedia of Ecology, 2nd edn, pp. 414–418. Elsevier (2019). https://doi.org/10.1016/B978-0-12-409548-9.10751-1

58. Badr, N.G.: Guidelines for health IT addressing the quality of data in EHR information systems. In: 12th International Joint Conference on Biomedical Engineering Systems and Technologies, 22–24 February 2019, Prague, Czech Republic (2019). http://insticc.org/node/TechnicalProgram/biostec/presentationDetails/69410
59. Wang, R.Y., Strong, D.M.: Beyond accuracy: what data quality means to data consumers. J. Manag. Inf. Syst. **12**(4), 5–33 (1996)

Benchmarking the Metabolism of European Union Countries to Promote the Continuous Improvement of Service Ecosystems

Ana Camanho$^{(\boxtimes)}$, Mafalda C. Silva, Isabel M. Horta, and Flávia Barbosa

Faculdade de Engenharia da Universidade do Porto, Porto, Portugal
acamanho@fe.up.pt

Abstract. In recent decades, the concept of urban metabolism has been widely applied at different scales. This paper proposes an optimization model, based on Data Envelopment Analysis, for the evaluation and benchmarking of countries' metabolism. The EU-28 countries are analyzed based on economic and environmental indicators, including the resources consumed (energy and materials) and environmental pressures (GHG emissions and waste) associated with the value-added from the economic activities. The empirical results produced a ranking of countries' based on their metabolic performance underlying the creation of wealth, along with the targets for the countries with lower metabolic performance. This new metabolic approach is a contribution to the design of policies for the promotion of sustainable and resilient services.

Keywords: Countries metabolism · Benchmarking · Data Envelopment Analysis · Directional distance function · Sustainable services

1 Introduction

The importance of efficient, fair and equitable distribution of resources amongst nations has long been emphasized in the international sustainability discourse (WCDE 1987). In the European Union (EU), key sustainable development concerns include the "the fight against climate change and promoting a low-carbon, knowledge-based, resource-efficient economy" (CEC 2009, p.2).

In a time when boosting the European economy is particularly relevant, it is important to safeguard that the economic development does not come at the expense of ecological and social assets, and vice-versa. The achievement of environmental targets should not burden developing economies, and so the efforts need to be fairly allocated. Setting fair goals at a national level is a challenging task. European member states should cut greenhouse gas (GHG) emissions by 20% by 2020 (EC, 2010), while a decrease of 80–95% below reference levels is

© Springer Nature Switzerland AG 2020
H. Nóvoa et al. (Eds.): IESS 2020, LNBIP 377, pp. 319–333, 2020.
https://doi.org/10.1007/978-3-030-38724-2_23

expected by 2050 (EC, 2012). The Effort Sharing Decision sets national targets for cutting non-EU Trading Scheme GHG emissions, accounting for wealth differences throughout EU member states. With regard to waste, targets are equally ambitious. By 2020, 50% of household and similar waste and 70% of construction and demolition waste should be prepared for re-use and recycling (EC, 2008), an effort common to all member states.

It is indisputable that sustainable development depends on resource conservation and efficiency. It is here made the point that setting national targets and commitments should take into account a number of input-output flows on national economies, along with the specific socioeconomic characteristics of countries. For this, we advocate that the social metabolism metaphor could provide a basis for a more integrated and equitable assessment of national economies. In the context of service science, it is essential to be able to measure the utility of resources' consumption in terms of the contribution to economic achievements, and the corresponding environmental impacts. This research contributes to service design and innovation in complex ecosystem, that often face challenges related to the increasing demand for resources. We show that the application of Data Envelopment Analysis could provide useful insights for comparing nations' environmental performance and for assessing the burden-sharing targets defined by existing international environmental policies. This approach bridges the gap between two fields of knowledge that have been kept apart in the scientific literature: the socioeconomic metabolism of territories and the development of sustainable and resilient service ecosystems.

Section 2 reviews the literature, providing the background to introduce the model proposed. Section 3 details the methodology applied in this research. Section 4 presents and discusses the results of the application of the model to EU-28 countries. Section 5 concludes and presents future research directions.

2   A Review of Performance Assessment of Territories

Assessing the performance of territories is a complex theme, covering several scientific areas (e.g., economics, management science, urban planning and environmental studies). This resulted in a diversity of approaches of analysis. This section explores the literature relating to social metabolism and eco-efficiency, as our methodological approach brings together these two research lines.

The Ecological Footprint [23] emerged as an attempt to characterize the human impact on the environment. This was translated in terms of area needed to generate the resources used by a person or community. It raised the concern that wealthy countries used more than a 'fair' share of the planet's carrying capacity. Along with Environmental Footprints, Blanc and Friot [5] suggested three more types of environmental assessment methods that could be applied to territories: Life Cycle Assessments (LCA), Material Flow Analysis (MFA) and Extended Environmental Input-Output Analysis (EEIO). LCA, however, is more often applied to smaller systems like companies or buildings, due to its complexity and data-intensive nature. MFA has been applied at different scales. There is a greater emphasis on the national level, at which data is typically available (e.g. Adriaanse et al. 1997). Nevertheless, applications at regional/local level are also available [10]. Finally, the EEIO is generally applied at country scale and is mostly based on the construction of input-output tables that aim to describe production processes and transactions within a given economy. More recently, Pauliuk et al. [20] proposed a general system structure with a graphic character for describing socioeconomic metabolic flows, comprising existing accounting methods.

The urban metabolism concept emerged from Abel Wolman's study of a hypothetical US-American city with one million inhabitants. He calculated resource and waste flows based on average national production and consumption data [29]. Urban metabolism interprets a city as a black box, requiring inputs to work in order to maintain the well-functioning of vital activities. This process originates outputs which can be either economic value or waste. The urban system is seen under a stocks and flows perspective, requiring a substantial portion of flows to be built and to function (see Schremmer et al. (2011) for an overview of this topic). It is acknowledged that the society should have a circular or closed metabolism, rather than a linear one [1]; (EC, 2015).

Studies of urban metabolism usually aim at analyzing the relative size of flows in and out of a given region [13,24], or to explore the metabolic implications on sustainable development [2]; [3,11]. Newman [19] extended the metabolism concept to include a stronger emphasis on a social component called 'livability'. More recently, urban metabolism has been argued to constitute an urban design tool, with a strategic dimension [12]. The methods most frequently used in these studies include LCA and MFA [26], which enable a detailed evaluation of the material balances (inputs, outputs and throughputs), but provide few insights on the relative performance of urban areas. As argued by Barles [3], it is important to move from this type of analysis by developing new methods that enable comparisons between different geographies. Also, understanding the causes and patterns behind metabolic behavior is a topic deserving further attention.

In parallel, the scope of the metabolic studies has been extended at a national level. Designations range from industrial metabolism (concerning industrial societies) to a wider concept of societal metabolism [28]. Metabolic assessments have ranged from analysis of material throughputs for individual countries [15] to world regions [25]. These studies have either focused on specific resources, like energy [9], or broader sets of resources [8]. Some authors point that the patterns behind material stocks (versus material throughputs) should be carefully considered in sustainability assessments. Whatever the focus of the analysis, methodological approaches adopted at an aggregated level are similar to the ones applied at urban scale. Accounting frameworks are usually adopted, whilst performance comparisons are disregarded due to the difficulties in obtaining comparable data. In Krausmann et al. [14] a global dataset including 175 nations was grouped into six clusters according to their economic development and population density, calling attention to the importance of conducting comparative studies. s the country GDP, and the environmental impacts considered were GHG emissions.

From a technical point of view, the first eco-efficiency studies, based on standard DEA models, investigated the potential for reductions of pollutants and consumption of environmental resources, whilst keeping the levels of outputs (economic value added) unchanged. Later studies enlarged the analysis by considering other directions for improvement. However, none of these studies considered all aspects of the WBCSD [27] definition, which suggests taking into account the resources consumed, including both environmental aspects (e.g. materials, water, energy) and economic aspects. The few studies that attempted to bridge the gap between the urban metabolism and eco-efficiency literatures were conducted by Liu et al. [17,18]. Liu et al. [18] conducted studies that attempted to bridge the gap between the urban metabolism and eco-efficiency literatures. It was explored the evolution of the efficiency of the Xiamen, China from 1985 to 2007. The inputs were water consumption, energy consumption, food requirements and land exploration. The desirable outputs were GDP and social welfare, whereas the environmental impacts considered emissions of SO2, industrial soot emission and industrial wastewater discharge. This was further developed by Liu et al. [17], which studied 30 Chinese urban areas. These studies treated the environmental pollutants as undesirable outputs after taking their reciprocals, and used an input-oriented DEA model to evaluate the potential for reducing resources consumption.

Herein, we propose an enhanced model that considers resource flows, which constitute the basis of urban metabolism studies, alongside economic features of the territories under assessment. The environmental dimension considers two types of variables: inputs flows associated with the resources consumed (energy and materials), and output flows corresponding to the by-products generated (e.g. GHG emissions and waste). The economic dimension is represented by the production value. The metabolic flows and economic development are all measured per capita, such that countries are compared taking into account the demographic features underlying the economic and environmental flows. The optimization model used for benchmarking the performance of territories is based on a directional distance function, such that specific improvement paths can be explored. Our study is primarily focused on the optimization of metabolic flows, involving the reduction of consumption of environmental resources and emission of pollutants.

3   Methodology

The model proposed to measure the metabolism of territories is based on the directional distance function (DDF) model first proposed by Chambers et al. [6]. The DDF model allows choosing the direction of projection of inefficient DMUs towards the frontier. The method applies the input-output framework from metabolic studies to evaluate how efficiently the territories transform the use of resources into economic value, whilst minimizing the environmental impacts. The original formulation of the DDF model is presented in 1.

$$\text{Maximize} \quad \beta \tag{1}$$

$$\text{s.t.} \quad \sum_{j=1}^{n} x_{ij}\lambda_j \leq x_{ij_0} - \beta g_x \quad i = 1,\ldots,m$$

$$\sum_{j=1}^{n} y_{rj}\lambda_j \geq y_{rj_0} + \beta g_y \quad r = 1,\ldots,q$$

$$\lambda_j \geq 0 \quad j = 1,\ldots,n$$

In model (1), x_{ij} ($i = 1,\ldots,m$) are the inputs consumed by unit j ($j = 1,\ldots,n$), and y_{rj} ($r = 1,\ldots,q$) are the outputs produced. Unit j_0 is the unit under assessment (e.g., country, company, school), where x_{ij_0} and y_{rj_0} correspond to the input and the output values of the unit j_0. The components of vector $g = (-g_x, g_y)$ indicate the direction of change for the inputs and outputs, respectively. Positive values of the components of the directional vector g are related to expansions of the original output values and negative values are associated with contractions of the input values. The factor β indicates the extent of units' inefficiency. An inefficiency score equal to zero corresponds to the best performance observed in the sample, whereas values greater than zero signal the existence of performance levels inferior to those observed in peers within the sample. The λ_j ($j = 1,\ldots,n$) are the intensity variables. This optimization model allows constructing a frontier of best practices against which all units are evaluated. The units that define the location of this frontier are considered the benchmarks of the sample. For inefficient units, i.e. those located inside the space enveloped by the frontier, the model identifies the peers that can be used in the search for best practices. These are assigned a value of λ_j greater than zero at the optimal solution of model (1) applied to the unit j_0 under assessment. The identification of targets is determined through a linear combination of the peers identified. These correspond to the left-hand side of the restrictions of model (1) for each indicator considered in the model. This information on the targets can be of utmost importance to managers and administrative authorities for supporting the design of sustainable development policies. As formulation (1) does not specify the values of slacks between the left-hand side and right-hand side of the restrictions, only the radial adjustments suggested by the β values are subject to optimization, and there is no incentive for non-radial improvements beyond these radial adjustments. In our study, we divided the indicators into environmental and economic dimensions, and specified a directional distance function model focusing on improvements to the environmental variables, without worsening the outputs corresponding to the economic dimension. The model specified involves a similar treatment for all indicators that should be minimized, irrespectively of their intrinsic nature being an input (x_{ij}) (e.g., use of resources) or an undesirable output (b_{sj}) (e.g. emissions), as in Färe et al. [7]. Furthermore, through a second stage optimization, we sought for non-radial improvements to the environmental indicators, which involves the incorporation of slacks for these variables in the objective function. The resulting model is shown in (2).

$$\text{Maximize} \quad \beta + \epsilon\left(\sum_{i=1}^{2} s_i + \sum_{s=1}^{2} s_s\right) \tag{2}$$

$$\text{s.t.} \quad \sum_{j=1}^{n} x_{ij}\lambda_j \leq x_{ij_0} - \beta g_x - s_i \quad i = 1, 2$$

$$\sum_{j=1}^{n} y_j\lambda_j \geq y_{j_0}$$

$$\sum_{j=1}^{n} b_{sj}\lambda_j \geq b_{sj_0} - \beta g_b - s_s \quad s = 1, 2$$

$$\lambda_j, s_i, s_s \geq 0 \quad \forall j, i, s$$

In model (2), x_{ij} $(i = 1, 2)$ are the resources consumed (materials per capita and energy consumption per capita), y_j is the production value per capita (output), and b_{sj} $(s = 1, 2)$ are waste production per capita and GHG emissions per capita (undesirable outputs) for the countries analysed $(j = 1, \ldots, n)$. Note that model assumes that the frontier exhibits constant returns to scale following the previous studies on environmental performance assessment [21]; [16]. For this empirical assessment, we specified a directional vector that searches for improvements in the environmental dimension whilst maintaining the economic levels observed in each country. This involves reductions to the use of resources consumed (materials per capita, energy consumption per capita) and levels of by-products produced (waste per capita, emissions per capita). The components of the directional vector were specified as being equal to the values of the resources and undesirable outputs of the country under assessment, i.e. $g = (-g_x, g_y, -g_b) = (-x_{ij_0}, 0, -b_{sj_0})$, in order to enable a proportional interpretation of the potential for improvement in the indicators of countries metabolism. Finally, the value of β constitutes a measure of metabolic performance, with zero corresponding to the best levels and positive scores representing the magnitude of proportional adjustments to the environmental indicators that would be required to reach the best-practice frontier of metabolic performance. Additional information obtained from the model relates to the slack variables $(s_i$ and $s_s)$. These indicate the extent to which the resources consumed, GHG emissions and waste produced could be improved beyond the radial adjustment β required to achieve metabolic performance. If $s_i^* > 0$ or $s_s^* > 0$ for some input i or undesirable output s, the unit under assessment is projected on an inefficient segment of the frontier of best practices, and thus these non-radial adjustments should also be considered in the target setting process.

4 Empirical Analysis of Metabolic Performance

4.1 Variables and Data

The data used in this study refers to EU-28 countries. It was collected from the Eurostat database for the year 2012. In order to illustrate how the metabolism of countries can be compared, it was considered an economic indicator (production value per capita) and four metabolic indicators (materials and energy used, GHG emissions and waste produced, all per capita). The consumption of materials is given by the direct material input (DMI) which corresponds to total material input into the country economy. Energy is expressed by the gross inland energy consumption, reflecting the total demand for primary energy of a country or region. Waste production corresponds to the total solid waste generated by all activities and households. GHG emissions account for all GHG emissions generated, excluding aviation. All these variables were divided by country population. Data on water and wastewater is not available for all EU-28 countries, so this variable could not be included in the model. Concerning the economic indicator, the national production value measures the value actually produced, based on sales (excluding financial and insurance activities). This is a standard variable used to measure the economic development of societies. The values of the variables considered in the assessment are reported in Table 1. Although the data is presented per capita, it is possible to observe that the EU-28 countries are quite heterogeneous concerning the variables analyzed.

4.2 Results and Discussion

This section reports the results obtained in the assessment of the metabolism of the EU-28 countries. The metabolic score (β) measures the proportion by which each country can reduce its metabolic flows (i.e. the use of energy and materials, and the production of waste and emissions) while maintaining its economic status (i.e. the production value with at least the level observed). A score of zero signals an efficient country, in terms of the metabolic components addressed. Table 2 reports the metabolic score (β) and the percentage of reduction for each undesirable output that inefficient countries should pursue (calculated as the target value minus the observed value divided by the observed value). The countries are ordered by their metabolic performance, from the best ($\beta = 0$) to the worst performance levels. The four efficient countries identified are Italy, Sweden, U.K. and Denmark. These countries provide evidence of having the most appropriate levels of metabolic flows (use of resources and environmental impacts), given their economic status.

The metabolically efficient countries are predominantly from Northern latitudes, except for Italy, and correspond to mature economies. These results confirm previous findings from Schandl and West [25], who claim that the best efficiencies are found in more developed contexts, and also from Beltrán-Esteve et al. [4] and Picazo-Tadeo et al. [22] who concluded that newer EU member states are more distant from the efficient frontier.

Table 1. Input data for the DEA model

Country	Materials consumption per capita (thousand tones/population)	Energy consumption per capita (thousand toe/population)	Waste production capita(thousand tones/population)	GHG Emissions per capita (thousand tones/population)	Production value per capita (millions of euros/(population)
Austria	0.0287	0.0040	0.004	0.0095	0.0502
Belgium	0.0311	0.0051	0.0061	0.0105	0.0589
Bulgaria	0.0211	0.0025	0.022	0.0083	0.0086
Croatia	0.0127	0.0019	0.0008	0.0062	0.0121
Cyprus	0.0210	0.0029	0.0024	0.0107	0.0189
Czech Rep.	0.0213	0.0041	0.0022	0.0125	0.0299
Denmark	0.0286	0.0033	0.0029	0.0093	0.0618
Estonia	0.0388	0.0046	0.0166	0.0145	0.0214
Finland	0.0414	0.0063	0.0170	0.0113	0.0480
France	0.0150	0.0040	0.0053	0.0075	0.0339
Germany	0.0212	0.0040	0.0046	0.0117	0.0488
Greece	0.0162	0.0025	0.0065	0.0100	0.0132
Hungary	0.0122	0.0024	0.0016	0.0062	0.0147
Ireland	0.0274	0.0030	0.0029	0.0128	0.0503
Italy	0.0120	0.0027	0.0027	0.0077	0.0365
Latvia	0.0290	0.0022	0.0011	0.0054	0.0138
Lithuania	0.0211	0.0024	0.0019	0.0072	0.0142
Luxembourg	0.0384	0.0085	0.0160	0.0226	0.0908
Malta	0.0151	0.0020	0.0035	0.0075	0.0238
Netherlands	0.0316	0.0049	0.0074	0.0115	0.0520
Poland	0.0205	0.0026	0.0043	0.0105	0.0158
Portugal	0.0191	0.0021	0.0013	0.0065	0.0192
Romania	0.0234	0.0018	0.0133	0.0059	0.0081
Slovakia	0.0175	0.0031	0.0016	0.0079	0.0219
Slovenia	0.0189	0.0034	0.0022	0.0092	0.0245
Spain	0.0120	0.0027	0.0025	0.0073	0.0236
Sweden	0.0318	0.0053	0.0165	0.0061	0.0594
UK	0.0118	0.0032	0.0038	0.0091	0.0372
Mean	0.0228	0.0035	0.0062	0.0095	0.0326
St. deviation	0.0086	0.0015	0.0060	0.0035	0.0205

Whereas the efficiency of UK and Italy seems to have a straightforward explanation - these countries are widely acknowledged world powers - the metabolic efficiency of Denmark and Sweden is likely due to the innovative and competitive nature of their economies, focused on modern technology industries (see for instance the case of Ericsson, a global leader in the ICT market), and to the high share of renewable energy in their energy mix. Conversely, the results from the DDF model indicate that less developed economies, notably those from Eastern Europe, are more material and carbon-intensive, i.e. are using a higher share of resources, with a resulting higher environmental impact in terms of undesirable outputs, in relation the economic value-added achieved. This suggests that as countries development progresses, a proportion of the revenues is reinvested in metabolic performance and environmental efficiency. As such, it is expected that future investments in "greener" technologies will result in important improvements in the metabolic performance of the countries currently considered inefficient. In order to understand the effort that will be required to achieve the best-practice metabolic levels in the different EU countries, Fig. 1 depicts the amounts that should be reduced for each of the four metabolic variables, in tones per capita (i.e. the difference between the current levels observed for each indicator and the targets corresponding to efficient metabolism in each country).

Table 2. Metabolic performance of EU-28 countries

Countries	Metabolic performance (β)	Potential reductions (%)			
		Waste	Materials	Energy	Emissions
Italy	0.0000	0	0	0	0
Sweden	0.0000	0	0	0	0
U.K.	0.0000	0	0	0	0
Denmark	0.0000	0	0	0	0
Ireland	0.1129	19%	16%	11%	41%
Belgium	0.1407	14%	14%	28%	14%
France	0.1413	23%	14%	33%	14%
Germany	0.1516	28%	15%	15%	17%
Luxembourg	0.1728	48%	17%	18%	17%
Austria	0.2031	20%	20%	29%	20%
Netherlands	0.2780	28%	28%	29%	28%
Croatia	0.2984	30%	56%	66%	71%
Spain	0.3074	31%	31%	35%	32%
Malta	0.3082	64%	31%	31%	48%
Portugal	0.3164	32%	54%	52%	56%
Czech Rep.	0.3541	35%	35%	61%	64%
Slovakia	0.3569	36%	42%	62%	58%
Latvia	0.4047	40%	78%	66%	61%
Slovenia	0.4170	42%	42%	58%	56%
Finland	0.4350	52%	43%	43%	43%
Hungary	0.4808	48%	48%	63%	59%
Cyprus	0.5987	60%	60%	63%	72%
Poland	0.6538	81%	65%	65%	76%
Lithuania	0.6553	66%	69%	69%	71%
Greece	0.6742	88%	67%	67%	77%
Estonia	0.7531	94%	75%	75%	78%
Romania	0.7634	97%	84%	76%	80%
Bulgaria	0.8054	98%	81%	81%	83%

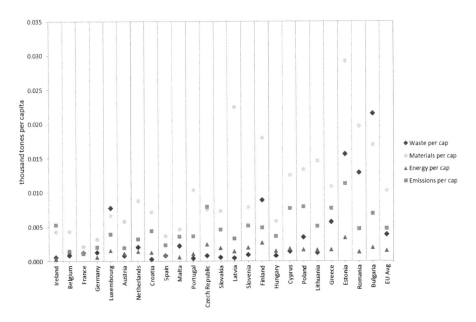

Fig. 1. Reduction required for each metabolic component for inefficient EU countries

Figure 1 shows that even considering per capita values, there is a significant variation between the magnitudes of the potential improvements among the inefficient countries.

Figure 2 provides a complementary insight to Fig. 1. In addition to highlighting, for the different metabolic components, the distance between the current position of each country in relation to the corresponding efficiency targets, it also shows the magnitude (in tons per capita) of resources used and undesirable outputs produced. For instance, comparing the Netherlands to Belgium in term of materials consumption, we can see that the Netherlands needs to reduce a larger amount of materials than Belgium. Nevertheless, the magnitude of the efficient target regarding materials use per capita is larger for Belgium than for the Netherlands. In addition, while some countries require similar reductions for the four components (e.g. Spain), in other cases, the reductions needed for the four metabolic components vary significantly (e.g. Poland).

Finally, Fig. 3 shows the proportion of reduction of metabolic flows for inefficient countries, highlighting the effort that inefficient countries should make in each metabolic dimension. Note that, for each country, the black part of the bars has the same magnitude for the four indicators of metabolism (Fig. 3), and represents the value of inefficiency (value of β obtained using model 2).

The colored parts of the bars represent the amount of slack associated with each indicator (waste, materials, energy and emissions), which represent the non-proportional improvements that are required, beyond the inefficiency score, to operate in the best-practice frontier constructed from what was observed in

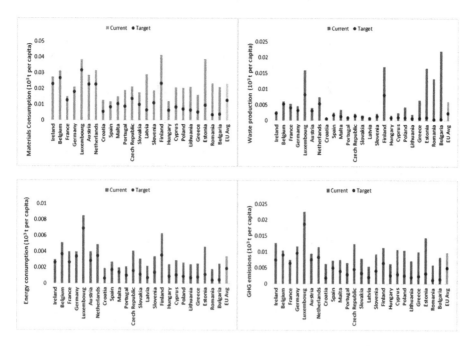

Fig. 2. Current standpoint for each metabolic component and the corresponding target: materials (upper left), waste (upper right), energy (lower left) and GHG emissions (lower right).

benchmark countries. This visualization of metabolic inefficiencies enables to identify where the potential for improvement lies, as well as to prioritize action.

From Fig. 3 it is evident the link between the metabolic input-output pair of energy and GHG emissions. The improvement effort that is required for energy in specific countries is, overall, quite similar to the improvement effort needed for GHG emissions. Nevertheless, this link is not as evident for the metabolic pair materials-waste. In order to further explore potential explanations for the metabolic profiles obtained for the EU countries, additional variables were considered in order to check for correlations between such variables and the metabolic inefficiency (beta). The additional variables considered were GDP per capita, the country population, the share of industry in the value added, and the human development index (HDI). Whereas the GDP returned a statistically significant correlation ($r = -0.620$, p-value 0.0010), the remaining variables evidenced no significant explanatory power. This result indicates that countries with more developed economies (with higher GDP per capita) correspond to those with lower metabolic inefficiencies. The country population does not have a significant effect on metabolic inefficiency, meaning that, at the national level, the metabolism is not influenced by the size of the country. Finally, it seems that the nature of the economic activities of the country (proxied by the industrial share indicator) and the levels of development in terms of

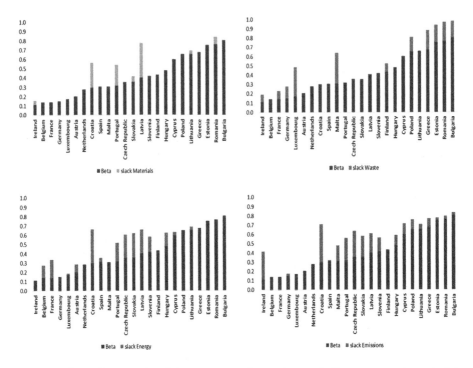

Fig. 3. Effort of proportional improvement for each metabolic component: materials (upper left), waste (upper right), energy (lower left) and GHG emissions (lower right). (Color figure online)

economic wellbeing, health services and education (proxied by the HDI) are not drivers of enhanced metabolic performance. Given these results, it is expected that the existing differences are a consequence of differences in lifestyles, regulatory conditions or environmental awareness leading to the take-up of more efficient technologies at national level. The model presented in this paper enables the comparison and benchmarking of countries metabolism. It accounts for the economic characteristics of a nation when determining the levels of inputs an outputs that would be appropriate for efficient metabolism levels. The results obtained are encouraging, and the relevance of the methodology proposed for decision-support in international policy was verified. This study is a pioneer exploratory application of DEA to the social metabolism field, whose results are aligned with previous research, confirming that the economic development leads to more efficient metabolism.

5   Conclusions

This study offers a new benchmarking approach that enables the assessment of the metabolic performance of different countries using a relatively low-time and low-data consuming process. It advances the state of the art by bringing to the social metabolism field, often grounded on descriptive and qualitative assessments, an analytical tool for service design and innovation tailored for complex ecosystems.

This article contributes to the literature in two ways. First, it extends previous national eco-efficiency studies by including a set of indicators aligned with the metaphor of social metabolism. It is targeted at minimizing both the amount of resources used and the pollutants produced, while simultaneously considering economic flows. Second, it provides a sense of performance, which has been lacking in metabolic approaches (either at urban and country scale). It represents an alternative method of analysis to conventional MFAs, with a strong strategic nature.

The empirical analysis of the metabolic performance of the EU-28 countries included indicators measuring inputs (resources) and output flows (production value, GHG emissions and waste) and has proven to be successful towards relative performance assessments inspired in eco-efficiency studies. This procedure contributes to the sustainable development of service ecosystems by promoting continuous improvement through learning from best-practices. The modelling results provide insights on possible resource inefficiencies, given the specific economic conditions of each country. Nevertheless, the causes for inefficiencies have to be carefully explored. They may be attributed to differences in lifestyles or take-up of 'greener' technologies. In addition, the empirical analysis provided a quantification of the potential for improvement, which may contribute to guide the process of setting international targets for specific metabolic flows.

Further research is needed for consolidating the potential of DEA to assess the metabolism of nations, as well as guiding, in an integrated and equitable way, the definition of targets for specific metabolic flows. Since targets are an important policy instrument for keeping track of the sustainable development agenda, this is certainly a field of research deserving further attention that could contribute to fine-tune policy-making in the future. Future research pathways could include other world regions, promoting learning from cross comparisons, or considering additional indicators to extend the assessment of social metabolism. It would also be interesting to test the suitability of this model at urban scale, although data availability could be a constraint.

DEA can provide a comprehensive measure of relative performance, which can contribute to the design of enhanced public policies. It can also be used as a tool for monitoring the compliance of countries in terms of environmental impact levels agreed in international forums. However, due to the non-parametric nature of DEA models, measurement error in the data sets is not accounted for in the efficiency analysis. Therefore, the results are sensitive to the quality of input and output data collected from international databases. Future research can extend the approach proposed in this paper, developing robust analysis using partial frontiers and conditional models for the evaluation of performance in ecosystem services.

References

1. Ayies, R.U., Ayres, L.W., Klöpffer, W.: Industrial ecology: towards closing the material cycle. Int. J. Life Cycle Assess. **2**(3), 154 (1997)
2. Baccini, P.: A city's metabolism: towards the sustainable development of urban systems. J. Urban Technol. **4**(2), 27–39 (1997). https://doi.org/10.1080/10630739708724555
3. Barles, S.: Society, energy and materials: the contribution of urban metabolism studies to sustainable urban development issues. J. Environ. Plan. Manag. **53**(4), 439–455 (2010). https://doi.org/10.1080/09640561003703772
4. Beltrán-Esteve, M., Gómez-Limón, J.A., Picazo-Tadeo, A.J., Reig-Martínez, E.: A metafrontier directional distance function approach to assessing eco-efficiency. J. Prod. Anal. **41**(1), 69–83 (2014). https://doi.org/10.1007/s11123-012-0334-7
5. Blanc, I., Friot, D.: Evaluation of Environmental Accounting Methodologies for the assessment of global environmental impacts of traded goods and services., Ph.D. thesis, Mines ParisTech (2010)
6. Chambers, R.G., Chung, Y., Färe, R.: Benefit and distance functions. J. Econ. Theory **70**(2), 407–419 (1996). http://www.sciencedirect.com/science/article/pii/S0022053196900964
7. Färe, R., Grosskopf, S., Pasurka, C.A.: Potential gains from trading bad outputs: the case of u.s. electric power plants. Resource Energy Econ. **36**(1), 99–112 (2014). http://www.sciencedirect.com/science/article/pii/S0928765513000791
8. Haas, W., Krausmann, F., Wiedenhofer, D., Heinz, M.: How circular is the global economy?: an assessment of material flows, waste production, and recycling in the european union and the world in 2005. J. Ind. Ecol. **19**(5), 765–777 (2015). https://onlinelibrary.wiley.com/doi/abs/10.1111/jiec.12244
9. Haberl, H., et al.: The energetic metabolism of the european union and the united states: decadal energy input time-series with an emphasis on biomass. J. Ind. Ecol. **10**(4), 151–171 (2006). https://onlinelibrary.wiley.com/doi/abs/10.1162/jiec.2006.10.4.151
10. Hendriks, C., Obernosterer, R., Müller, D., Kytzia, S., Baccini, P., Brunner, P.H.: Material flow analysis: a tool to support environmental policy decision making. case-studies on the city of vienna and the swiss lowlands. Local Environ. **5**(3), 311–328 (2000). https://doi.org/10.1080/13549830050134257
11. Kennedy, C., Cuddihy, J., Engel-Yan, J.: The changing metabolism of cities. J. Ind. Ecol. **11**(2), 43–59 (2007). https://onlinelibrary.wiley.com/doi/abs/10.1162/jie.2007.1107
12. Kennedy, C., Pincetl, S., Bunje, P.: The study of urban metabolism and its applications to urban planning and design. Environ. Poll. 159(8), 1965–1973 (2011). Selected papers from the conference Urban Environmental Pollution: Overcoming Obstacles to Sustainability and Quality of Life (UEP2010), 20–23 June 2010, Boston, USA. http://www.sciencedirect.com/science/article/pii/S0269749110004781
13. Kimberley Warren-Rhodes, A.K.: Escalating trends in the urban metabolism of hong kong 1971–1997. AMBIO: J. Hum. Environ. **30**(7), 429–438-10 (2001)
14. Krausmann, F., Fischer-Kowalski, M., Schandl, H., Eisenmenger, N.: The global sociometabolic transition. J. Ind. Ecol. **12**(5–6), 637–656 (2008). https://onlinelibrary.wiley.com/doi/abs/10.1111/j.1530-9290.2008.00065.x
15. Krausmann, F., Gingrich, S., Nourbakhch-Sabet, R.: The metabolic transition in japan. J. Ind. Ecol. **15**(6), 877–892 (2011). https://onlinelibrary.wiley.com/doi/abs/10.1111/j.1530-9290.2011.00376.x

16. Kuosmanen, T., Kortelainen, M.: Measuring eco-efficiency of production with data envelopment analysis. J. Ind. Ecol. **9**(4), 59–72 (2005). https://onlinelibrary. wiley.com/doi/abs/10.1162/108819805775247846

17. Liu, Y., Song, Y., Arp, H.P.: Examination of the relationship between urban form and urban eco-efficiency in china. Habitat Int. **36**(1), 171–177 (2012). http://www.sciencedirect.com/science/article/pii/S0197397511000609

18. Liu, Y., Wang, W., Li, X., Zhang, G.: Eco-efficiency of urban material metabolism: a case study in xiamen, china. Int. J. Sustain. Dev. World Ecol. **17**(2), 142–148 (2010). https://doi.org/10.1080/13504501003603223

19. Newman, P. W. G.: Sustainability and cities: extending the metabolism model (1999)

20. Pauliuk, S., Majeau-Bettez, G., Müller, D.B.: A general system structure and accounting framework for socioeconomic metabolism. J. Ind. Ecol. **19**(5), 728–741 (2015). https://onlinelibrary.wiley.com/doi/abs/10.1111/jiec.12306

21. Picazo-Tadeo, A.J., Beltrán-Esteve, M., Gómez-Limón, J.A.: Assessing eco-efficiency with directional distance functions. Eur. J. Oper. Res. **220**(3), 798–809 (2012). http://www.sciencedirect.com/science/article/pii/S0377221712001579

22. Picazo-Tadeo, A.J., Castillo-Giménez, J., Beltrán-Esteve, M.: An intertemporal approach to measuring environmental performance with directional distance functions: Greenhouse gas emissions in the european union. Ecol. Econ. **100**, 173–182 (2014). http://www.sciencedirect.com/science/article/pii/S0921800914000408

23. Rees, W.E.: Ecological footprints and appropriated carrying capacity: what urban economics leaves out. Environ. Urban. **4**(2), 121–130 (1992). https://doi.org/10. 1177/095624789200400212

24. Rosado, L., Niza, S., Ferrão, P.: A material flow accounting case study of the lisbon metropolitan area using the urban metabolism analyst model. J. Ind. Ecol. **18**(1), 84–101 (2014). https://onlinelibrary.wiley.com/doi/abs/10.1111/jiec.12083

25. Schandl, H., West, J.: Resource use and resource efficiency in the asia-pacific region. Glob. Environ. Change **20**(4), 636–647 (2010). 20th Anniversary Special Issue. http://www.sciencedirect.com/science/article/pii/S0959378010000592

26. Suh, S.: Handbook of Input-Output Economics in Industrial Ecology, vol. 23. Springer, Heidelberg (2009). https://doi.org/10.1007/978-1-4020-5737-3

27. WBCSD: Eco-efficiency: Creating More Value with Less Impact, 1st ed. World Business Council for Sustainable Development, Geneva (2000)

28. Weisz, H., Schandl, H.: Materials use across world regions: inevitable pasts and possible futures. J. Ind. Ecol. **12**(5–6), 629–636 (2008)

29. Wolman, A.: The metabolism of cities. Sci. Am. **213**(3), 178–193 (1965)

Ten Years Exploring Service Science: Looking Back to Move Forward

Jorge Grenha Teixeira[✉], Vera Miguéis, Marta Campos Ferreira,
Henriqueta Nóvoa, and João Falcão e Cunha

Faculty of Engineering, University of Porto and INESC TEC,
Rua Dr. Roberto Frias, s/n, 4200-465 Porto, Portugal
jteixeira@fe.up.pt

Abstract. In celebration of the 10th anniversary of the International Conference on Exploring Service Science (IESS), this paper takes a historical look at the papers that have been published in the IESS proceedings. The analysis is focused on the development and evolution of the IESS community and of the main research topics covered by the published papers over time. The IESS community is portrayed in terms of authors, their affiliations and co-authoring network, while the topics are analyzed according to the papers' keywords. Moreover, this paper analyzes the impact of the papers published in this decade, in terms of citations. These results are then discussed in light of the observed trends and of the evolution of the service science field, to guide the future development of the IESS conference and of research on service science.

Keywords: Service science · IESS conference · Research agenda

1 Introduction

Service science has had a remarkable development since initial calls to action for its creation, more than ten years ago [1]. These calls emphasized the importance of services in the modern economies and the comparative lack of academic, business, and policy interest in service, with research efforts being dispersed in isolated silos [1, 2]. Since then, the service science field started to take shape by building a significant body of research, defining its main constructs, developing dedicated research outlets and organizing a service science research community.

From a conceptual point-of-view, service science is now well-established and defined. Service science is defined as a combination of organization and human understanding with business and technology understanding to study service systems, thus creating a basis for systematic service innovation [3, 4]. Service science key constructs are service-dominant logic [5, 6] and service systems [7, 8] that are combined with theories and methods drawn from service-related fields such as, marketing, management, engineering, design, operations, computer science, psychology, among others [3, 9]. Service-dominant logic provides the perspective and vocabulary for service science, and the service system construct provides a basis for modeling interactions among entities [3], being defined as a "dynamic value-cocreation configuration of resources, including people, organizations, shared information (language, laws,

© Springer Nature Switzerland AG 2020
H. Nóvoa et al. (Eds.): IESS 2020, LNBIP 377, pp. 334–346, 2020.
https://doi.org/10.1007/978-3-030-38724-2_24

measures, methods), and technology, all connected internally and externally to other service systems by value propositions" [7].

Along with the conceptual development of the field, building a research community was identified early on as a priority to ensure service science sustainability as a research field [1]. Several efforts have been successful regarding the creation and development of such a community. The Service Science journal recently celebrated its tenth anniversary, having published more than 200 research papers in a broad range of topics related to service science and getting increasingly more submissions and impact [10]. The Handbook of Service Science, first published in 2010 [11] is a cornerstone for the field by clarifying the definition, role, and future of service science from early on. The second volume of this publication further consolidates and evolves the field by emphasizing the human side of service systems, the networked nature of service experience and the broad scope of service ecosystem [12]. Moreover, the Handbook of Service Science is part of a series on "*Service Science: Research and Innovations in the Service Economy*" that regularly publishes research on service science-related topics. Furthermore, service science is frequently discussed and published in other service research journals [13–15]. However, the development of a research field also requires dedicated forums, where the community can meet regularly to share and discuss their latest research developments. One of such forums has accompanied the development of service science almost since its inception: the International Conference on Exploring Service Science (IESS).

At the time of this publication, the IESS conference is celebrating its tenth anniversary, as the first conference occurred in 2010 in Geneva. This memorable mark enables the always fruitful exercise of looking back and analyzing the history and contribution of the IESS conference for deriving prospective paths and trends for future research.

First, it is important to briefly understand the origins and the vision that guided the creation of the IESS conference. In fact, the initial meetings organized by IBM, nearly 15 years ago, sponsoring the emergent field of service science, gathered an important number of researchers from a large span of universities and research institutions across the world. Particularly, the University of Geneva, the Public Research Centre Henri Tudor from Luxembourg and the University of Porto immediately understood the need to proactively work together in order to build a coherent set of skills for this emergent field, as well as developing appropriate learning paths for professionals. This awareness led to the creation of a master program in Service Engineering and Management by the University of Porto, a pioneer project back in 2007.

Other important outcomes of this initial stimulus consubstantiated in two major initiatives: (1) the European project DELLIISS (DEsigning Lifelong Learning for Innovation in Information Services Science, www.delliiss.eu), followed by the (2) IESS conference, that started immediately as the project ended. The DELLIISS project was funded through the ERASMUS Lifelong Learning Program run by the EU Education and Culture Directorate General (DG). DELLIISS, as the other projects of this Program, had the objective of designing new education curricula aiming at filling the skills gap identified by enterprises [16]. In particular, the main goal of the project was to establish a one-year full-time equivalent Executive Master degree in Innovative Service Systems (EMISS) targeting professional people. The DELLIISS project enabled an important impetus for the Service Science field in Europe, involving a systematic and exhaustive process of identifying industry requirements, based on Think Tanks held around Europe during one year, as well as the development of a knowledge map, that structured the state-of-the-art on ICT, service science, and innovation scientific domains at the time. This jointed effort played an important role to gather and foster a critical mass of service science researchers that afterwards led to a second initiative, the IESS conference.

This paper reviews the first ten years of the IESS conference, analyzing the 258 papers that were published in the conference proceedings [17–25] aiming at characterizing this service science community as well as identifying the main topics addressed by published papers and their evolution, and seeking to understand trends and emerging areas that can guide future research on service science.

In the next sections the paper methodology is described, followed by the results of the analysis. Based on these results, a research agenda for service science is proposed in the final section of the paper.

2 Methodology

From 2010 and 2018, nine editions of the IESS conference have been organized, across five different European countries (Switzerland, Portugal, Romania, Italy and Germany), and 258 papers have been published in the conference proceedings. Table 1 shows the number of papers published per year and the country where the conference was organized. The average number of papers published by edition is 29, although there are some editions whose number of papers is significantly different, i.e. 57 in 2016 and 10 in 2014.

Table 1. Number of papers published in IESS conference proceedings per year.

Year	Organizing country	Nr of papers published
2010	Switzerland	27
2011	Switzerland	19
2012	Switzerland	28
2013	Portugal	28
2014	Switzerland	10
2015	Portugal	27
2016	Romania	57
2017	Italy	33
2018	Germany	29

To analyze the 258 papers published in the IESS conference proceedings since 2010, several steps were performed. The first phase of this process consisted on collecting detailed information about each paper, namely: title, authors, affiliations and country, abstract, keywords, and number of citations in Google scholar. Afterwards, this data was processed in Excel, RapidMiner, and R software, which enabled a descriptive analysis of the data gathered.

This analysis covered three aspects: authors, research topics and impact. First, regarding the authors of the papers published, the analysis involved a ranking of the most prolific authors in the IESS conference, the distribution by countries of origin and the development of a network graph of the service science community which has published in IESS over the years. The representation of the IESS community network was based on the existence of papers in co-authorship. This means that a link was established among those who are co-authors of at least one paper.

Second, the analysis of the topics covered by all the papers was sustained by the keywords provided in each one. A word cloud representing these keywords was mapped for each IESS edition. Only the keywords appearing at least twice are considered and the keyword "service science" is omitted due to its ubiquity.

Finally, regarding impact analysis, the most cited papers of the IESS conference were identified, and the distribution of citations across IESS editions was analysed.

3   Results

This section includes the results of the analysis performed. Initially we present some indicators regarding the authors. This is followed by the presentation of the main topics covered by the papers over time. In the end, the results are focused on the impact of the papers published in IESS conference.

3.1   Authors

In the last ten years some authors stand out from the vast number of authors published in the IESS proceedings (507 in total), due to the substantial and continuous contribution over the years. Figure 1 shows the number of papers published by the authors who have published five or more papers in the IESS proceedings. These represent a list of experts in the field who made significant contributions to the IESS conference over this decade (Table 2).

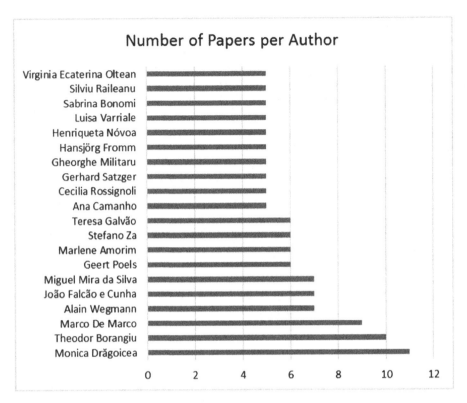

Fig. 1. Number of papers published in IESS conference proceedings by the authors with 5 or more papers.

Finally, the analysis of the data collected also enabled the development of the network graph presented in Fig. 2, based on the co-authorship of the papers published. In order to simplify the graph, we have decided to exclude from the plot those sub-networks which contained only 5 or less authors. This means that for example, a group of 4 authors, who have shared some work but did not collaborate with other IESS authors was disregarded. From a total of 115 sub-groups, we ended up with 17 larger sub-groups. There is one group with 80 authors and another with 40, while the remaining are smaller, including 6-11 authors. The largest group (in the central upper area of Fig. 2) includes authors from Portugal, Romania, and Italy, who published together. The second-largest group (in the upper left side of the graph) includes almost exclusively authors from Germany. Despite these larger groups, the network graph shows a varied, but dispersed community (Table 3).

Table 2. Authors' affiliations per country (2% or more).

Country	Authors' affiliations (number)	Authors' affiliations (%)
Portugal	129	17%
Germany	112	15%
Italy	111	15%
Romania	110	15%
Switzerland	47	6%
Luxemburg	24	3%
Belgium	20	3%
Spain	18	2%
France	17	2%
Austria	15	2%
Netherlands	15	2%
Canada	13	2%
Philippines	12	2%

Table 3. Contribution of the top 5 countries per year.

Country	2010	2011	2012	2013	2014	2015	2016	2017	2018
Portugal	1%	**17%**	10%	**51%**	8%	23%	14%	12%	14%
Germany	**14%**	0	1%	14%	**21%**	14%	4%	19%	**51%**
Italy	4%	0	**23%**	0	11%	**24%**	21%	**26%**	7%
Romania	0	0	0	9%	**21%**	14%	**33%**	20%	8%
Switzerl.	13%	13%	17%	4%	5%	1%	0	11%	3%
Total	32%	30%	51%	78%	66%	76%	72%	88%	83%
Organiz. country	Swit.	Swit.	Swit.	Port.	Swit.	Port.	Rom.	Italy	Ger.

3.2 Topics Covered

The topics covered by the papers published in the IESS conference proceedings were explored through the analysis of the keywords using text analytics. The keywords of each paper were identified and then the number of times each keyword appeared in each year edition was calculated. A word cloud was then generated for each year, as shown in Fig. 3. The larger the word size, the more often it appears in relation to other words. These word clouds include the words which were used at least in two different papers. It is relevant to highlight that, in 2014, only one keyword, "service system", was repeated. As shown in Fig. 1, this edition had only 10 papers published.

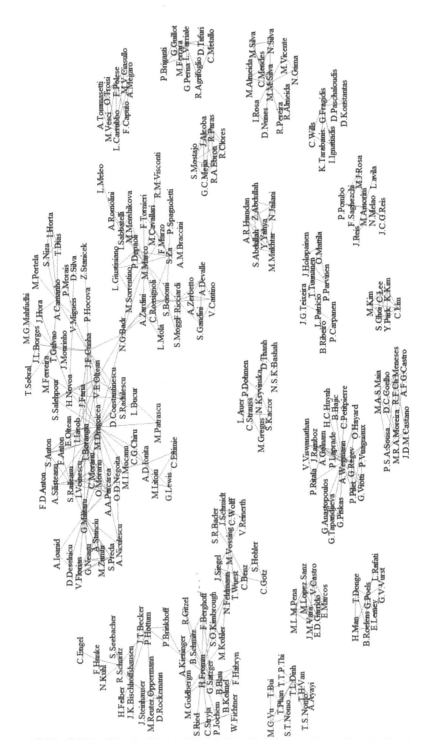

Fig. 2. Network graph of the authors who published in the IESS conference.

Fig. 3. Word clouds of the keywords in IESS proceedings.

The word "service science" was also excluded as it is at the heart of almost every edition of the conference and obscured the other keywords. The overwhelming prevalence of this keyword shows a good alignment between the published papers and the conference research field.

When comparing these word clouds, it is possible to infer some patterns and trends. "Service design" and "service system" have also been a constant in almost every edition. "Service innovation" had a strong presence in the early years of the conference, having regained prominence again in 2016 and 2017.

This conference has also been marked by a strong information systems component, having evolved from concepts such as "Web Services" (2010 and 2011), and "Social Network" (2010 and 2017) to concepts such as "Big Data" (2016 onwards). The concept of "Value Co-creation" [26] starts appearing in 2015 onwards, while "Service-Dominant Logic" [5, 6, 27], somewhat unexpectedly, only shows up in two editions (2012, 2016). The word clouds also highlight the importance of several service sectors such as healthcare (from 2015 onwards), education (2012, 2016 and 2017), transportation (2016) and creative industries (2017).

3.3 Impact

One of the most commonly used, though imperfect, metric to infer the impact of a paper is the number of citations that a paper receives [28]. This may be an imperfect measure given the time lag between publication date and citations. Longer published papers tend to have more citations simply because they are available for a longer period. The analysis of the impact of the IESS conference consisted of collecting the Google Scholar citations for each paper to uncover the most impactful IESS papers and editions. Google Scholar was chosen because it is the multidisciplinary database with the broader coverage [28].

The impact of the papers overtime is shown in Fig. 4. The median of the number of citations of the first three IESS editions papers is the same, i.e. 4 citations. The papers from 2013 and 2014 present a slight increase in the median number of citations, which was followed by a gradual decrease, for those published in the following years. There is some variance on the number of citations of the papers published in each edition, especially regarding the papers published in 2013, year in which the third and fourth quartiles are the highest. Concerning the outliers, 2012 is the year with the highest number of papers presenting an atypical number of citations. Three of them present more than 22 citations. In 2016 and 2017, two papers are considered outliers in terms of number of citations. It is worth to highlight that one of these outliers, published in 2017, is the most cited paper of all time on IESS, with 71 citations. In contrast, in all IESS editions, with the exception of the 2014 edition, there are also papers without any citation. In 2016 and 2018 the number of cited papers that were not cited is particularly high, which may be a result of how recent is this publication.

Finally, Table 4 shows the ten most cited papers in recent years. Although, the most cited papers are the ones from earlier editions, the most cited paper is quite recent (2017).

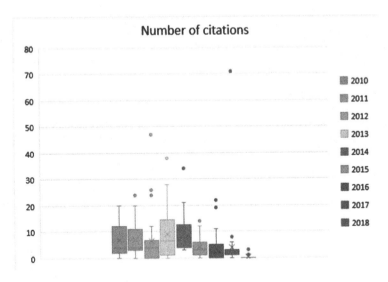

Fig. 4. Number of citations per paper in each IESS edition.

Table 4. Ten most cited papers.

Year	Number of citations	Authors	Title
2017	71	Stefan Seebacher and Ronny Schüritz	Blockchain Technology as an Enabler of Service Systems: A Structured Literature Review
2012	47	Sybren de Kinderen, Khaled Gaaloul and H. A. Erik Proper	Integrating Value Modelling into ArchiMate
2013	38	Rafael Almeida, Rúben Pereira and Miguel Mira da Silva	IT Governance Mechanisms: A Literature Review
2014	34	Muriel Foulonneau, Sébastien Martin and Slim Turki	How Open Data Are Turned into Services?
2013	28	Rui Alves and Nuno Jardim Nunes	Towards a Taxonomy of Service Design Methods and Tools
2012	26	Roberta Ferrario and Nicola Guarino	Commitment-Based Modeling of Service Systems
2012	24	Francesca Ricciardi and Marco De Marco	The Challenge of Service Oriented Performances for Chief Information Officers
2011	24	Natalia Kryvinska, Lukas Auer and Christine Strauss	An Approach to Extract the Business Value from SOA Services
2013	23	Axel Kieninger, Florian Berghoff, Hansjörg Fromm and Gerhard Satzger	Simulation-based Quantification of Business Impacts Caused by Service Incidents
2016	22	Maria Conceição Portela and Ana S. Camanho	Performance Assessment of Portuguese Secondary Schools

4 Discussion and Future Research Agenda

The analysis of the 258 papers published on the first ten years of the IESS conference opens the ground for several reflections and sets the path for the future development of the service science field. Following the structure of the results, this discussion covers three main aspects: authors, topics and impact. While the first of these aspects, authors, addresses more specifically the IESS conference, the other two, topics and impact, can have broader implications to the service science field.

Regarding the authors, results have shown a core of contributing authors and research teams from five different countries (Portugal, Italy, Germany, Romania and Switzerland). These are also the five countries that hosted the conference so far. These results highlight the dedicated community built around IESS that constitute a major driving force for the conference success throughout the years. However, while such a cohesive core of contributors helps the resilience of the conference, the results also show an increasing concentration on the five countries mentioned. Also, a great majority of published papers are historically from European origin. As such, there is here the opportunity to leverage the dedicated core of contributors and expand the conference to other geographies, from Europe and beyond. For example, Scandinavian countries and the United States have a strong service research tradition, with several service research centers [14], that can be tapped to enrich IESS. Furthermore, Fig. 2 shows once again strong collaboration between research teams from Portugal, Italy and Romania, and within Germany, but scarce to non-existent interaction with the other research teams. One path to expand the range of contributions to the IESS conference can be for the established research hubs to establish further connections with these more dispersed research teams.

Regarding the analysis of the topics, it is first important to highlight how the IESS conference is strongly anchored on core service science related constructs, with service science being prevalent throughout the years, and service system being highly relevant in most editions. The analysis of the topics also positions service science and the IESS conference as an improvement-oriented, problem-solving or solution-oriented field, with keywords such as "service innovation", "innovation", or "service design" being relevant. From a service-dominant logic perspective, service innovation is defined as the rebundling of diverse resources that are beneficial to some actors, or to a network of actors [29]. Service innovation can also be defined as the creation of new and/or improved service offerings, service processes, and service business models through complex adaptive combinations of people, technology, processes, and information [15]. Finally, building on Herbert Simon's definition of design [30], i.e. devising courses of action aimed at changing existing situations into preferred ones, service design is the field that brings service strategy and innovative service ideas to life [15]. This improvement orientation makes service science well-positioned to tackle highly relevant service research priorities [14] such as stimulating service innovation, leveraging service design and improving societal well-being through transformative service research [31]. The analysis of the topics also shows a strong connection between service science and technology and information systems related topics, such as "social networks" or "big data". Once again, this characteristic is highly aligned with service research priorities such as leveraging technology and big data to advance service. With this analysis it can be argued that the improvement and technology-orientation of service science makes it well positioned to address many relevant service research priorities. Service science researchers can further leverage these topics by using dedicated problem-solving methodologies coming from information systems, such as Design Science Research (DSR) [32, 33], or Action Design Research [34]. DSR is dedicated to understanding and solving real-world classes of problems through building and evaluating artifacts [32, 35]. DSR is increasingly used in service research and service design [36–39] making it suitable for future service science research. Furthermore, several technology-related service research topics are arising, such as artificial intelligence [40], service robots [41, 42] and the Internet-of-Things [43] that seem to hold relevant research opportunities for future research.

Finally, regarding impact, while a more robust analysis did not offer strong enough results, the list of the top ten published papers (Table 4) highlights the strong impact of literature reviews and related efforts (paper 1, 3 and 5). Prescriptive-driven research (paper 2, 4, 6 and 8), that shows how to achieve a certain outcome, also seems to have high citation potential. As can also be seen in Table 4, technology, information technology and information systems topics seem to attract significant interest. As service science expands and consolidates its position as a research field, systematic ways of organizing literature are increasingly relevant. On the other hand, and bridging with the topics discussed, prescriptive-driven can be paired with improvement-oriented research and methodologies such as DSR, to develop knowledge that explains how to achieve a certain outcome. With the world facing important societal challenges, this can increase service science research impact.

Overall, as IESS reaches its tenth edition, this paper shows that the conference has a solid body of contributing authors that produce impactful research, oriented to change and improve the world. Hopefully, this paper can also guide future research for ten more years of exploring service science.

References

1. Chesbrough, H., Spohrer, J.C.: A research manifesto for services science. Commun. ACM **49**(7), 35–40 (2006)
2. ifM and IBM: Succeeding through Service Innovation: A Discussion Paper. University of Cambridge Institute for Manufacturing, Cambridge, UK (2008)
3. Maglio, P.P., Kieliszewski, Cheryl A., Spohrer, J.C., Lyons, K., Patrício, L., Sawatani, Y.: Introduction: why another handbook? In: Maglio, P.P., Kieliszewski, Cheryl A., Spohrer, J. C., Lyons, K., Patrício, L., Sawatani, Y. (eds.) Handbook of Service Science, Volume II. SSRISE, pp. 1–9. Springer, Cham (2019). https://doi.org/10.1007/978-3-319-98512-1_1
4. Maglio, P.P., Spohrer, J.: Fundamentals of service science. J. Acad. Mark. Sci. **36**(1), 18–20 (2008)
5. Vargo, S.L., Lusch, R.F.: Service-dominant logic: continuing the evolution. J. Acad. Mark. Sci. **36**(1), 1–10 (2008)
6. Vargo, S.L., Lusch, R.F.: Evolving to a new dominant logic for marketing. J. Mark. **68**(1), 1–17 (2004)
7. Maglio, P.P., et al.: The service system is the basic abstraction of service science. Inf. Syst. E-Bus. Manag. **7**, 395–406 (2009)
8. Spohrer, J.C., et al.: Steps toward a science of service systems. IEEE Comput. **40**(1), 71–77 (2007)
9. Spohrer, J., Kwan, S.K.: Service science, management, engineering, and design (SSMED): an emerging discipline–outline and references. Int. J. Inf. Syst. Serv. Sect. **1**(3), 1 (2009)
10. Maglio, P.P.: Toward ten years of service science. Serv. Sci. **9**(4), iii–iv (2017)
11. Maglio, P.P., Kieliszewski, C.A., Spohrer, J.C.: Handbook of Service Science (2010)
12. Maglio, P.P., et al.: Handbook of Service Science, vol. II. Springer, Heidelberg (2018)
13. Mahr, D., Kalogeras, N., Odekerken-Schröder, G.: A service science approach for improving healthy food experiences. J. Serv. Manag. **24**(4), 435–471 (2013)
14. Ostrom, A.L., et al.: Service research priorities in a rapidly changing context. J. Serv. Res. **18** (2), 127–159 (2015)
15. Ostrom, A.L., et al.: Moving forward and making a difference: research priorities for the science of service. J. Serv. Res. **13**(1), 4–36 (2010)
16. Dubois, E., Cunha, J.F.e., Leonard, M.: Towards an executive master degree for the new job profile of a service systems innovation architect. In: 2011 Annual SRII Global Conference (2011)
17. Morin, J.-H., Ralyté, J., Snene, M. (eds.): IESS 2010. LNBIP, vol. 53. Springer, Heidelberg (2010). https://doi.org/10.1007/978-3-642-14319-9
18. Snene, M., Ralyté, J., Morin, J.-H. (eds.): IESS 2011. LNBIP, vol. 82. Springer, Heidelberg (2011). https://doi.org/10.1007/978-3-642-21547-6

19. Snene, M. (ed.): IESS 2012. LNBIP, vol. 103. Springer, Heidelberg (2012). https://doi.org/10.1007/978-3-642-28227-0

20. Falcão e Cunha, J., Snene, M., Nóvoa, H. (eds.): IESS 2013. LNBIP, vol. 143. Springer, Heidelberg (2013). https://doi.org/10.1007/978-3-642-36356-6

21. Snene, M., Leonard, M. (eds.): IESS 2014. LNBIP, vol. 169. Springer, Cham (2014). https://doi.org/10.1007/978-3-319-04810-9

22. Nóvoa, H., Drăgoicea, M. (eds.): IESS 2015. LNBIP, vol. 201. Springer, Cham (2015). https://doi.org/10.1007/978-3-319-14980-6

23. Borangiu, T., Drăgoicea, M., Nóvoa, H. (eds.): IESS 2016. LNBIP, vol. 247. Springer, Cham (2016). https://doi.org/10.1007/978-3-319-32689-4

24. Za, S., Drăgoicea, M., Cavallari, M. (eds.): IESS 2017. LNBIP, vol. 279. Springer, Cham (2017). https://doi.org/10.1007/978-3-319-56925-3

25. Satzger, G., Patrício, L., Zaki, M., Kühl, N., Hottum, P. (eds.): IESS 2018. LNBIP, vol. 331. Springer, Cham (2018). https://doi.org/10.1007/978-3-030-00713-3

26. Vargo, S.L., Maglio, P.P., Akaka, M.A.: On value and value co-creation: a service systems and service logic perspective. Eur. Manag. J. **26**, 145–152 (2008)

27. Vargo, S.L., Lusch, R.F.: Institutions and axioms: an extension and update of service-dominant logic. J. Acad. Mark. Sci. **44**(1), 5–23 (2016)

28. Waltman, L.: A review of the literature on citation impact indicators. J. Inform. **10**(2), 365–391 (2016)

29. Lusch, R.F., Nambisan, S.: Service innovation: a service-dominant logic perspective. MIS Q. **39**(1), 155–175 (2015)

30. Simon, H.A.: The Sciences of the Artificial. MIT Press, Cambridge (1969)

31. Anderson, L., et al.: Transformative service research: an agenda for the future. J. Bus. Res. **66**(8), 1203–1210 (2013)

32. Hevner, A.R., et al.: Design science in information systems research. MIS Q. **28**(1), 75–105 (2004)

33. Peffers, K., et al.: A design science research methodology for information systems research. J. Manag. Inf. Syst. **24**(3), 45–77 (2007)

34. Sein, M.K., et al.: Action design research. MIS Q. **35**(1), 37–56 (2011)

35. Van Aken, J.: Management research based on the paradigm of the design sciences: the quest for field-tested and grounded technologicla rules. J. Manag. Stud. **41**(2), 219–246 (2004)

36. Teixeira, J.G., et al.: The MINDS method: integrating management and interaction design perspectives for service design. J. Serv. Res. **20**(3), 240–258 (2017)

37. Patrício, L., et al.: Service design for value networks: enabling value cocreation interactions in health care. Serv. Sci. **10**(1), 76–97 (2018)

38. McColl-Kennedy, J.R., et al.: Gaining customer experience insights that matter. J. Serv. Res. **22**(1), 8–26 (2019)

39. Grenha Teixeira, J., Patrício, L., Tuunanen, T.: Advancing service design research with design science research. J. Serv. Manag. (2019). in press

40. Huang, M.-H., Rust, R.T.: Artificial intelligence in service. J. Serv. Res. **21**(2), 155–172 (2018)

41. Wirtz, J., et al.: Brave new world: service robots in the frontline. J. Serv. Manag. **29**(5), 907–931 (2018)

42. Čaić, M., Odekerken-Schröder, G., Mahr, D.: Service robots: value co-creation and co-destruction in elderly care networks. J. Serv. Manag. **29**(2), 178–205 (2018)

43. Ng, I.C.L., Wakenshaw, S.Y.L.: The Internet-of-Things: review and research directions. Int. J. Res. Mark. **34**(1), 3–21 (2017)

The Digital Twin as a Service Enabler: From the Service Ecosystem to the Simulation Model

Jürg Meierhofer[1,3(✉)], Shaun West[2,3], Mario Rapaccini[4], and Cosimo Barbieri[4]

[1] School of Engineering, ZHAW Zurich University of Applied Sciences, Winterthur, Switzerland
juerg.meierhofer@zhaw.ch
[2] School of Engineering and Architecture, HSLU Lucerne University of Applied Sciences and Arts, Lucerne, Switzerland
shaun.west@hslu.ch
[3] Swiss Alliance for Data-Intensive Services, Expert Group Smart Services, Thun, Switzerland
[4] School of Engineering, University of Florence, Florence, Italy
{mario.rapaccini, cosimo.barbieri}@unifi.it

Abstract. This paper investigates the concept of the digital twin as an enabler for smart services in the context of the servitization of manufacturing. In particular, a concept is developed and proposed for the derivation of appropriate simulation models starting from the model of the service ecosystem. To do so, smart industrial services are analyzed from the point of view of their value proposition. Next, the role of the digital twin as an enabler for these services is analyzed and structured in a multi-layer architecture. Hybrid simulation approaches are identified as suitable for building simulation models for this architecture. Finally, a procedural end-to-end approach for developing a simulation based digital twin departing from the service ecosystem is proposed.

Keywords: Smart Services · Servitization of manufacturing · Digital twin

1 Introduction

The goal of this paper is to elaborate a concept for creating service value using a digital twin. The aim is to design and engineer services derived from the customer needs that leverage the potential of the digital twin. The approach discussed in this paper starts with the concept of "Service-Dominant Logic" (SDL). With the transition from products to services, the economy moves from the concept of "Goods-Dominant Logic" (GDL) to SDL. In SDL, service is considered the fundamental purpose of economic exchange (foundational premise 1, FP 1). The focus of value creation is moved from the manufacturer as creator to co-creation through customer interaction [1]. Here the quality of the service is determined by the customer's perception, rather than by the engineering on the side of the provider: hence, the value is always co-created by the customer (FP 6). Value is deployed over a period of time which exceeds

© Springer Nature Switzerland AG 2020
H. Nóvoa et al. (Eds.): IESS 2020, LNBIP 377, pp. 347–359, 2020.
https://doi.org/10.1007/978-3-030-38724-2_25

the discrete moment of sales and distribution and is created in ecosystems through actor-generated institutions and institutional arrangements (FP 11).

SDL also states that operant resources - knowledge and skills - are the fundamental source of competitive advantage for the actors in the ecosystem (FP 4). Service providers apply their knowledge and skills for the benefit of another entity or the entity itself [1]. In the context of industrial services, the ability to use a digital twin based on data-based models and analytics represents an operant resource. The concepts of SDL are translated into practical procedures by, e.g., service design approaches. According to [2], service design [3] can be considered an operationalization of SDL. When designing a new service, it is essential to first define and understand the target customers and to explore their needs for service in the specific context. The benefit for the customer depends strongly on the customer himself and on his individual situation and context (FP 10).

2   Smart Industrial Services

The service sector is continuously growing and makes up a substantial part of employment and the gross domestic product [4]. On the transition from products to services, companies start to move from the concept of GDL to SDL, and the concept of industrial companies as service providers has emerged [5]. The focus of firms' value creation, thus, shifts from the manufacturer to co-creation by firm and customer [1].

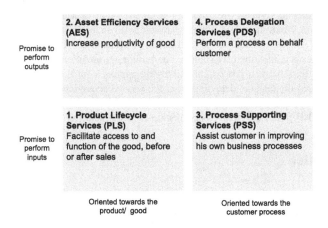

Fig. 1. Input- vs. output-based services (adapted from [7])

The shift to services is driven by saturated markets and high competitive intensity [6], as well as by customer demand for the values and benefits provided by services [7]. In particular, there is an evolution for customers to demand and pay for some agreed performance output instead of the resource inputs the provider offers. Therefore, the transition from goods to services and the addition of services to products is considered essential for manufacturing firms [8]. For the development of this service economy, the omnipresence of information and communications technology is a major driving force [9].

The literature provides a classification of industrial services based on the value provided to the customer, who is guaranteed either an input or some output performance [7, 10]). There is a differentiation between services that are oriented towards the supplier's goods or towards the customer's processes (Fig. 1). According to the figure, traditional service models are located in the PLS (product life-cycle services) quadrant. Examples are the installation of new equipment, maintenance, repair or spare parts delivery. When the provider moves to new service models around its products, the PLS services are complemented or replaced by output-oriented asset efficiency services (AES), such as customization, condition monitoring, predictive maintenance, performance optimization, or consulting for the customer along the end-to-end journey. New service models that focus on output performance are also referred to as "advanced services" [8]. This is in line with the concept of moving from value in exchange to value in use with the good becoming a distribution mechanism for the service value (FP 3). Assessing and quantifying the fluctuations and risks inherent to the output provided as well as the production costs to achieve the promised level of output quality, therefore, becomes a key capability for a provider when moving to output-based advanced services.

The provider needs to understand and manage the risk of failing to provide the guaranteed performance. It is evident that the ability to manage, process, and analyze data of the installed base is essential to do this. Therefore, with the increasing degree of servitization of manufacturing and in order to move to advanced services, leveraging data for the development and the provision of services becomes a key prerequisite, and at the same time it is a key challenge for the providers. In this perspective, the ability to process and analyze data is considered an operant resource according to SDL. The intention to move from basic services in the area of product life-cycle services to advanced asset efficiency services was confirmed in case studies with several manufacturing companies [11, 12].

3 The Digital Twin as an Implementation Form of Smart Services

3.1 Actors' Jobs and Needs Over the Product Lifecycle

Each actor within a firm's business ecosystem will have different jobs, pains, and gains and these are dependent on the situation in question; in effect, the actor's problem. The digital twin needs to match closely with "the actor's problem" in a systematic approach, so the resulting services are centered around the specific needs of the actor in that particular situation. This fosters a close relationship and co-creation of value between the different actors (e.g., the providers and the customers), which contributes to long-term relationships and loyalty. There is no single applicable value proposition but rather, many value propositions that are based around individual actors and their situational problems. In particular, the digital twin lends itself to contribute to the value propositions by supporting all of the actors around the product service system, in particular by relieving the pains and increasing the gains of the actors. As discussed before, the digital twin is based on a combination of data, analytics, and the visualization of the insights to support decision-making by providing advice. Depending on the phase of the product lifecycle (beginning, middle, end of life, i.e., BOL, MOL, EOL), different components of the twin and different data are required.

When designing and applying the digital twin, the provider applies its knowledge and skills in analyzing the needs of the different actors in the ecosystem and conceiving appropriate service value propositions based on data, modelling the physical equipment and combining this with domain specific knowledge to provide a service benefit. Therefore, the approach discussed in this paper considers the digital twin as a data-driven operant resource for the design and provision of services. The twin is structured according to technical and business hierarchies. The product, which is the carrier of the industrial service, on the one hand consists of sub-components and on the other hand is part of a larger system. Therefore, the objects considered in these layers may range from physical components and integrated machines, up to shop floors, factories, and systems of factories [13].

The research question in this study requires framing within the context of data-driven product service systems so that the complexities of the B2B (business to business) environment and digital twins can be investigated in the industrial world. Around this framing the research question is: "In a complex data-driven product service system, what service value can the digital twin provide, who is the provider, who is the beneficiary, and how can this service value be implemented by data and simulation models?"

3.2   Layered Approach for Service Value Creation Using the Digital Twin

In this section, we discuss a structure of the digital twin that lends itself to the design of industrial services along the product lifecycle. From the perspective of the product life-cycle management phases (BOL, MOL and EOL), systematic service design delivers customer value in all these phases (Fig. 2). The actors in the ecosystem, i.e., the diverse actors along the product lifecycle, require value propositions that fit their jobs, pains, and gains. These diverse instances of value propositions are generated by the family of digital twins, consisting of several sub-twins each serving a specific sub-set of jobs, pains, and gains, and using specific elements of data and models.

According to [14], value creation based on data and analytics is very similar to that based on the digital twin. Data and analytics can be considered as an enabling part of the digital twin. Although technical discussions of the digital twin predominate in literature, there are also sources that illuminate the object from the perspective of services and value contribution. According to [15], the integration of the digital twin and service represents a promising research direction which should be addressed in future paradigms. [16] as well as [17] and [15] discuss the application of the digital twin along the product lifecycle. It can provide value in the form of services in product design and engineering, in the design and optimization of the shop floor, in product operations and usage monitoring, and in after-sales services and prognostics and health monitoring of the product. [18] discusses the digital twin as a basis of a decision support system in manufacturing.

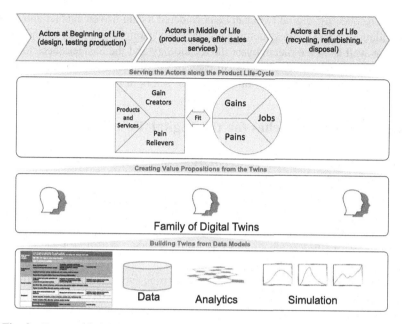

Fig. 2. Data enabled value creation framework of the simulation-based digital twin

Given the wide range of interpretations of the term "digital twin", it becomes evident that further dimensions are required to get a differentiated perspective on the value contribution. A hypothesis for a layered approach adapted from [19] is shown in Fig. 3.

Many factories / business ecosystem	Asset management optimization within market conditions (e.g., opening new factories/lines, upgrading existing lines, mothballing lines, closing lines/factories)		
	ERP, CRM, SCM, logistics, market relevant indicators		
	System Dynamics (SD) Modelling		
Production line/ factory	Sizing of production lines Lesson for next generation factory / product	Forecasting, production optimization, maintenance integration, line configuration, capacity management, customer management	Factory/line CMU
	Capacity of machines, volumes, production mix, order process, customer behavior		
	Discrete Event Simulation (DES) or Agent Based Simulation (ABS) Modelling		
Product / machine	Design product and module, optimization for application Lessons for next generation machines	Production line optimization, machine performance, manage machine health, machine configuration	Refurbish, recycle, upgrade components, extent life, design improvements
	Operational data, business processes, machine status, static and live machine parameters, location		
	Physical simulation (FEM, differential equations, machine learning)		
Component	Design of the component based on past experience	Manage health of component, maintenance	Refurbish, recycle, upgrade components, extend life, design improvements
	Material properties, dimensions, environmental data, operation data, maintenance data		
	Physical simulation (FEM, differential equations, machine learning)		
	Beginning of life (BOL)	Middle of life (MOL)	End of life (EOL)

Fig. 3. Architecture of the digital twin in several layers with design objective, data used, and simulation technique applied

In this respect, the digital twin can be considered a data-driven enabler and support for providing services. The structuring of the twin along the phases of the product lifecycle as well as in the hierarchy of the technology and the business decisions leads to the concept of the family of twins. In each of the phases BOL, MOL, and EOL, a digital twin can provide service value to specific actors in the product service system.

4 Simulation Approaches for the Digital Twin

4.1 Literature Review for Simulation Approaches for Multi-layer Manufacturing Systems

Implementing a family of digital twins for the architecture shown in Fig. 3 requires the development of different elements of simulation models on a different level of granularity, which are then combined. The family of twins in a specific case may comprise elements covering different phases of BOL, MOL, or EOL, as well as different layers of the architecture, e.g., a machine layer as well as a factory layer.

Simulation across several layers in manufacturing environments is referred to as "hybrid simulation" in the literature [21–25]. "Hybrid simulation" is also methodologically further elaborated in [26–28]. Under this term, the three modelling approaches "agent-based simulation" (ABS), "discrete event simulation" (DES), and "system dynamics" (SD) are considered in a systematic and combined approach. According to [21], simulation in manufacturing systems today mainly focuses on DES, while the potential of ABS and SD are rather under-utilized. In [23], a comprehensive overview of eight issues in production planning and control is provided, together with an indication which of the simulation modelling approaches of ABS, DES, or SD is suited to each. These eight issues are: facility resource planning, capacity planning, job planning, process planning, shop floor scheduling, inventory management, production and process design, purchase and supply management. In [25] a hybrid approach of ABS and DES is elaborated for the modelling of a product-service-system and compared to an approach based only on DES. The study reveals that the hybrid approach is more useful for modelling the customer interaction in a service system. [24] elaborates on the specific challenges imposed by the high variability inherent in services systems. This is due to the co-creative involvement of a variety of customers as well as the provider resources in the process, which may be a cause of the "service paradox", i.e., companies suffering from poor return on their service business. Both [24] and [29] argue that simulation approaches are a suitable means to cope with these challenges of increased variability. They compare different hybrid approaches of ABS, DES, and SD, and conclude that the combination of ABS and DES is well suited for the modelling of manufacturing issues, where ABS makes modelling of customer variability easier and also allows the detection of phenomena as they emerge in the system. SD lends itself to, e.g., modelling sustainable cause-effects loops.

4.2   Hybrid Simulation Concept for the Family of Digital Twins

Given the simulation concepts available in the literature, it becomes obvious that a hybrid simulation approach comprising SD, DES, and ABS is adequate to support decision-making by the family of digital twins in industrial environments. Thus, the simulation models represent an operant resource that is integrated by the actors for providing a benefit for decision support.

System Dynamics (SD) Modeling:

- SD modeling is focused on the observation on the system level and used for strategic-level modeling. SD modeling uses the laws of physics to describe and investigate the dynamics of economic and social systems, whereby individual entities are not specifically modelled, but are represented as a continuous quantity in a stock. The mechanisms behind DS modeling are feedback loops and stock and flow models (e.g., for stocks of products or jobs or for flows of purchases).
- The target objects of the models in manufacturing systems are, for example: demand forecasting (on an aggregated level), estimation of total cost and revenues, network and supply chain planning, logistics and inventory planning, ecology modeling, supply, corporate planning, etc. [21, 23].

Discrete Event Simulation (DES) Modeling:

- DES modeling is used for process-oriented modelling of the behaviour and impact of actors or items in their ecosystem on an operational-tactical level. DES models can be considered as an interacting set of entities that evolve through different states while internal or external events happen. The system is frequently modeled as a process, i.e., a sequence of operations performed across entities and resources.
- Discrete event simulation is the most common simulation methodology used for manufacturing systems. Target objects of the models are, for example, capacity planning, quantity planning, job-shop management, shop floor scheduling, order time planning etc. [21, 23].

Agent-Based Simulation (ABS) Modeling:

- ABS is a bottom up approach. The modelers do not need to know the complex structure and operation of the overarching system. Modeling starts with designing individual agents, defining their characteristics and behaviors. In the execution of the simulation models, they interact together their environment. Thus, agents in ABS are autonomous, heterogeneous and possibly intelligent entities. The evolution of the agents over time can be modeled by state charts inside agents. The global system behavior does not need to be defined. It emerges as a result of the evolution of the individual agents, each following its own rules [24].
- Agents may be, for instance, customers, machines, capacities, or goods flowing across a system. Therefore, in manufacturing contexts, ABS modeling is applied to job planning, equipment planning, process sequence planning, shop floor scheduling etc. or, e.g., cooperation and communication within the supply chain [21, 23].

Physical Modeling:

- In specific cases, the input data for feeding the digital twin may stem from physical sensors, e.g., vibrations, temperatures etc. This data can be used in a physical model that is combined with the other models, typically based on differential equations/FEM (e.g., [20]).
- However, in the context of decision support services in manufacturing environments, physical data can be used for models based on data science applications. The underlying methodologies are described in, e.g., [30], with the application for data-driven services discussed in [31].
- This can be used, for example, to assess or predict the status of machines by classification, based on physical input data. The discrete variables describing this status can then be fed into agents of ABS to factor into the decision-making process.

To conclude, as discussed in the literature about hybrid simulation models, the combined models of ABS and DES are most frequently applied in manufacturing systems. DES is used to model the work processes on the shop floor, in which raw-materials and semi-finished products are worked on by humans and machines. Throughput or waiting times based on queuing systems can be evaluated and used for the decision-making process supported by the digital twin. ABS can enhance the DES model by modulating the behavior of the actors (e.g., customers can change their needs based on contextual information, which is typical in service systems). Equally, human workers or machines can change their behavior over time. ABS can also be used for modeling autonomous material or products moving across the system.

In the architecture of the digital twin according to Fig. 3, SD simulation will be added at the top of the scheme if the behavior on the system level needs to be understood for decision support services. This may be, e.g., strategic evolutions at the supply chain level, or the impact of the production schemes on ecological variables. Additionally, where required, physical models can be supplemented at the bottom of the scheme in a targeted manner to integrate physical measurement points into the model.

5   Integration of Service Ecosystem and Simulation Model

When hybrid simulation models according to Sect. 4.2 and based on the architecture of Fig. 3 are implemented, it is important not to start building the entire system in a GDL approach, which would be followed by traditional simulation approaches (i.e., building a simulation model around a product or a physical system and then trying to get value out of this). According to [16], "The simulation models making-up the Digital Twin are specific for their intended use and apply the suitable fidelity for the problem to be solved." [28] claims that the goal of a simulation modeling process is not to build a simulation model, but to build a decision support system. Taking the SDL perspective, the needs of the actors in the ecosystem and the value in use to be provided need to be modeled first. Only then, the application of simulation models to provide the value in use may be considered.

In the service design perspective, this means that first, the ecosystem of actors and their jobs, pains, and gains need to be modeled. Digital twins – in particular simulation-based ones – are only considered if they can add value to the service value proposition for supporting the actors in their decision processes. In practical cases, it often turns out that simulation-based digital twins are appropriate only for a specific subset of the customer jobs. Substantial components of the service are then based on other – not data or simulation based - components of knowledge and skill. This also means that the architecture of Fig. 3 is not implemented in its entirety, but specifically in individual elements to achieve the targeted service benefit. These elements then make up the members of the family of twins shown in Fig. 2.

Given this SDL perspective, we therefore elaborate and propose a procedure in six steps for the development of the digital twin starting from modelling the service ecosystem and leading to the integration of the technical resources (i.e., the simulation-based members of the family of twins).

1. **First, the ecosystem of actors in the system is analyzed and documented:**

   - In agreement with the concept of SDL and in particular of service ecosystems, the boundaries of the system need to be determined by including those actors that are directly relevant for the use case and that may influence the decision to be taken.
   - Service ecosystems are typically drawn in the form of a graph with the actors being the nodes and the connecting edges representing the transactions and the value flows [2] (Fig. 4).

Fig. 4. Mapping the existing service ecosystem with different actors being interconnected by value exchanges that can be described by the transactions.

2. **Next, the problems of the actors are described using service design tools:**

   - In particular, this means conducting contextual customer insight research and understanding the jobs to be done [3].
   - This leads to documented actors'jobs, pains, and gains (see also Fig. 2), their customer journey, and their position in the service ecosystem.

3. **Next, the moments of truth for decision support are identified:**

   – As discussed in Sect. 4, hybrid simulation-based models (like physical modelling, SD, DES, or ABS) lend themselves to provide value in the form of decision support to the actors in the ecosystem. For deciding which simulation elements (i.e., members of the family of twins) provide most value, the most critical decision points of the actors are identified.
   – In terms of the service design methodology, these are the moments of truth (MoT, [3]).

4. **For these identified decisions, their factors and impacts are analyzed:**

   – What is driving the decision? Are there capacity constraints (e.g., not enough resources or agents)? Or are there capability constraints (e.g., not the right resources)? Are there other driving factors?
   – Which information is required to assist the actor in the decision? How does it have to be presented to the actor in order to provide a service value?
   – Which is the relevant time horizon for the service value? I.e., in which time horizon does the decision have an impact? E.g., an actor responsible for the maintenance of equipment may need to know how it will perform in the next days or weeks or months, depending on the context.

5. **For the required information for decision support, the integrability of the technical resources is checked:**

   – Which data and other technical resources (operant resources) are required to provide the required value to the actor?
   – Are the resources available and can they be integrated? If not, can they be made available? Can they be simulated or emulated for estimating the potential of the new service?

6. **Build the models for the family members of the digital twin:**

   – In order to provide the actor-specific service value, the technical models according to the concepts described in Sect. 4 are built, thus making up the family of twins.
   – These models then are integrated as resources in service value propositions, i.e., combined with other resources (that are typically not data- or simulation-based).
   – Of course, according to the service design procedures (e.g., [3]), the development of the service value proposition follows an iterative approach, i.e., the service value is developed in a continuous cycle of "design – test – improve" steps until it provides the relevant value. This also means that the family of twins undergoes a continuous evolution.

If decision support for several actors in the system is developed, this means elaborating several family members of the twin, which together constitute the digital twin.

6 Discussion and Conclusions

In this conceptual paper we derived a concept for modeling industrial service ecosystems based on the digital twin. We apply SDL to the digital twin concept, which means that the development process starts with understanding the business system and the problem of (all) the actors (e.g., people, things and processes) in the system. Transitions are identified between actors and then value propositions are derived from this and translated in appropriate members of the "family of twins". This results in a multi-layer approach for the required simulation models. The concepts around the topic "hybrid simulation" comprising ABS, DES, and SD described in the literature lend themselves to implementing parts of this architecture in simulation models.

The application of "agency" is important as the simulation results should be viewed as advisory and limited by the system boundaries. There are many instances where the digital twin will have key management information missing. It should provide consequences of the advice and provide alternative options. In effect, we are suggesting that the data-driven models ("artificial intelligence", AI) do not hold all the answers but rather they are supportive to the decision making within the ecosystem. The AI here is made up of many digital twins and it is possible (or expected) that it may provide conflicting or unclear advice as it has used different underlying simulation models. A simple example of a digital twin providing advisory are the widely used navigation system apps and their travel services: options are provided both in terms of the mode of transport and routes – the digital twin here provides the consequences (e.g., travel times and complexity of journey). Finally, a digital helper then instructs the traveler.

The simulation based digital twin can be considered as a service agent when viewed through the lens of SDL. It should provide advice within the constraints of its boundaries as well as options allowing the human counterparties to take the action as the knowledge of the digital twin is limited to its bodies. It needs the use cases to be well understood and described and the sources of data and the beneficiaries of information to be defined. Based on this, we conclude that however the system is simulated, it should be brought all to the same level: people, machines and other objects as well as digital twins (including digital helpers) are all actors within the ecosystem.

In the ongoing and future research by the authors, a multiple case study is conducted to establish which of the modeling methodologies is most suitable to create value in which specific type of service situation. It is evaluated whether there are typical patterns and blueprints for selecting this methodology. Additionally, the amount and quality of the available data may limit the benefit of the services provided by the family of twins, which needs to be investigated. These investigations need to be conducted against the background of the business objectives of the actors in the ecosystem. Furthermore, it remains to be verified whether and how simulation agents in ABS lend themselves to the modelling of non-human actors in service ecosystems, which opens up new research questions in the area of the role of autonomous actors in service ecosystems [32].

References

1. Vargo, S., Maglio, P., Akaka, M.A.: On value and value co-creation: a service systems and service logic perspective. Eur. Manag. J. **26**(3), 145–152 (2008)
2. Vargo, S.L., Lusch, R.F.: The SAGE Handbook of Service-dominant Logic. SAGE, Los Angeles (2019)
3. Stickdorn, M., Hormess, M., Lawrence, A., Schneider, J.: This is Service Design Doing: Using Research and Customer Journey Maps to Create Successful Services. O'Reilly UK Ltd. (2017)
4. Kindström, D., Kowalkowski, C.: Service innovation in product-centric firms: a multidimensional business model perspective. J. Bus. Ind. Mark. **29**(2), 96–111 (2014)
5. Lay, G.: Introduction. In: Lay, G. (ed.) Servitization in Industry. Springer, Cham (2014)
6. Gebauer, H., Ren, G.-J., Valtakoski, A., Reynoso, J.: Service-driven manufacturing, provision, evolution and financial impact of services in industrial firms. J. Serv. Manag. **23**(1), 120–136 (2012)
7. Kowalkowski, C., Ulaga, W.: Service Strategy in Action: A Practical Guide for Growing Your B2B Service and Solution Business. Service Strategy Press (2017)
8. Baines, T., Lightfoot, H.W.: Servitization of the manufacturing firm. Int. J. Oper. Prod. Manag. **34**(1), 2–35 (2013)
9. Chen, Y.G., Hsu, C.M., Chen, Z.H.: The service design strategy of manufacturing service industry. In: PICMET 2010 Technology Management for Global Economic Growth, pp. 1–6 (2010)
10. Ulaga, W., Reinartz, W.J.: Hybrid offerings: how manufacturing firms combine goods and services successfully. J. Mark. **75**(6), 5–23 (2011)
11. Meierhofer, J.: Data driven servitization for SMEs in manufacturing. In: Spring Servitization Conference - Driving Competition through Servitization, pp. 101–108, Aston University, Birmingham (2018)
12. Meierhofer, J., Kugler, P., Etschmann, R.: Challenges and approaches with data-driven services for SMEs: insights from a field study. In: Spring Servitization Conference: Delivering Services Growth in the Digital Era, Lin-köping, Sweden, 13–15 May 2019, pp. 39–49. Aston University, Birmingham (2019)
13. Porter, M.E., Heppelmann, J.E.: How smart, connected products are transforming competition. Harv. Bus. Rev. **92**(11), 64–88 (2014)
14. Qi, Q., Tao, F.: Digital twin and big data towards smart manufacturing and industry 4.0: 360 degree comparison. IEEE Access **6**, 3585–3593 (2018)
15. Tao, F., Zhang, M.: Digital twin shop-floor: a new shop-floor paradigm towards smart manufacturing. IEEE Access **5**, 20418–20427 (2017)
16. Boschert, S., Rosen, R.: Digital Twin—The Simulation Aspect. In: Hehenberger, P., Bradley, D. (eds.) Mechatronic Futures, pp. 59–74. Springer, Cham (2016). https://doi.org/10.1007/978-3-319-32156-1_5
17. Qi, Q., Tao, F., Zuo, Y., Zhao, D.: Digital twin service towards smart manufacturing. Procedia CIRP **72**, 237–242 (2018)
18. Kunath, M., Winkler, H.: Integrating the digital twin of the manufacturing system into a decision support system for improving the order management process. Procedia CIRP **72**, 225–231 (2018)
19. Meierhofer, J., West, S.: Service value creation using a digital twin. In: Naples Forum on Service, Service Dominant Logic, Network & Systems Theory and Service Science: Integrating Three Perspectives for a New Service Agenda, Ischia, 4–7 June 2019

20. Thiede, S., Schönemann, M., Kurle, D., Herrmann, C.: Multi-level simulation in manufacturing companies: The water-energy nexus case. J. Clean. Prod. **139**, 1118–1127 (2016)
21. Scheidegger, A.P.G., Pereira, T.F., de Oliveira, M.L.M., Banerjee, A., Montevechi, J.A.B.: An introductory guide for hybrid simulation modelers on the primary simulation methods in industrial engineering identified through a systematic review of the literature. Comput. Ind. Eng. **124**, 474–492 (2018)
22. Chandra, C., Grabis, J.: Simulation modeling and hybrid approaches. Supply Chain Configuration, pp. 173–195. Springer, New York (2016). https://doi.org/10.1007/978-1-4939-3557-4_9
23. Min Jeon, S., Kim, G.: A survey of simulation modeling techniques in production planning and control (PPC). Prod. Plan. Control **27**(5), 360–377 (2016)
24. Rondini, A., Tornese, F., Gnoni, M.G., Pezzotta, G., Pinto, R.: Hybrid simulation modelling as a supporting tool for sustainable product service systems: a critical analysis. Int. J. Prod. Res. **55**(23), 6932–6945 (2017)
25. Rondini, A., Tornese, F., Gnoni, M.G., Pezzotta, G., Pinto, R.: Business process simulation for the design of sustainable product service systems (PSS). In: Umeda, S., Nakano, M., Mizuyama, H., Hibino, H., Kiritsis, D., von Cieminski, G. (eds.) APMS 2015. IAICT, vol. 460, pp. 646–653. Springer, Cham (2015). https://doi.org/10.1007/978-3-319-22759-7_74
26. Eldabi, T., et al.: Hybrid simulation: historical lessons, present challenges and futures. In: 2016 Winter Simulation Conference (WSC), pp. 1388–1403, Washington (2016)
27. Brailsford, S.C., Eldabi, T., Kunc, M., Mustafee, N., Osorio, A.F.: Hybrid simulation modelling in operational research: a state-of-the-art review. Eur. J. Oper. Res. **278**(3), 721–737 (2019)
28. Lättilä, L., Hilletofth, P., Lin, B.: Hybrid simulation models – When, Why, How? Expert Syst. Appl. **37**(12), 7969–7975 (2010)
29. Pezzotta, G., Rondini, A., Pirola, F., Pinto, R.: Evaluation of discrete event simulation software to design and assess service delivery processes. In: Choi, T.-S. (ed.) Service Supply Chain Systems, A Systems Engineering Approach. Kogan Page Ltd., London (2018
30. Provost, F.P., Fawcett, T.: Data Science for Business. O'Reilly, Sebastopol (2013)
31. Meierhofer, J., Meier, K.: From data science to value creation. In: Za, S., Drăgoicea, M., Cavallari, M. (eds.) IESS 2017. LNBIP, vol. 279, pp. 173–181. Springer, Cham (2017). https://doi.org/10.1007/978-3-319-56925-3_14
32. Maglio, P., Lim, Ch.: On the impact of autonomous technologies on human-centered service systems. In: Vargo, S.L., Lusch, R.F.: The SAGE Handbook of Service-Dominant Logic, pp. 689–699. SAGE, Los Angeles (2019)

Service Management

Is There a Relationship of Interdependence Between Resilience, Viability and Competitiveness? Ditron Ltd. Case-Study

Luca Carrubbo[1]([✉]), Monica Drăgoicea[2] [iD], Xhimi Hysa[3],
Antonietta Megaro[1], and Besjon Zenelaj[3]

[1] University of Salerno, Via Giovanni Paolo II, 132, 84084 Fisciano, SA, Italy
`{lcarrubbo,amegaro}@unisa.it`
[2] Faculty of Automation and Computers, University Politehnica of Bucharest,
Splaiul Independenței 313, 060042 Bucharest, Romania
`monica.dragoicea@upb.ro`
[3] Department of Business Administration, Epoka University, Tirana, Albania
`{xhhysa,bzenelaj}@epoka.edu.al`

Abstract. The present work aims to deepen the concept of resilience in order to better understand, in terms of viability, the determinants of the competitiveness of companies that operate in conditions of uncertainty and change. A systems view and some of the fundamental concepts (FCs) of the Viable Systems Approach (VSA) have been taken into consideration. To evaluate the findings with practical evidences, a case study has been conducted to evaluate resilience in a managerial context and to apply the proposed systems approach. Even though this study considers resilience in a systems way only, using a specific theoretical framework, several practical implications have been derived. They refer to the possibility of analysing resilience in a systems perspective, also as a distinctive feature of the viability (PN.1). In this approach, resilient behaviours can be interpreted as badges of acting viable, because they are provided with a precise meaning linked to the purpose of the person who implements them (PN.2). The determinants that favour resilience represent rather subset of the determinants that pertain to the competitiveness of a viable system company (PN.3). As well, findings show that the competitiveness of a business system is a matter of viability (PN.4).

Keywords: Resilience · Viability · Viable system · Survival · Viable Systems Approach

1 Introduction

The term resilience, in the transition to the common language, has a sense more similar to the "ability to sustain impacts without breaking", a particularly congenial aspect in the study of materials in Physics. Over time, resilience has been brought closer to the characteristics of flexibility and resistance, and for this it seen as an important quality of an organism (individual or organization) useful for facing contingencies and stresses over time.

© Springer Nature Switzerland AG 2020
H. Nóvoa et al. (Eds.): IESS 2020, LNBIP 377, pp. 363–376, 2020.
https://doi.org/10.1007/978-3-030-38724-2_26

The present work aims to deepen the concept of resilience of individuals and organizations in order to better understand, in terms of viability, the determinants of the competitiveness of those companies operating in conditions of uncertainty and change. We tried to understand why the ability to be viable differentiates organizations that survive from others regardless of their resilience (whether structure or system). Therefore, we have tried here to formulate some specific interpretative propositions (PNs) to respond promptly to the following research questions (RQs):

Is resilience a distinctive trait of viability?
Is resilience a basic element for business competitiveness?

The structure of the document provides at first an excursus on the concept of Resilience in literature. Then the comparison with the viability concept and with the business competitiveness is presented. Finally, the description of the case study "Ditron Ltd" is proposed.

2 Methodology

The work in this paper is conceptual and this analysis is based on the systems thinking. Particularly, the Viable Systems Approach (VSA) has been assumed as theoretical framework [1] for using some of the fundamental concepts (FCs) [2], for understanding the linkage, and for making a comparison in a systems perspective, between "viability" of an organization understood as a system and "resilience" (RQ1). The excursus on the concept of Resilience in literature is based especially on international literature with respect to the topic of business organization.

Subsequently, the work comes up with some considerations concerning systems determinants of viable systems of business competitiveness that can be associated with being resilient (RQ2).

A case study has been carried out for supporting this paper's findings with practical evidences. To conclude the presented ideas on Resilience, a case study is addressed, the Ditron Ltd., where the empirical evidences are analyzed based on the insights (IN) derived and developed within the present contribution. The last part of this contribution presents managerial practical implications and some conclusive considerations.

3 Resilience, an Evolving Multi-disciplinary Concept

3.1 Nudges from Literature

Resilience is a multi-disciplinary concept and has been developed in different fields of knowledge, including psychology [3], *supply-chain management* [4], corporate strategies [5], organizational models [6], *disaster management* [7], human resource management [8].

In the early 2000s, two different perspectives of resilience analysis were theorized. The first one concerns *high reliability organizations* (HROs) that operate in extreme conditions in an attempt to reduce errors and inefficiencies [9]. The second perspective concerns the resilience and reconstitution ('*restoration*') of the abilities and resources of an organization [10].

Resilience has been also evaluated as the ability to absorb, collect and metabolize negative surprises that can affect the survival of organizations by effectively overcoming destructive *shocks* and debilitating consequences [11].

Hamel and Välikangas [4] believe that resilience can help to reinvent business models and organizational strategies. Lengnick-Hall and Beck [10, 11] think that resilience derives from a mix of abilities, routines, practices and processes integrating resources, able to orient behavior in an adaptive way. According to Freeman et al. [12] it is important, in this sense, to consider a decided orientation to results, to strong values, to a genuine vision and to a certain property of language (understood as the ability to interface with others).

3.2   The Resilience of Organizations 'Read' from a Systems Perspective

In the social sciences, resilience is associated with different themes, also addressed by Systems Thinking. This is not always explicitly presented, especially if we refer to the concepts of system, context, resources, self-regulation, adaptation, viability.

Resilience can be linked to the concept of system. For Hollnagel et al. [13] resilience is the intrinsic ability of an organization (understood as a system) to dynamically maintain a stable equilibrium (IN.1) and implement growth actions even in the presence of continuous stress. Resilience is a function of the context in which it operates. Block and Block [14] believe that resilience is the dynamic ability to modify its organizational model according to the changing characteristics of the reference context, emphasizing the negative aspect of this variability. Resilience is close to the themes of autopoiesis and self-regulation. Vogus and Sutcliffe [15] argue that resilience is the capacity of organizations to preserve themselves and always recover despite the adversities.

Resilience is linked to the availability of resources. Bhamra et al. [6] associate resilience with the dynamics that create or maintain resources (cognitive, emotional, relational or structural) in a form that is sufficiently flexible, memorable, convertible and malleable. This allows organizations, their units and individuals to behave sufficiently adaptive to cope with uncertainty, turbulence and discontinuities. Glassop [16] summarizes in 3R the elements that influence resilience, *requisite variety*, *redundancy*, *resources*, typical factors of complex environments.

Resilience is about being able to be adaptive. Weick and Sutcliffe [9] speak of resilience as the collective ability to implement adaptive behaviors (IN.2) able to reduce the stress condition deriving from the contingencies that appear more or less suddenly on the path of development. Mallak [17] has considered resilience as the ability to anticipate, respond or adapt quickly in response to catastrophic and destructive events.

Resilience is also an expression of viability. Wildavsky [18] believed that resilience was the ability to overcome potential unexpected harmful situations (IN.3).

Dalziell and McManus [19] identify resilience as a function of the vulnerability of a system and its effective adaptability. They talk about *situation awareness* and *resilience ethos* (referring to an intrinsic instinctive self-preservation approach) (IN.4).

As a consequence, for resilient organizations, being adaptive appears to be a necessity and no longer simply a strategic choice.

In summary, based on the analysis of literature, that we reread in a systems key, the answer to RQ1 is Yes. This is because resilience can be connected to the concept of system, it is a function of the context in which it operates, it is close to the themes of autopoiesis and self-regulation. As well, it is linked to the availability of resources, it concerns the ability to be adaptive, and it is an expression of viability. The 'Proposition' n.1 which derives from this is the following.

PN.1: Resilience can be analyzed in a systems perspective, also as a distinctive feature of the viability.

3.3   Between Resilience and Viability, the Contribution of VSA FCs

Studying companies as systems also makes it possible to understand the value of relations with the outside world and the need to continue to grow and innovate for continually adapting to changes. In this way, it is possible to achieve a sustainable balance for survival [20, 21]. According to the VSA, the company is considered as an "open, aimed, organic, autopoietic, cognitive, cybernetic, equifinal" system [22]. According to the VSA, a *System* represents a (dynamic) entity that emerges from a specific (static) structure, defined as a set of individual elements with assigned roles, activities, responsibilities and tasks to be performed in compliance with specific shared rules and constraints [23, 24].

The viable system, understood in the VSA, is an entity adaptable to an environment in continuous evolution, able to survive regardless of the onset of a given event [22, 25]. Every *viable system company* is characterized by: (i) numerous tangible and intangible sub-components; (ii) interdependence and communication between these sub-components; (iii) the need to activate the relationships in order to achieve the systems purpose [20].

A systems interpretation of resilient behaviors is made using specific reference to some FCs of the VSA and precisely to: FC5 (self-regulation), FC6 (structure/system), FC7 (consonance/resonance), FC8 (viability) [2]. First, the VSA studies have deepened the theme of systems autopoiesis and self-regulation aimed at allowing an organization to self-organize itself. These studies are useful in order to find a condition of equilibrium in contexts (stable) or turbulent situations (occasional) (FC5), trying to align internal and external complexity to reach the expected results in the most sustainable way possible [26].

Furthermore, according to the dichotomy structure/system (FC6), the set of endogenous 'structural' components operates through a virtuous interaction, in an equifinal manner and passing through different evolutionary paths [27, 28]. The behavioral and dynamic aspect, typical of the 'systems' component of every organization, must take into account the evolution of the surrounding situation to favor the right exchange of resources [29, 30] (IN.5).

The contextual analysis highlights the relationships between actors, between those potentially consonant with them, so potentially able of acting together in a harmonious way, and therefore predisposed to be resonant when these harmonious interactions take place (FC7). The systems viability (FC8), as mentioned, regards the capacity of each organization to sustain its activity over time in an effective and sustainable way [22, 25]. This leads us to reflect on what it really means "survive" and therefore successfully overcome the contingent difficulties and at the same time plan the future, even in difficult conditions of uncertainty and complexity.

The viability is the final result of all the operations in progress (including learning and experience), obtained through the fusion of many contributions from all parts of the same viable system [31, 32] and certainly appears close to the concept of resilience.

Compared to what has been said, even after the analysis carried out through the FCs, the answer to the RQ1 is Yes, again. The resilience represents a distinctive feature of the viability of organizations understood as systems, although mainly referred to the reactivity in moments of difficulty and therefore not considered for all stages of growth of an organization. This concept is summarized in the following 'Proposition' n.2.

PN.2: The systems approach allows us to interpret resilient behaviors as badges of acting viable.

Martinelli and Tagliazucchi [33] have recently made a long and careful *excursus* on the interdisciplinary theme of resilience and have approached it directly and explicitly to entrepreneurs (especially of SMEs). The authors always associate the concept of resilience with the ability to react to critical conditions creating new opportunities (organizational renewal) (IN.6) and not only reacting to something else (organizational *recovery*) [34, 35].

Sabatino [36] also try to associate resilience with the ability of entrepreneurs to remain competitive over time. In particular, he compares the vulnerability factors with 7 defined determinants of resilience (of which, here, we recall the decision-making speed (IN.7), the sense of identity and *customer centricity*).

All this is combined with the conviction that in contingent and critical situations it is also possible to adopt a *bouncing forward* attitude [15, 17] (IN.8) rather than simply bouncing back, that is, orienting oneself towards growth and moving forward rather than restoring only the starting *status quo*. Therefore, resilience can be also understood as an attitude of positive challenge [37]. Martinelli and Tagliazucchi [33] identify resilient behaviors as the willingness of entrepreneurs to get out of their *comfort zone* and enter a sort of *challenge zone* (p.40) and complete the reasoning summarizing some of the structural and behavioral characteristics of a resilient enterprise, i.e. solidity (*hardness*), *optimism* and above all the ability to regenerate resources (*resourcefulness*) (p.41) (IN.9).

When actors are looking for new interactions with other actors operating in the same context due to continuous new stimuli, they are conceptually close to the meaning of being resilient. This appears intrinsic and inherent in the existence of organizations for resilient behaviors, which create robustness in response to uncertainty and organizational flexibility in response to changes (IN.10).

Today, for companies operating in conditions of uncertainty and complexity it is much more frequent to try to update. The intent to change configuration is almost like a habit (IN.11), a natural aspect. It is no longer an occasional or temporary event.

To be effective and able to adapt in a pro-active way and survive over time, companies must stimulate a continuous confrontation with the market. They must always stimulate an active involvement of people who will have to judge the value of their offer, this regardless of the extemporaneousness of the events [38–41].

Typically, resilience is studied and observed in literature as a characteristic capable of absorbing almost entirely the negative effects of occasional and *contingent* events. This often concerns critical and sometimes *pathological* moments in the life of organizations. Sometimes that involves the development of adaptive processes. The adaptation of an organization, understood as a system (especially of business organizations), on the other hand, must be expressed in a more *continuous*, almost *physiological* way, pro-actively (IN.12), predictive, abductive, and aimed at achieving one's ultimate goal, which is the survival (IN.13).

The survival of an entity implies the persistence of identity, which does not exclude such a change [42]. In fact, we do not change to modify our identity, we change rather to try to stabilize it over time (IN.14). Given the interest in managing this type of change, we need to learn and interpret the surrounding conditions and their complexity, in order to understand how to change, when to change and why [43].

4   Reflections on the Competitiveness of Businesses, a Question of Viability

For being resilient and viable, it is important to dynamically maintain a stable balance, to modify one's organizational model according to the changing characteristics of the context (as it is personally perceived), showing awareness, self-preservation approach, decision-making speed, sense of identity, and customer centricity. It is necessary to have the ability to regenerate resources (*resourcefulness*), implement adaptive behaviors, overcome potential harmful situations and adversity (also and especially the unexpected ones). It is necessary to imagine organizational *renewal* (and not just organizational *recovery*), favor interactions and assume *bouncing forward* attitudes instead of simply *bouncing back.*

Nevertheless, being viable, unlike being resilient, is something more. It also means updating oneself as a habit and not only temporarily setting one's existence in a relational way (IN.15). It means not suffering change processes, but trying to impose them (IN.16), know how to 'read', filter and use information regardless of occasional events (IN.17), identify the most propitious moment to propose something new as an enlightened and coherent anticipation with the future.

In this sense, resilience turns out to be a component of the business viability, an element to be considered in order to aim for survival, but it is not in itself nor sufficient to achieve lasting success, as it is particularly linked to impromptu events (usually adverse). At the same time, however, it helps us to have a greater awareness, to motivate us, to stimulate us, to make us think differently (*think resilience, be resilient*) as a necessary condition for the development and growth of our activities. In essence, being resilient is not enough, but it helps.

In summary, the answer to the RQ2 therefore is No. Indeed, resilience is an expression of viability, but it is not completely superimposable on it, it rather represents a sub-component of it, as reported in the following 'Proposition' n.3.

PN.3: The determinants favoring resilience represent a contained sub-set of the determinants relating to the competitiveness of a viable system.

The competitiveness of a company (intended as viable system) is the result of far-sighted awareness (IN.18). It suggests that it is not possible to perpetrate the same behaviors (or strategies) too long.

The competitiveness of a viable system company derives from the contribution of knowledge, the application of skills, the ability to configure and re-configure [44] (IN.19), the desire to weave long-term relationships with entities considered strategic, which all represent elements of a systems and not necessarily resilient way of growing.

The competitiveness of a viable system company depends on the ability to preserve the value of its offer. It does not necessarily imply a form of resilient reaction, rather it means updating appropriately proposing a new *concept* of supply, production and use of the product offered. Companies able to maintain a sustainable value proposal over time (IN.20) show an approach that is less resilient and more oriented towards the viability.

The answer to the RQ2 therefore is again No, resilience cannot be considered as a basic element for the competitiveness of companies, especially for viable systems. Being good at not having to manage a crisis or a moment of difficulty, because you have managed to anticipate it, allows you to set and pursue new strategies and policies with greater awareness, you are competitive because you are viable, it is not enough to be resilient, as summarized for the occasion in the following 'Proposition' n. 4.

PN.4: The competitiveness of a viable system business is a matter of viability.

5   Ditron Ltd. Case-Study. Methodology and Results

In an attempt to consolidate what has been deduced so far, a case study was conducted [45, 46]. The use of a case study seemed particularly interesting to propose a more practical study respecting to the topic discussed. An only theoretical approach, in an attempt to contribute to the literature on that issue, would have been ineffective [47]. It is believed to be more useful to test a theory by looking for its dynamic details within an empirical and effective context [48].

The company studied is Ditron Ltd., an Italian multinational holding company, national leader in the design and production of cash registers and scales. Its targets are DO and GDO since 2008, and it is a clear leader in the retail market with a market share of over 50%.

In recent years, Ditron has maximized investments, always looking for innovative solutions. In fact now it develops integrated hardware and software and systems of integration towards cloud platforms. The case study was carried out through a collection and analysis of data. For the data collection, 40 interviews were carried out addressed to some key informants who were the ownership of the company. The interviews were carried out in the form of semi-structured questionnaires, moving on the basis of some pre-defined key concepts to guide the conversation. This type of interview has allowed the spontaneous emergence of further questions to clarify unclear meanings or with respect to unplanned topics and which were therefore worthy of further study, so as to be able to improve the quality of the result. The interviews lasted about an hour each and each interview was recorded and transcribed.

The primary data detected with this method have been analyzed with a logical inductive approach: the reading of the transcribed interview made it possible to refer to some specific topics and theoretical framework used to move from the particular to the general, such as the observation practice be able contribute to the validation of the theory. The data-coding took place using some basic assumptions of the theory of Systems Thinking and VSA. Threats arising from market turbulence, evolving needs and legislative changes have forced the company to manage a condition of complexity.

Ditron understood that he had to organize an articulated, integrated and functional value proposition and increase his perception of quality, resources and potential; through a transit that leads from a product oriented concept to one aimed at customer satisfaction, aiming at customer loyalty and the establishment of a lasting relationship not only with its own customers, but with all relevant stakeholders.

The key issues carried out are as follow:

Flexibility - "We have managed all issues to be competitive, because we make decisions in front of the coffee machine."

Ditron understood that its long-standing historical action needed adaptation and transformation processes to be able to maintain and improve its market positions or intercept new ones. Although its focus is on innovation, it was understood that, in order for it to materialize, a process of structural reorganization of the company within its operational context was also necessary. Over the years has begun to organize itself in a fragmented way, on several business units, which has made it more flexible and faster, able to anticipate market trends.

Over time, the company as a machine supplier has begun to propose itself to its interlocutors as a supplier of devices able to provide solutions, a set of integrated objects in order to solve a problem and satisfy a need. The owner understood the variability of market needs and he divided his organization into different individual business units, each aimed at its own reference market. The breakdown has made it possible to obtain a more agile structure, with a faster decision process and an ever wider and more varied range of services offered, which can be integrated into a single, complete and holistic value proposition.

The ability to offer integrated and complete solutions allowed Ditron to be more and more efficient compared to the expectations of its customers whose approach today is always more like "I don't want problems you do".

Dematerialization - "We have completely changed our vision for 3 years and have begun to propose ourselves as solution sellers. From cash register providers, we realized that we had become data providers and therefore a potential new service".

Technology is useful because it allows actors to be faster in answering to the market expectations. Reactions must be supported by the application of technologies within the decision support systems.

In order to reduce complexity and generate innovation, they need to be able to valorize and share knowledge, which today means first of all to collect data, detect information flows, and then convey them to those who can actually make that knowledge a value for the company. The cash register, through the collection and processing of data, would thus have made it possible to establish a direct relationship with customers and focus on emotions. Not only that, the data would have allowed retailers to obtain information also on the capacity of their own structure, on the most effective physical areas of the store and on the less visible or captivating ones, to acquire information on employee performance. All levers able to allow traders more and more weighted decisions in terms of effectiveness, but also to improve the efficiency of their structure.

Strategic collaborations - "Thanks to the collaboration with Repas Lunch Coupon Ltd., we are able to collect personalized data, so as to be always more efficient than the customers' expectations".

The new technology implemented has prompted the establishment of a new partnership, with Repas Lunch Coupon Ltd., a meal voucher company for certain categories of workers, through which retailers would have been able to acquire, through pos payments and vouchers, personal data of customers. Not only that, the data would have allowed retailers to obtain information also on the capacity of their own structure, on the most effective physical areas of the store and on the less visible or captivating ones, to acquire information on employee performance. Thanks to the new technology and this partnership with Repas, it has begun to carry out customized data collection activities. The possible processing of such data would allow customers (retailers) to adapt their proposal to the needs of specific and known consumers and offer them an increasingly personalized and personal offer and evaluate the loyalty of each. The introduction of the new technology has allowed Ditron to collaborate with new players and forge alliances, able to improve performance towards its customers.

5.1 Discussion of the Results in Terms of Resilience, Competitiveness and Viability

The results obtained through the guided observation of this case study, show how the company Ditron Ltd., although unconsciously, has positioned itself as an innovator and precursor of new co-creative logics within its own context. These results are summarized in Table 1.

Table 1. Ditron Ltd. is intended as viable system on the results of the case study

Results	VSA fundamental concepts	Implications
S.C.	FC8: The viability of the system is described by its ability to perform harmonious behavior, with relevant actors, through consonant and resonant relationships	The introduction of the new technology enabled Ditron to collaborate with new players and forge alliances, which can improve its performance towards customers
F.	FC9: The dynamics and the viable of companies are related to the continuous dynamic structural and systemic changes aimed at aligning internal structural potentials with external systemic needs	Ditron understood that in order to best adapt to market expectations, it was necessary to be fast and responsive and to have a "fluid" structure that can be quickly modeled on the basis of contextual needs
D.	FC10: Viable systems must continually look for an alignment between internal and external complexity in order to better manage the changes that affect its viable behavior	Ditron understood that in order to better manage external complexity, there was the need for a functional reconfiguration based on redefining its proposal to the market

Source: Elaboration from authors

The data emerged from the interviews regarding the advantages encountered in having a fragmented structure, useful in terms of flexibility, allows us to consider Ditron as a resilient reality. If resilience manifests itself when contextual conditions impose to react and remodel its own decision-making apparatus, in the case of Ditron, the new tax legislation with which the traceability of payments to retailers was imposed, has determined a reaction in terms of re-qualification of its own value proposition and reconfiguration of its own structure.

Through an implementation of the new technology, Ditron managed to preserve its balance (IN.1) while maintaining its market share stable. It has succeeded in activating, in response to the emergency, adaptive behaviors (IN.2) by reading the change in terms of possible codification of new consciousness (IN.9, IN.19) and useful knowledge to be able to face, based on experience, new solicitations from the context (IN.11), even functional for the configuration of new markets (IN.16; IN.18) with direct effects also on the actors connected to it (*dematerialization*). With this attitude, Ditron proves to be strongly oriented towards the long term, and therefore interested not only in the management of the current and contextual condition, but in the activation of a growth process (IN.8) essential in this way for Ditron to overcome the emergency, potentially harmful (IN.3), turning it into an opportunity through an approach to improve one's own proposal based on learning and oriented towards a viable attitude (IN.20). This substantial reconfiguration of the proposal has allowed Ditron to preserve its market share also thanks to its leadership position held until then (IN.4; IN.14). The new value proposition, characterized by the introduction of new technology, has defined for Ditron the arise of new alliances. It established new relationships within its context, for example with software house for satisfying the need for new specific resources (IN.5). He has begun a collaboration or with Repas Ltd., to overcome the concept of pathological emergency, for moving towards proactivity and physiological survival for better fit with the market needs (IN.12; IN.15; IN.17; IN 18) (*strategic collaborations*).

Ditron is an evident example of a 'flexible' enterprise. It has been able to anticipate important changes coming in its reference context, 'significantly' transforming its 'specific structure' (IN.6; IN.10) which allowed it to being able to be more agile and faster in finding solutions (IN.7) (*flexibility*).

The commentary of the results allows us to understand how Ditron has understood "when" and "how" to change effectively, in a way that more than simply being a resilient reality, as a "viable" reality, also maintaining and increasing its competitiveness.

To understand how much Ditron has managed being not only resilient, but viable, it is necessary to consider the three conditions of a viable system. It substantially satisfies three main systemic conditions: *openness*, represented by the ability to selectively exchange resources (with relevant actors); *dynamism*, understood as a coherent development of a structure with emerging changes; *contextualization*, or search for viability via interaction with contextual conditions [49].

6 Final Remarks

The concept of resilience does not concern (if not marginally) the structural characteristics but rather the system capacity of react to negative situations to find solutions
and to convert an apparent difficulty into an opportunity to co-create value [50]. The
decision maker of a viable system company must put in place all the necessary actions
not only to demonstrate its resilient abilities, but also to be able to survive over time.
The resilient behavior is therefore an element of support for a viable process of decision
making [51].

In the Service Science field, studies on smart service systems (SSS) and business
behavior related to the complexity of contexts [52, 53], in the attempt to pursue
conditions of viability [54], are very relevant. This consideration introduces the need to
understand the most appropriate behaviors for companies in complex contexts, and
therefore to analyze their possible resilient behaviors [55]. As suggested by the Service
Science, a univocal interpretative lens, a common language and a multidisciplinary
approach to this topic are necessary [56].

In sum, there is a great link between the value co-creation process and viability in
SSS. Nevertheless, this is a starting point. In the next future we want to continue this
work by making a quantitative research on variables helpful to the scope.

References

1. Golinelli, G.M.: Viable Systems Approach (VSA). Governing Business Dynamics. Kluwer
   Cedam, Padova (2010)
2. Barile, S., Polese, F.: Smart service systems and viable service systems. Serv. Sci. 2(1/2),
   21–40 (2010)
3. Norris, F.H., Stevens, S.P., Pfefferbaum, B., Wyche, K.F., Pfefferbaum, R.L.: Community
   resilience as a metaphor, theory, set of capacities, and strategy for disaster readiness. Am.
   J. Commun. Psychol. 41(1–2), 127–150 (2008)
4. Sheffi, Y.: The Resilient Enterprise. MIT Press, Cambridge (2007)
5. Hamel, G., Välikangas, L.: The Quest for Resilience. Harvard Bus. Rev. 9(81), 52–63 (2003)
6. Bhamra, R., Dani, S., Burnard, K.: Resilience: the concept, a literature review and future
   directions. Int. J. Prod. Res. 49(18), 5375–5393 (2011)
7. Paton, D., Smith, L., Violanti, J.: Disaster response: risk, vulnerability and resilience.
   Disaster Prev. Manag. 9(3), 173–180 (2000)
8. Nilakant, V., Walker, B., Rochford, K., Van Heugten, K.: Leading in a post-disaster setting:
   a guide for human resource practitioners. New Zealand J. Employ. Relat. 38(1), 1–14 (2013)
9. Weick, K.E., Sutcliffe, K.M.: Managing the Unexpected: Assuring High Performance in an
   Age of Complexity. Jossey-Bass, San Francisco (2001)
10. Lengnick-Hall, C.A., Beck, T.E.: Adaptive fit versus robust transformation: how organi-
    zations respond to environmental change. J. Manag. 31(5), 738–757 (2005)
11. Lengnick-Hall, C.A., Beck, T.E.: Resilience capacity and strategic agility: prerequisites for
    thriving in a dynamic environment. In: Nemeth, C., Hollnagel, E., Dekker, S. (eds.)
    Resilience Engineering Perspectives, vol. 2. Ashgate Publishing, Aldershot (2009)

12. Freeman, S.F., Hirschhorn, L., Maltz, M.: Organization resilience and moral purpose: Sandler O'Neill and partners in the aftermath of 9/11/01. Paper presented at the National Academy of Management meetings, New Orleans, LA (2004)

13. Hollnagel, E., Woods, D.D., Leveson, N.: Resilience Engineering: Concepts and Precepts. ASHGATE publishing, Farnham (2006)

14. Block, J.H., Block, J.: The role of ego-control and ego-resiliency in the organization of behavior. In: Collins, W.A. (ed.) Development of Cognition, Affect, and Social Relations: Minnesota (1980)

15. Vogus, T.J., Sutcliffe, K.M.: Organizational resilience: towards a theory and research agenda. Paper presented at the IEEE International Conference on Systems, Man and Cybernetics (2007)

16. Glassop, L.: The three R's of resilience: redundancy, requisite variety and resources in building and sustaining resilience in complex organization. J. Purchasing Supply Chain Manag. **16**(1), 17–26 (2007)

17. Mallak, L.A.: Putting organizational resilience to work. Ind. Manag. **40**(6), 8–13 (1998)

18. Wildavsky, A.: Searching for Safety. Transaction Books, New Brunswick (1988)

19. Dalziell, E.P., McManus, S.T.: Resilience, vulnerability, and adaptive capacity: implications for system performance. International Forum for Engineering Decision Making (IFED), University of Canterbury, Christchurch (2004)

20. Barile, S.: Management sistemico vitale. Giappichelli, Torino (2009)

21. Barile, S., Carrubbo, L., Iandolo, F., Caputo, F.: From 'EGO' to 'ECO' in B2B relationships. J. Bus. Market Manag. **6**(4), 228–253 (2013)

22. Golinelli, G.M.: L'approccio sistemico al governo dell'impresa. L'impresa sistema vitale, CEDAM, Padova (2005)

23. Parsons, T.: The System of Modern Societies. Prentice-Hall, Englewood Cliffs (1971)

24. Golinelli, G.M., Pastore, A., Gatti, M., Massaroni, E., Vagnani, G.: The firmas a viable system: managing inter-organisational relationships. Sinergie **58**, 65–98 (2002)

25. Barile, S.: (a cura di), L'impresa come sistema. Contributi sull'approccio sistemico vitale, II ed., Giappichelli, Torino (2008)

26. Barile, S., Polese, F.: Linking the viable system and many-to-many network approaches to service-dominant logic and service science. Int. J. Qual. Serv. Sci. **2**(1), 23–42 (2010)

27. Wieland, H., Polese, F., Vargo, S., Lusch, R.: Toward a service (eco)systems perspective on value creation. Int. J. Serv. Sci. Manag. Eng. Technol. **3**(3), 12–25 (2012)

28. Polese, F., Di Nauta, P.: A viable systems approach to relationship management in S-D logic and service science. Bus. Adm. Rev. Schäffer-Poeschel **73**(2), 113–129 (2013)

29. Gatti, C., Dezi, L.: Un modello di analisi delle traiettorie evolutive del sistema impresa. Struttura e governance, Esperienze d'Impresa, vol. 1, no. 1, pp. 7–27 (2000)

30. Parente, R., Petrone, M.: Strategie di co-evoluzione nei sistemi locali innovativi. Sinergie Italian J. Manag. **83**, 31–52 (2011)

31. Barile, S., Pels, J., Polese, F., Saviano, M.: An introduction to the viable systems approach and its contribution to marketing. J. Bus. Market Manag. **5**(2), 54–78 (2012)

32. Polese, F., Carrubbo, L.: Gli eco-sistemi di servizio in Sanità. Giappichelli, Torino (2016)

33. Martinelli, E., Tagliazucchi, G.: Resilienza e Impresa, l'impatto dei disastri naturali sulle piccole imprese commerciali al dettaglio. Franco Angeli, Milano (2018)

34. Kantur, D.: İşeri say: organizational resilience: conceptual integrative framework. J. Manag. Organ. **18**(6), 762–773 (2012)

35. Teixeira, E.O., Werther Jr., W.B.: Resilience: continuous renewal of competitive advantages. Bus. Horizons **3**(56), 333–342 (2013)

36. Sabatino, M.: Economic crisis and resilience: resilient capacity and competitiveness of the enterprises. J. Bus. Res. **69**(5), 1924–1927 (2016)

37. Cantoni, F.: La resilienza come competenza dinamica e volitiva. Giappichelli editore, Torino (2014)
38. Barile, S., Gatti, M.: Corporate governance e creazione di valore nella prospettiva sistemico-viable. Sinergie **73–74**, 151–168 (2007)
39. Barile, S., Polese, F., Calabrese, M., Iandolo, F., Carrubbo, L.: A theoretical framework for measuring value creation based on Viable Systems Approach (VSA). In: Barile, S. (a cura di), Contributions to Theoretical and Practical Advances in Management, Viable Systems Approach, ARACNE Ed., Roma (2013)
40. Bonfanti, A., D'allura, G.: L'incidenza della professionalità sul valore percepito dal cliente e sulla sua soddisfazione durante il service encounter (The impact of professionalism on the value perceived by the client and on his satisfaction during the service encounter. Sinergie **95**, 99–120 (2014)
41. Tommasetti, A., Troisi, O., Vesci, M.: Measuring customer value co-creation behavior: developing a conceptual model based on service-dominant logic. J. Serv. Theory Practice **27**(5), 930–950 (2017)
42. Schein, E.H.: Organizational culture. Am. Psychol. Assoc. **45**(2), 109–119 (1990)
43. Barile, S., Polese, F., Carrubbo, L.: Il Cambiamento quale Fattore Strategico per la Sopravvivenza delle Organizzazioni Imprenditoriali. In: Barile, S., Polese, F., Saviano, M. (a cura di) Immaginare l'innovazione, e-book Giappichelli, Torino. Editore, Torino, pp. 2–32 (2012)
44. Polese, F., Carrubbo, L., Bruni, R., Maione, G.: The viable system perspective of actors in eco-systems. TQM J. **29**(6), 783–799 (2017)
45. Eisenhardt, K.M.: Building theories from case study research. Acad. Manag. Rev. **14**(4), 532–550 (1989)
46. Baxter, P., Jack, S.: Qualitative case study methodology: Study design and implementation for novice researchers. Qual. Rep. **13**(4), 544–559 (2008)
47. Flyvbjerg, B.: Five misunderstandings about case-study research. Qual. Inquiry **12**(2), 219–245 (2006)
48. Stake, R.E.: The Art of Case Study Research. SAGE Publications Ltd., Thousand Oaks (1995)
49. Polese, F.: Management sanitario in ottica sistemico vitale, vol. 57. G Giappichelli Editore (2013)
50. Ciasullo, M.V., Polese, F., Troisi, O., Carrubbo, L.: How service innovation contributes to co-create value in service networks. In: Borangiu, T., Drăgoicea, M., Nóvoa, H. (eds.) IESS 2016. LNBIP, vol. 247, pp. 170–183. Springer, Cham (2016). https://doi.org/10.1007/978-3-319-32689-4_13
51. Polese, F., Tommasetti, A., Vesci, M., Carrubbo, L., Troisi, O.: Decision-making in smart service systems: a viable systems approach contribution to service science advances. In: Borangiu, T., Drăgoicea, M., Nóvoa, H. (eds.) IESS 2016. LNBIP, vol. 247, pp. 3–14. Springer, Cham (2016). https://doi.org/10.1007/978-3-319-32689-4_1
52. Barile, S., Polese, F., Carrubbo, L., Caputo, F., Wallerzky, L.: Determinants for value co-creation and collaborative paths in complex service systems: a focus on (smart) cities. Serv. Sci. **10**(4), 379–477 (2018)
53. Polese, F., Carrubbo, L., Bruni, R., Caputo, F.: Enabling actors' viable behaviour: reflections upon the link between viability and complexity within smart service system. Int. J. Markets Bus. Syst. **3**(2), 111–119 (2018)
54. Polese, F., Carrubbo, L., Caputo, F., Megaro, A.: Co-creation in action: an acid test of smart service systems viability. In: Satzger, G., Patrício, L., Zaki, M., Kühl, N., Hottum, P. (eds.) IESS 2018. LNBIP, vol. 331, pp. 151–164. Springer, Cham (2018). https://doi.org/10.1007/978-3-030-00713-3_12

55. Barile, S., Polese, F., Carrubbo L.: La resilienza come elemento base per la competitività d'impresa? No, è una questione di vitalità! Proceedings Convegno Sinergie-SIMA "Management and sustainability: creating shared value in the digital era. Roma (2019)

56. Polese, F., Barile, S., Loia, V., Carrubbo, L.: The demolition of Service Scientists' cultural-boundaries. In: Maglio, P., Kieliszewski, C., Spohrer, J., Lyons, K., Patrício, L., Sawatani, Y. (eds.) Handbook of Service Science. SSRI, pp. 773–784. Springer, Heidelberg (2018). https://doi.org/10.1007/978-3-319-98512-1_34

Modelling Service Processes as Discrete Event Systems with ARTI-Type Holonic Control Architecture

Theodor Borangiu$^{(\boxtimes)}$, Ecaterina Virginia Oltean, Silviu Răileanu,
Iulia Iacob, Silvia Anton, and Florin Anton

Department of Automation and Industrial Informatics,
University Politehnica of Bucharest, Bucharest, Romania
{theodor.borangiu, silviu, iulia.iacob, silvia.anton,
florin.anton}@cimr.pub.ro, ecaterina.oltean@aii.pub.ro

Abstract. Starting from the generic, activity-oriented Service System lifecycle model, the paper considers the service as a flow of connected activities having a discrete event nature and being formalized as discrete event systems. The service activity flow is optimized off-line by a discrete supervisor and monitored at delivery by a logical controller in a 4-layer embedded digital twin architecture. The supervisor is developed using an ARTI-type holonic architecture in which operant resources (skills, knowledge) are considered intelligent agents acting on operand resources virtualized by their associated digital twins. Because they offer reality-awareness, these virtualized entities form the "intelligent beings" layer of a holonic supervisor with distributed intelligence. The holonic approach offers new perspectives for the service management: optimization and reality awareness.

Keywords: Service System · Activity · Discrete Event System · Holon · Holarchy · ARTI reference architecture · Digital twin

1 Introduction

System theory approaches, concepts and principles (e.g., Discrete Event Systems, Viable System Approach) and Data science-based insights and tools (e.g., clustering: grouping of people, products, features; profiling: description of behaviours, detecting deviations; similarity matching: identification of similar items, features; prediction of relationships between people or items; reduction of data sets; classification: mapping of people, items, features on defined groups; causal modelling: identification of influence between events or actions) are used today to design, develop, test, deploy and monitor complex services in information-based Service Systems [1]. While the new discipline of Service Science proposes a novel perspective on creating value in a continuous interaction between the service provider and service consumer [2], more complex IT tools and methodologies are used to approach the processes of modelling, simulating, developing and tracking optimized, consumer-in-the-loop services [25, 26].

© Springer Nature Switzerland AG 2020
H. Nóvoa et al. (Eds.): IESS 2020, LNBIP 377, pp. 377–390, 2020.
https://doi.org/10.1007/978-3-030-38724-2_27

One of the first techniques used in service design was service blueprinting, a visual tool used in the Service Engineering process for service innovation and improvement [3]; it is a customer-focused approach that helps visualizing working processes associated to service development, highlighting points of customer contact in service provision.

More recent references introduce Service Systems Engineering as a methodology to approach service design that addresses a service system from a lifecycle, cybernetic and customer perspective [4]. In this approach, a service is seen as the outcome generated in a service system that has to fulfil customer expectations [5]. The Service System Engineering approach is intended to produce both good service outcome well perceived by customers, and optimized and robust service delivery process in a viable and sustainable service system [6].

The above mentioned perspective on service systems engineering opens a new line of thought over the design of information-based service systems that allow optimizing and customizing complex services, configured and validated before deploying by help of digital models and simulation techniques featuring reality awareness. These digital models can be also used in observability and controllability scenarios to track service delivery processes in real time [7]. Business Process Modelling (BPM) and Multi-Agent System (MAS) techniques were proposed as means to foster understanding the integration of different components of services and service system entities to define the activity, event, information and perception flows required for the governance, provisioning, management, tracking and follow-up of complex services [8, 9].

Formalising process configurations of service design, delivery and tracking systems fostering collaboration between service provider, customer, competitors and authority aligned with organisational practice led to a new class of IT-enabled service systems driven by customer requirements, collaboration between actors and pervasive instrumenting of resources and processes for dynamic reconfiguring: (i) service features (at customer request) or (ii) service delivery processes (at disturbances) [10, 11].

The attempts of modelling service systems from different perspectives are still in early stages of development. For example, in [12] different aspects of service modelling are presented, taking into consideration a component model, a resource model, and a process model. The modelling framework SEAM is investigated in [13] to be applied to the design and analysis of viability in service systems.

A generic, activity-oriented Service System (SSyst) lifecycle model is proposed in [14]; it has four components: Customer Order Management (COM); Service Management (SM), Service Operations Management (SOM), and Service Taxation and Invoicing (STI), that are mapped onto core service activities: COM and SM onto (1) *design and development*, and SOM onto (2) *delivery* and (3) *operations management*. STI acts as regulator onto (1), (2), (3) and service marketing. The viability of complex services is the key factor for a sustainable host Service System, and can be reached, according to the activity-based lifecycle service model, through:

1. *Optimization* of the mixed planning and scheduling of service activities and resource allocation, relative to the customer's requirements and provider's capacity and operating costs, during a Service Configuring and Set-up (SCSU) off-line stage of SM. The service level agreement (SLA) is reached in the COM through an

iterative and interactive process involving the service provider and customer; during this process, SCSU is repeatedly run, priced and its results negotiated. SCSU optimization is redone whenever the customer changes a service specification or the provider changes the operating costs or delivery conditions. An aggregated service model is needed that includes all operational, timing and logistics conditions.

2. *Reality-awareness* by activity monitoring (instrumenting equipment and processes, communication with personnel) and information transmission during service delivery. Two types of informations are generated at the occurrence of abnormal events in service delivery: (a) Exceptions with respect to normal service schedule (delayed completion of current activity, unavailable resources for next activities); (b) Alarms rose upon an interruption or failure in performing the current activity. Such events are fed to a service progress report that generates a Service Reconfiguring Request (for new resource allocation or activity redefinition) to the SCSU in real time [15].

In the perspective of information-based distributed service systems, digital twins may be used as models of discrete event services that mirror parts of the reality like "intelligent beings" [15]. They can be deployed in digitalized service systems using the ARTI-type high abstraction level holonic reference architecture with multi-agent system implementing framework.

This paper proposes a discrete event approach of the service management process using the discrete event control theory of Ramadge and Wonham [16], and defines physical operant and operand resources as "intelligent beings" modelled as digital twins, respectively operant resources as "intelligent agents" of the holonic reference architecture ARTI (Activity-Resource-Type-Instance) [15]. The control of the activities flow is exerted by a discrete event supervisor capable of reality-awareness, which forbids undesired events to occur, thus preventing integrated physical operant (people) and operand (technology) resources, initially optimized, to reach undesired states. Intelligent agents are modelled in the activity-oriented SSyst approach by two classes of holons that encapsulate decision making rules (expertise type) respectively technology (expertise instance or implementing software). The holarchy describes the way in which holons are integrated and generated starting from basic service resources.

The paper is structured as follows. Section 2 introduces the basic concepts of discrete event control theory applied to event-driven service processes. Section 3 integrates the service system entities, defined according to the first foundational premise of service science, in the ARTI-type holonic reference architecture for optimization and reality awareness purpose. Section 4 transposes the discrete service model in the embedded digital twin for deployment in service optimization and reconfiguring strategies. The association of service management and control functions with the holonic paradigm is justified in Sect. 5, followed by concluding remarks.

2 Discrete Event Control and Service Processes

Service activities designed in the SCSU off-line SM stage, respectively monitored and managed during delivery in the SOM stage are event-triggered; they can be represented as discrete event systems (DES). The two defining characteristics of a DES are: (1) the state space Q is *discrete* and (2) the dynamics are *event-driven*, as opposed to time-driven. Consider that Q is a finite set. Denote $\sum$ the finite alphabet of possible events and $\sum^*$ the set of finite strings of events from $\sum$, including the *empty string* ε. A string $\omega = \sigma_1\sigma_2...\sigma_i \in \sum^*$ is a (partial) events path [16].

In the theory of Ramadge and Wonham, a *process* is a logical DES modelled as a *generator* $P = (Q, \Sigma, f, q_0)$, where $f : Q \times \Sigma \to Q$ is the state transition function and q_0 is the initial state. The process behaviour is given by the set of *physically possible strings of events* from $\sum$, which generates o formal language over $\sum$, denoted $L(P)$. Once arrived in a state $q \in Q$, the (service) process can generate events from $\Sigma(q) \subseteq \Sigma$. The ordering index of the events from $\sum$ is denoted by the integer variable k, which is also called *logical moment*.

In supervisory control, it is considered that some of the uncontrolled process behaviour is *illegal* or undesired, and thus must be prevented to occur under control. A controlled process in a state $q \in Q$ can be viewed as DES that receives a *list of forbidden events* Φ and outputs an event from $\Sigma(q)\backslash\Phi$. The actions of the *supervisor* are limited, in the sense that it can prevent to occur only *controllable* events from $\sum$ and it can detect only *observable* events from $\sum$. From the controllability viewpoint, denote $\sum_c$ the *set of controllable events* and consider the partition $\Sigma = \Sigma_c \cup \Sigma_u$, where $\sum_u$ is the set of *uncontrollable events* of the process.

In service-dominant (S-D) logic, the service is viewed as the application of competencies for the benefit of another party, and operant resources (skills and knowledge) act upon operand resources (computers, control information) [17]. Thus the service can be regarded as a controlled flow of activities performed by operant resources. Activities have essentially a discrete event nature; they can be described as sequences of discrete event systems. In an approach based on discrete event control theory, the operant resources in a service system can be modelled as DES with events triggering begin and end of activities as units of work that can be abstracted as sequences of discrete events: a state transition diagram for an activity instance a may generate, for example, the sequence $\omega = i_a e_a b_a t_a$, where i_a is the event *initialize*, e_a is the event *enable*, b_a is the event *begin* and t_a is the event *terminate*.

Considering the operant resources of a service system as agents that perform atomic or composed activities, one can model the operant resources as discrete event systems which evolve generating strings of events. These strings describe the flows of activities performed by agents. The flows of activities may be controlled by distributed control patterns, like in the business process approach [18], or the control may be centralized, as in the DES supervisory theory. The operand resources are modelled as logic automata and their integrated behaviour, i.e. the flow of activities defining the service, is controlled by a discrete event supervisor.

This leads to the feedback scheme in Fig. 1, where P is the **service process** under the control of the **supervisor** S. The control action of the supervisor at the logical moment k is the list of forbidden events $\Phi^{(k)} \subset \Sigma_c$. The supervised service process $P(\Phi^{(k)})$ is in state q and, at the logical moment $(k + 1)$, it will generate an *allowed event* $\sigma^{(k+1)} \in (\Sigma(q)\backslash\Phi^{(k)})$. The occurrence of $\sigma^{(k+1)}$ may determine a transition of the supervisor into a new state, and instantly S will output a new list of forbidden events, $\Phi^{(k+1)}$ which may be eventually distinct from the previous one, and so on.

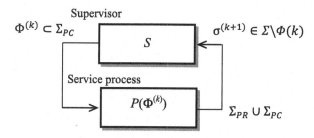

Supervisor

$$\Phi^{(k)} \subset \Sigma_{PC}$$

S

$$\sigma^{(k+1)} \in \Sigma \backslash \Phi(k)$$

Service process

$$P(\Phi^{(k)})$$

$$\Sigma_{PR} \cup \Sigma_{PC}$$

Fig. 1. The feedback loop of supervisory service process control (k is some logical moment and $\Phi(k)$ is the current list of forbidden events).

Two facts are notable: (*i*) the observation of the service process P by the supervisor S is *asynchronous* and (*ii*) since S can only prevent some events to occur, it follows that it can only *restrict* the evolution of the service process P. Between two successive events occurrences, $\sigma^{(k)}$ and $\sigma^{(k+1)}$, the supervisor output $\Phi^{(k)}$ remains unchanged.

The *supervision language* $L(S)$, generated by S, is the set of all events strings, which are *authorized* by S. The *closed loop behaviour* is given by the evolution of the service process coupled to the supervisor, i.e. the evolution of the controlled service process. According to the feedback scheme depicted in Fig. 1, a string of events ω can be generated in closed loop, if it can be generated by the uncontrolled service process ($\omega \in L(P)$) and if it is authorized by the supervisor ($\omega \in L(S)$). Denote S/P the automaton representing the process P coupled to the supervisor S. The associated language is $L(S/P) = L(P) \cap L(S)$ and it defines *the closed loop behaviour*.

For a given service process P and its language $L(P)$, the necessary and sufficient condition for the existence of a supervisor is formulated as follows: for a prefix-closed and non-empty language $L_D \subseteq L(P)$, there exists a supervisor S such that $L(S/P) = L_D$ if and only if L_D is *controllable with respect* to $L(P)$. The concept of language controllability reflects the fact that only controllable events can be prevented to occur by supervision.

3 Service System Entities: Holons in the ARTI-Type Holarchy

As previously declared, the objectives for the design of the information-based SSyst are double: (i) *optimization* of multi-activity services in the off-line service configuring (SCSU) stage; (ii) *reality awareness* in the service delivery stage. The first objective involves mixed service process planning, activity scheduling and resource allocation in the global activity context of the provider's organization while the second one relies on instrumenting operand resources, tracking their state and behaviour, detecting changes/unexpected events in the service delivery process and resource breakdowns.

A supervisory service process control (S in Fig. 1) is therefore needed. The holonic paradigm is proposed to formalize this control process, because holons are well suited to represent highly abstracted SSyst entities (service activities, resources) in distributed, closed control loops of service processes, for agentified ICT implementing.

A holon has the dual capacity to preserve and assert its individuality as quasi autonomous whole, and to function as an integrated part of a larger whole. This polarity between the *self-assertive* and *integrative* tendencies is inherent in the concept of hierarchic order - a universal characteristic of life (Koestler [19]). Holons are included in other holons in vertical structures with nested hierarchical order [20] called holarchies. A holarchy is a hierarchy of self-regulating holons that function first as autonomous wholes in supra-ordination to their parts, secondly as dependent parts in sub-ordination to controls on higher levels, and thirdly in coordination with their local environment.

The holonic paradigm is applied to supervised control of service processes in ICT-based service systems in two steps:

1. Establishing which parts of the reality of interest (e.g., service processes and activities) will be modelled by DES represented software as digital twins, and how these virtual models will be networked, aggregated and embedded at logical control and supervision level.
2. Building up the holarchy and its components (holons) that model the stages of the SSyst (e.g., service design, configuring, delivery, monitoring); for this purpose the ARTI (Activity-Resource-Type-Instance) holonic reference architecture [15] is used and further implemented in multi-agent system (MAS) framework.

ARTI builds on PROSA [21] and upgrades it by: (i) *Generalisation*: the names of the PROSA holons are replaced by generic ones, applicable beyond manufacturing to other classes of discrete, multi-resource service activities like those in the SCSU and SOM stages; (ii) *Reality awareness*: this architecture reduces the gap between the control software and the reality of interest defined by highly abstract elements (activities, resources and outcomes) mirrored by digital twins embedded in SSyst stages.

The more abstract and generic interpretation of ARTI turns some scientific laws of the artificial [22] into unavoidable implications of bounded rationality:

- Flexible time-variant aggregation hierarchies are mandatory for the adaptation to a dynamic environment and thus hierarchies are time-variant (e.g. mirroring corresponding reality).
- Autocatalytic sets that include human resources are crucial for viability in service systems.
- Intelligent beings are separated from intelligent agents, and decision making technologies from the deployment of the selected decision making mechanisms.
- Proactive behaviour and coordination requires the ability to include the impact of future interactions, i.e., short term forecasts from the intentions of activity instances when multiple actors operate in a shared working space.

The holarchy for optimized and robust at disturbances (reality-aware) service management in ICT-based SSyst includes three classes of basic holons: *service* (process) *holon*, (delivery) *order holon* and *resource* (people, technology) *holon*, and one class of *expertise* (supervisory) *holon*. These classes are derived from the highly-abstracted, generic entities: "activity", "resource", "type" and "instance" of the ARTI holonic reference architecture represented in Fig. 2, in the form of the coloured ARTI-cube, which focuses on the relationship between (material) reality and the ICT-based Service System organization.

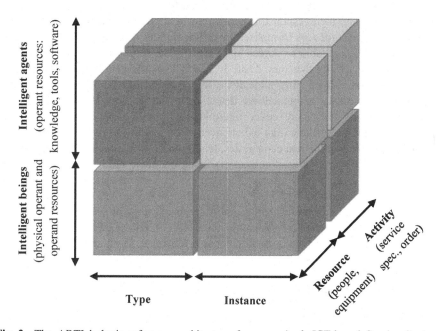

Fig. 2. The ARTI holonic reference architecture for supervised, ICT-based Service Systems (adapted from [15])

Service (process) holons contain the specifications (technical, logistics) reached in the SLA by the parties and are defined as *activity types*. Order holons become *activity instances* that are optimally calculated in the SCSU stage of service management (e.g. sequencing activities, allocating resources), while *resource* holons are subdivided into *types* (people, technology) and *instances* (e.g., assigned personnel, used equipment).

Decoupling decision making (i.e. the plug-ins for SCSU or SOM) from the service processes becomes a top-level feature of the architecture, separating *intelligent agents* from intelligent beings. The first agents are classified into *types* (e.g., optimized planning methods, management principles) and *instances* (i.e., the connection of these concepts and mechanisms to intelligent control plug-ins and software tools). Expertise holons reside in the green cubes; the quality of supervision they assure upon composite services depends on how the yellow components connect them to the physical operands of the SSyst. Thus, ARTI subdivides an ICT-based service system to assure flexibility and reality awareness.

The blue part of the ARTI holonic cube complies with the theory of flexibility (or design for the unexpected [22]). The green part is founded on basic principles of service-, system- and data science, e.g., mixed-integer programming, constrained optimization, recurrent neural network for resource state prediction, selected as needed in the SM and SOM stages: service process optimization in SCSU, service delivery monitoring in SOM, etc. The yellow part is represented by the control technologies (instrumenting resources, detecting events) and software tools that implement the green methods; the yellow parts introduce limitations and restrictions that do not exist in the world of interest and their responsibilities should be limited as much as possible.

Activity and resources are formalized by distinct holons such that an activity is free to search and utilise suitable resources in an appropriate sequence and the resource is able to serve any activity that may desire it. This allows the yellow software parts to optimally perform: (a) off line allocations in the SCSU stage, and (b) real time assignments at anomaly or breakdown detection in the delivery stage. Humans are activity performers; they may cover multiple cubes in the ARTI model without inducing the penalties to the extent experienced by the yellow software agents.

The holistic view of the real capabilities, features and states of resource and activity instances (service processes and resources) including their DES operating models, execution context, behaviour, and status in time can be encapsulated in *digital twins* (DT) that represent the blue ARTI parts – the intelligent beings. A DT is defined as an extended virtual model of a physical resource, service process or delivery order that is persistent even if its physical counterpart is not always on line/connected; this extended digital model can be shared in a cloud database with other SSyst entities [23].

4   Embedded Digital Twins for Service System Supervision

The supervised service process management and control (ICT-based configuring, set up and monitoring) can be implemented in the new ARTI holonic reference architecture based on the concepts of embedded and networked digital twins that provide a collective and predictive situation awareness thus bringing closer the software control to the real service process evolution during the delivery stage.

The 4-layer structure of the digital twin embedded in the supervised management and control of an aggregate service process is shown in Fig. 3. An example is suggested – public transport service; the transport process, the resources (busses) and activity performers (bus drivers) are defined as physical twins. There are multiple aggregations of DTs: (1) virtual twins of service processes (DES) and resources, and (2) predictive and decision-making twins projected in the supervisor for global service optimization and service process resilience. The DT layers are:

- *Data acquisition and transmission*: this layer defines the acquisition and eventually edge computing of data collected from service processes, technology resources and activity performers (humans operators), and how data streams are joined in time.
- *Virtual twins of service activities and resources*: DES models of individual service activity progress and resource operation, which are aggregated to provide in real time information about the evolution of a composite service. Resource virtualization with DT allows for secure communication with the SSyst supervisor.
- *Data analysis*: predictive twins directly fed with activity and resource data; using machine learning, they predict evolutions of service processes and resources and

detect anomalies. The output of the DT models is compared with sensor data of same type to detect deviations from normal operating or anomalies.

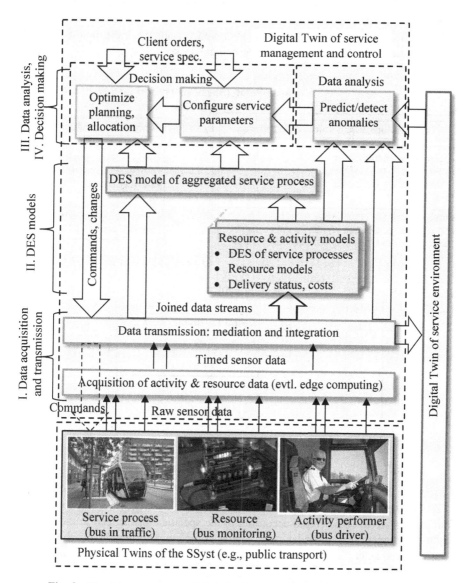

Fig. 3. The 4-layer aggregated digital twin embedded in the supervised SSyst

- *Decision making*: these are twin projections on the highest aggregation layer that apply the insights into the evolution of processes and operations of resources in supervised service monitoring and control. The role of the decision-making twins is twofold: optimally configuring service processes in the SCSU stage and triggering of appropriate corrective actions during the delivery stage at anomaly detection

5   Service Optimization and Control in the Holonic Approach

This section discusses the holarchy for the optimization, monitoring and reconfiguring of composite, multi-operation services, based on the ARTI reference architecture. Figure 4 shows how functions and data are mapped onto the basic holons classes, previously placed on the ARTI "Intelligent beings" layer.

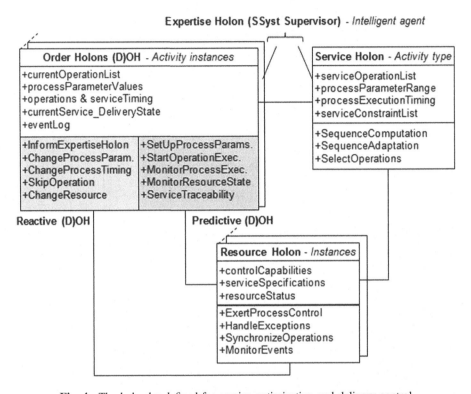

Fig. 4. The holarchy defined for service optimization and delivery control

The Service (process) holon, seen as a generic activity type, holds the knowledge about the service to be delivered according to the specifications agreed in the SLA by both the customer and provider. It contains complete information on the service operations, sequences of activities, permissions, timing and exceptions. As such, it contains the "service model" of the activity type, not the "service state model" of one instance being executed; it acts also as information server to optimization.

Resource holons classify: (a) types: human (performers) and equipment (technology) characterized by roles, skills, operating styles, respectively technical characteristics, operating modes and exploiting rules embedded in dynamic models of digital twins; (b) instances of these two type classes that are selected to deliver the contracted service.

Order (delivery) holons are activity instances of a service process type; each instance contains the necessary information to implement service delivery in terms of: resources, timing and delivery conditions, and process parameterization for client orders optimally grouped and planned off-line by the Expertize holon (Supervisor).

Order holons send to the resource holons the sequence and parameters of the service activities, start service delivery according to the timing optimally computed by the expertise holon (the SSyst supervisor), and monitor the occurrence of predicted service process events (maintaining an event log); these functions are performed by the *predictive part* of the order holon, see Fig. 4.

The *reactive part* of order holons is enabled whenever unexpected events occurring at service delivery time are detected by the Data analysis twin upon receiving specific information from the resources' virtual twins. If significant deviations from the optimized activities sequence are identified or anomalies are predicted by the centralized expertise holon, new reactive order holons are created (if possible) by the intelligent supervisor agent. New service process specifications are calculated, operations rescheduled and resources reallocated by the aggregate Decision-making twin of the supervisor; these changes are applied by the new (delivery) order holon (D) OH.

For each contracted service, a logical control is created in the ICT-based SSyst at delivery time. In the DES model of a service deployed with digital twins of resources and operations, the events in Σ_{PR} model the information coming from sensors, while the events in Σ_{PC} represent the commands issued by the logical control which is included in P (see Fig. 1). We consider the outputs of the logical service control as *potentially controllable events*, i.e. $\Sigma_c \subset \Sigma_{PC}$, and the outputs of the service process as *uncontrollable events*, $\Sigma_{PR} \subseteq \Sigma_u$; depending on the corresponding enabled discrete state transitions, the latter may signal the transitions *to normal* or *pre-alarm* states in the delivery stage:

- If transitions to normal states are observed, then *no control-symbol is forbidden* and the service's logical control remains in charge. We designate these events signalling normal transitions in the service evolution as **predictable events**, Σ_{PR_p}.
- If a detected event signals a transition associated to a possible technical disturbance or an improper activity evolution, then the SSyst Supervisor *forbids those control-symbols* whose actions do *not* ensure a stable service delivery evolution and the service parameters (re)configuring unit sends new appropriate values to the logical service control. We designate these events that signal abnormal state transitions in the evolution of the process as **unexpected events**, Σ_{PR_u}.

The supervision task can be designed to restrict the evolution of the service delivery process in a sequence with cost optimization – obtained through mixed service planning, operation scheduling and resource allocation. The intelligent agent chosen to instantiate the generic optimization mechanism is the IBM ILOG CPLEX optimizer engine technology [24].

The service control task is thus decoupled from the supervision task, and hence the supervision restrictions can be changed even in real time, during service delivery. The programming technology used to implement the SSyst distributed supervisory control is object-oriented JADE that builds up the ARTI-type holarchy with clusters of multi-agent systems.

6   Conclusions

The paper proposes a discrete event approach of Service Systems within the discrete event control theory. Physical operant and operand resources are modelled as discrete event systems that evolve generating events and triggering activities. The flow of activities composing the service is monitored by a logical controller and optimized by a global supervisor. The discrete event SSyst supervisor is implemented software in a 4-layer embedded digital twin (DT) the first two lower layers of which are distributed in multi-agent clusters, whereas its two higher Data analysis and Decision making layers are centralized. Activity instances (executions of delivery orders) and resource allocation techniques are the DTs of the SSyst decision-making part; predictive situation awareness can be reached making available a collective global image through: (i) virtual execution of decisions at complete horizon of the composite service at a high frequency that exceeds its real-world execution, and (ii) prediction of the physical twins' evolution, e.g., of technology resources.

These embedded digital twins are involved in all activities that imply their physical twins – e.g. service set up and optimized configuring, delivery monitoring and update. The main benefits brought by the digital twin concept are: (a) *visibility*: DTs allow visibility in the operations of resources; (b) *prediction*: using various modelling techniques (DES – based), the DT model can be used to predict the future state of a service process or resource; (c) *interaction with the DES model*: simulate conditions that are impractical to create in real life through "what if" analysis; (d) *documenting*: mechanisms to understand and explain behaviours of individual or interconnected resources; (e) *integration*: the DT model can be used to connect with backend business applications to co-create value.

The holonic reference architecture ARTI is proposed to formalize the supervision of service processes. ARTI-type abstract entities are used to formalize the closed loop service control. The information part of the basic holons is represented by the low level virtual twins that provide in real time information about resource capabilities, behaviour and state, and about service operations work-in-process execution state; because they offer reality-awareness, these virtualized entities form the "intelligent beings" layer of ARTI. By help of predictive twins, unexpected events leading to anomalies can be detected, and future evolutions can be foreseen and influenced by decision making twins that influence logical control and supervision – implemented in ARTI as "intelligent agents" that form an expertise holon.

The research reported in this paper demonstrates that the holonic paradigm that was first applied to discrete manufacturing control can be used for composite, multi-stage service supervision in ICT-based Service Systems with the benefits that holonic organisations provide to living organisms: direct connectivity between physical operant and operand service resources and their virtual twins, cost optimization, robustness at disturbances, service process resilience and service system sustainability. The dual holonic ARTI supervising solution is under development at present for a public private transport service.

The reported research will be extended in the direction of after sales services running over the entire product's life cycle; a strategy is constructed to define digital twins that provide the necessary information for maintenance, repair and upgrade.

References

1. Borangiu, T., Polese, F.: Introduction to the special issue on exploring service science for data-driven service design and innovation. Serv. Sci. **9**(4), v–x (2017). https://doi.org/10. 1287/serv.2017.0195. Informs PubsOnLine
2. Spohrer, J.C., Demirkan, H., Krishna, V.: Service and Science, in Service Science: Research and Innovations in the Service Economy. Springer, Berlin (2011)
3. Bitner, M.J., Ostrom, A.L., Morgan, F.N.: Service blueprinting: a practical technique for service innovation. Calif. Manag. Rev. **50**(3), 66–94 (2008)
4. Lopes, A.J., Pineda, R.: Service systems engineering applications. Procedia Comput. Sci. **16**, 678–687 (2013)
5. Slack, N., Brandon-Jones, A.: Operations Management, 9th edn. Pearson, London (2019). ISBN 978-1292253961
6. Peters, C., Maglio, P., Badinelli, R., Harmon, R.R., Maull, R., Spohrer, J.C.: Emerging digital frontiers for service innovation. Commun. Assoc. Inf. Syst. **39**(1), 136–139 (2016). https://doi.org/10.17705/1CAIS.03908
7. Borangiu, T., Răileanu, S., Voinescu, I., Morariu, O.: Dynamic service capacity and demand matching in a holonic public transport system. In: Borangiu, T., Drăgoicea, M., Nóvoa, H. (eds.) IESS 2016. LNBIP, vol. 247, pp. 573–589. Springer, Cham (2016). https://doi.org/10. 1007/978-3-319-32689-4_44
8. Meierhofer, J., Meier, K.: From data science to value creation. In: Za, S., Drăgoicea, M., Cavallari, M. (eds.) IESS 2017. LNBIP, vol. 279, pp. 173–181. Springer, Cham (2017). https://doi.org/10.1007/978-3-319-56925-3_14
9. Wooldridge, M.: An Introduction to MultiAgent Systems. Wiley, UK (2011). ISBN 978-0-470-51946-2
10. Borangiu, T., Oltean, E., Răileanu, S., Anton, F., Anton, S., Iacob, I.: Embedded digital twin for ARTI-type control of semi-continuous production processes. In: Borangiu, T., Trentesaux, D., Leitão, P., Giret Boggino, A., Botti, V. (eds.) SOHOMA 2019. SCI, vol. 853, pp. 113–133. Springer, Cham (2020). https://doi.org/10.1007/978-3-030-27477-1_9
11. Peng, Y.: Modelling and designing IT-enabled service systems driven by requirements and collaboration. Ph. D. thesis, L'institut National des Sciences Appliquées de Lyon (2012)
12. Böttcher, M., Fähnrich, K.-P.: Service systems modelling: concepts, formalized meta-model and technical concretion. In: The Science of Service Systems, Service Science: Research and Innovations in the Service Economy, pp. 131–149 (2011). https://doi.org/10.1007/978-1-4419-8270-4_8
13. Golnam, A., Regev, G., Wegmann, A.: On viable service systems: developing a modeling framework for analysis of viability in service systems. In: Snene, M., Ralyté, J., Morin, J.-H. (eds.) IESS 2011. LNBIP, vol. 82, pp. 30–41. Springer, Heidelberg (2011). https://doi.org/10.1007/978-3-642-21547-6_3
14. Borangiu, T., Drăgoicea, M., Oltean, V.E., Iacob, I.: A generic service system activity model with event-driven operation reconfiguring capability. In: Borangiu, T., Trentesaux, D., Thomas, A. (eds.) Service Orientation in Holonic and Multi-Agent Manufacturing and Robotics. SCI, vol. 544, pp. 159–175. Springer, Cham (2014). https://doi.org/10.1007/978-3-319-04735-5_11

15. Valckenaers, P.: ARTI reference architecture – PROSA revisited. In: Borangiu, T., Trentesaux, D., Thomas, A., Cavalieri, S. (eds.) SOHOMA 2018. SCI, vol. 803, pp. 1–19. Springer, Cham (2019). https://doi.org/10.1007/978-3-030-03003-2_1

16. Ramadge, P.J., Wonham, W.M.: The control of discrete event systems. Proc. IEEE **77**(1), 81–98 (1989)

17. Maglio, P.P., Vargo, L.S., Caswell, N., Spohrer, J.: The service system is the basic abstraction of service science. Inf. Syst. e-Bus. Manag. **7**, 395–406 (2009)

18. Weske, M.: Business Process Management. Concepts, Languages, Architectures, 2nd edn. Springer, Heidelberg (2012). https://doi.org/10.1007/978-3-642-28616-2

19. Koestler, A.: The Ghost in the Machine. Arkana, London (1967)

20. Mella, P.: The Holonic Revolution. Holons, Holarchies and Holonic Networks. The Ghost in the Production Machine. Pavia University Press, Pavia (2009)

21. Van Brussel, H., Wyns, J., Valckenaers, P., Bongaerts, L., Peeters, P.: Reference architecture for holonic manufacturing systems: PROSA. Comput. Ind. **37**(3), 255–276 (1998)

22. Simon, H.A.: The Sciences of the Artificial. MIT Press, Cambridge (1996)

23. Oracle: Digital Twins for IoT Applications: A Comprehensive Approach to Implementing IoT Digital Twins (2017). http://www.oracle.com/us/solutions/internetofthings/digital-twins-for-iot-apps-wp-3491953.pdf. Oracle White Paper, January 2017

24. IBM: IBM ILOG CPLEX Optimization Studio v12.9 (CJ4Z5ML) (2018). https://www-01.ibm.com/software/info/ilog/

25. Barile, S., Polese, F.: Smart service systems and viable service systems: applying systems theory to service science. Serv. Sci. **2**(1/2), 21–40 (2010)

26. Borangiu, T.: Digital transformation of manufacturing. agent technology and service orientation. plenary talk at the digital transformation. pursuit of relevance, excellence, and leadership. In: IBM Academic Days 2016 Conference, Lyon, France, May 19–20 (2016). http://site.ibm-academicdays2016.com

The Role of Error Management Culture and Leadership on Failures and Recovery in Services

Teresa Proença[1,2]($\boxtimes$) (iD), João F. Proença[1,3] (iD), and Inês Teixeira[1]

[1] School of Economics and Management, University of Porto, Porto, Portugal
{tproenca, jproenca}@fep.up.pt,
inesiteixeira21@gmail.com
[2] CEF.UP, School of Economics and Management,
University of Porto, Porto, Portugal
[3] Advance/CGS, ISEG, University of Lisbon, Lisbon, Portugal

Abstract. This study aims to establish the relationship between the leader behavioral integrity and the service recovery of the employees in services, considering also the influence of the error management culture and job satisfaction. An online questionnaire was directed to Portuguese service workers and a sample of 142 responses was collected. Data were analyzed by means of the software tool SmartPLS 3.0, and structural equations analysis was conducted. The study verified four of the seven hypotheses of the research. An important contribution of this research was the conclusion that the error management culture is a mediator of the relationship between leader behavioral integrity and the service error recovery performance of the employees, a relationship that has not yet been tested in the literature.

Keywords: Service recovery performance · Leader behavior integrity · Job satisfaction · Error management culture · Service failure

1 Introduction

Failure and errors occur in all organizations and can result in numerous negative consequences such as lost time, defective products, quality losses, increased costs, lost revenue, decreased employee performance and morale, lost clients and even physical injuries [1–3]. In service we can find errors and more specifically failure and service recovery. Service recovery refers to the actions an organization takes to respond to a service failure [4–8]. Managing these failures effectively is crucial because well-executed recoveries improve customer satisfaction, and poorly performed recoveries lead to customer displeasure [7, 9–13]. Errors and failures may develop within the organization, not occurring within the customer service itself but within the organization in the course of a day's work and which likely end up having an impact on the overall quality of the service that is provided [1, 14–16].

Research analyzing such errors and organizational aspects related with service failure is scarce, and more specifically the role that a leader can play in the effectiveness

© Springer Nature Switzerland AG 2020
H. Nóvoa et al. (Eds.): IESS 2020, LNBIP 377, pp. 391–398, 2020.
https://doi.org/10.1007/978-3-030-38724-2_28

of service employees to measure and overcome error situations or service failure and recovery.

This paper investigates the relationship between leader behavioral integrity and service recovery performance discussing the mediating role of both job satisfaction and error management culture. The literature is presented in order to formulate several research hypotheses, which will be presented afterwards. A survey was passed to collect data from service employees working in several sectors. Then, structural equation analysis was conducted [18, 39]. The data were analyzed trough PLS-SEM Smart PLS 3.0 [17, 18] and discussed. The following sections present the research, our findings and implications and finally the paper contribution.

2 Literature Review

Errors or mistakes are unintentional deviations from goals, standards, codes of behavior, truth, or some true value [19]. Failure and errors occur in all organizations and can result in numerous negative consequences such as lost time, defective products, quality losses, increased costs, lost revenue, decreased employee performance and morale, lost clients and even physical injuries [1, 3].

In service we can find errors and more specifically failure and service recovery. Failure in services has been defined as the performance of a service below customer expectations [2, 20, 21]. Most service failure/recovery studies focus on customers and the importance of employees' role in minimizing errors in providing direct customer service.

However, errors and failures may develop within the organization, not occurring within the customer service itself but within the organization in the course of a day's work and which likely end up having an impact on the overall quality of the service that is provided [1, 14–16].

Error recovery performance is seen as a junction of service recovery performance with error management [14]. This new concept encompasses not only customer service failures but also internal errors. And, it is known that there is a relationship between the behavioral integrity of a leader [22] and error recovery performance [14]. The behavioral integrity of a leader is the alignment between a leader's words and deeds [22], and is one of the factors that influence a follower's decision-making process [23, 24]. Over the years, several researches have underpinned this premise [25, 26], proving that there is a strong relationship between a leader who acts on his words and the highest effective commitment of his/her followers to the organization.

It is interesting to research, which is a component of very specific service employee performance through the use of the concept of the error and failure management culture, which is transmitted to service employees by their leaders as the mediator of the concepts besides the leader influence, the error and failure management culture may also influence service employee performance. The underlying premise is that the front office employees' perceptions of how the organization trains, rewards, and manages its employees, result in feelings such as satisfaction and attachment to the organization, which in turn determine the service recovery performance [27, 28]. Moreover, organizational culture reflects in part its leader's behaviors and assumptions [29, 30].

This research intends to broaden the study by Guchait *et al.* [14] by researching not only the relationship between a leader's behavioral integrity and employee service recovery performance, but also the influence that error management culture has on that relationship. The perception of consistency between a leader's words and deeds sets the direction for his followers, promoting clear values and directions with which followers can identify [26]. As a result, the trust and credibility of leadership will have power over employees' adaptation to service delivery and a strong incentive for employees to create value-added initiatives.

As a result, the trust and credibility of leadership will have power over employees' adaptation to service delivery and a strong incentive for employees to create value-added initiatives. The relationship between service quality commitment management and service recovery performance was researched [27, 28], as well as the error management culture and its potential effects [31]. The error management culture encompasses organizational practices related to communicating errors and failures, sharing knowledge inherent to them, helping in situations of error and quick detection and treatment of errors and failures [31]. There is a positive correlation between common practices of error management culture and the performance of organizations by reducing the negative impacts of errors, learning, innovation and improving the quality of products, services and procedures [31]. Open communication about errors may enable employees to learn and handle faster the situation, reducing the possible negative consequences of errors and inhibition due to fear of making mistakes [32].

Finally, it was proposed the use of the job satisfaction as a mediator variable [14], which fell under the same logic as organizational commitment. The behavioral integrity of a leader leads employees to identify with the leader and the organization, since when the goals and expectations are clear it is conveyed a sense of security for employees about what will happen and as so, employees will be more satisfied [33].

Synthetizing, this research aims to study how the leader behavior integrity impacts over management culture and on service employees to be motivated and to believe more in what they do, improving their effectiveness, and then their service or error recovery tends to be greater and/or faster.

Accordingly, seven hypotheses were formulated based on the literature presented before, see Table 1, and a theoretical research model is proposed, see Fig. 1.

Table 1. Research hypotheses

H1	Leader's behavioral integrity has a positive impact on error recovery performance
H2	Leader's behavioral integrity has a positive impact on the error management culture
H3	The error management culture has a positive impact on error recovery performance
H4	The error management culture can mediate the relationship between the leader's behavioral integrity and error recovery performance
H5	Leader's behavioral integrity has a positive impact on job satisfaction
H6	Job satisfaction has a positive impact on error recovery performance
H7	Job satisfaction can mediate the relationship between the leader's behavioral integrity and error recovery performance

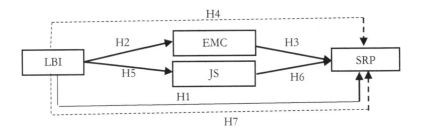

Fig. 1. Theoretical research model (Source: the authors)

3 Methodology

We research the relationship between leader behavioral integrity and service recovery performance discussing the mediating role of both job satisfaction and error management culture. Thus, the research will analyze and discuss the direct and indirect relationships between leader behavioral integrity and service recovery. The indirect relationship will be inquired through means of error management culture and job satisfaction (see Fig. 1).

The data were collected through an online survey from 184 service employees in different sectors between June and July 2018, and a total of 142 answers were validated.

The questionnaire was validated and based on the following researches and scales: behavior integrity (LBI) [23], error management culture (EMC) [31], job satisfaction (JS) [34], and service recovery performance (SRP) [26]. The questions require an answer using a Likert scale from 1 to 5 (strongly disagree with strongly agree).

Finally, structural equation analysis, using *Partial Least Squares Structural Equation Modelling* – PLS-SEM SmartPLS 3.0 [17, 18] was used to analyze and discuss the results.

4   Results and Findings

Adequate values were obtained for the measurement model: convergent validity and composite validity between 0.511–0.783 and 0.744–0.955, respectively. Discriminant validity cross loading test and the Fornell and Larcker criteria [35] also indicate adequate values.

The structural model reveals that the indicator Stone-Geisser Q^2 of the quality of accuracy of the adjusted model is adequate, and bigger than zero [18]. Regarding the Pearson's coefficients of determination (R^2), the model explains 34.3% of EMC, 22.5% of JS and 16.2% of SRP.

The significance and height of path coefficients [18] reveal the confirmation of H2, H3, H4 and H5. We found that EMC mediates the relationship between LBI and SRP, and thus LBI has an indirect effect in SRP (H4). We show that LBI has a positive impact on EMC (H2) and that EMC has a positive impact on SRP (H3). Contrary to the previous work [14], we found that:

(i)   LBI has not a positive direct impact on SRP (rejecting H1);
(ii)   JS has not been confirmed as a mediator between LBI and SRP (rejecting H7);
(iii) There was not found a positive relationship between JS and SRP (rejecting H6).

The Table 2 shows and summarizes our findings related with the hypotheses formulated.

Table 2. Validation of the research hypotheses

H1	Leader's behavioral integrity has a positive impact on error recovery performance	Rejected
H2	Leader's behavioral integrity has a positive impact on the error management culture	Confirmed
H3	The error management culture has a positive impact on error recovery performance	Confirmed
H4	The error management culture can mediate the relationship between the leader's behavioral integrity and error recovery performance	Confirmed
H5	Leader's behavioral integrity has a positive impact on job satisfaction	Confirmed
H6	Job satisfaction has a positive impact on error recovery performance	Rejected
H7	Job satisfaction can mediate the relationship between the leader's behavioral integrity and error recovery performance	Rejected

It is highlighted that our results do not suffer from the common method bias, as recommendations of the literature [39] were followed. Moreover after the Harman test only 32% of the total variance corresponds to a single factor [38].

5   Discussion

The use of EMC as a mediator of the relationship between LBI and SRP had not been previously tested, which is an important contribution of the present research.

Our research showed the positive relationship between LBI and JS (H5), which confirms previous work that have analyzed the relationship between leadership and employee satisfaction and trust with the leader [37]. The literature review showed that there is a link between leadership and employee satisfaction and trust with the leader [37]. Being behavioral integrity, the perception of the alignment between words and actions of a leader [22], it is natural that employees feel more satisfied with their leader and that this satisfaction has repercussions on their job satisfaction.

Though, the validation of the relationships researched, namely H2, H3 and H4 support the case for the formulation of the model presented before. This means that the formation of the culture that mirrors the shared and perceived practices and procedures [31] is positively influenced by LBI, which encourages greater sharing of error situations, positively influencing the error management culture in the organization. This greater openness to sharing allows individuals to develop a better and more sophisticated understanding of a specific situation that caused an error to occur [36] and positively influences SRP.

6   Conclusion

Our research shows the impact that the behavioral integrity of a leader has on the SRP, through its effect on error management culture. The paper contributes to the literature showing that the error management culture is a mediator between the behavioral integrity of the leader and the performance in the recovery of errors, a relationship that has not yet been tested in the literature.

This is an important conclusion because service failure and errors occurrence cannot be eradicated in organizations. However, errors can be minimized and in service organizations employees and organizations as a whole can best and evolve with them. In this sense, organizations, and more specifically the leaders of an organization, play a vital role in making employees feel safe and comfortable sharing their mistakes, reducing the likelihood of counterproductive behavior. Thus the present study contribution highlight the importance of perceiving the behavioral integrity of leaders in the eyes of their followers, as a way to boost openness to communication between the parties. In fact, the perception of behavioral integrity will have a positive impact on an organization's error management culture, largely justified by openness to communication between the parties, which subsequently influences employee performance, and then service recovery performance.

Meanwhile, some limitations of this research may be pointed out, which future research should try to overcome. First, the model used and tested in this dissertation could be evaluated through a different sample, independently of the original, so that the results could be doubly validated, as suggested by other authors [40]. Second, evaluation of error recovery performance was measured in this research based on employees' perceptions of how they handle error situations. This type of research design can lead to common method biases [39], because the measures derived from a single data source. Despite it, the Harman test demonstrated that there was no common variance error in the sample used [40]. However, future investigations can try to measure the employee's performance in recovering their errors through company information to assure stronger validity.

Acknowledgments. João F. Proença gratefully acknowledges financial support from FCT – Fundação para a Ciência e Tecnologia (Portugal), national funding through research grant UID/SOC/04521/2019.

Teresa Proença gratefully acknowledges financial support by Portuguese public funds through FCT - Fundação para a Ciência e a Tecnologia, I.P., in the framework of the project UID/ECO/04105/2019.

References

1. Homsma, G.J., Van Dyck, C., De Gilder, D., Koopman, P.L., Elfring, T.: Learning from error: the influence of error incident characteristics. J. Bus. Res. **62**(1), 115–122 (2009)
2. Shin, H., Ellinger, A.E., Mothersbaugh, D.L., Reynolds, K.E.: Employing proactive interaction for service failure prevention to improve customer service experiences. J. Serv. Theory Practice **27**(1), 164–186 (2017)
3. Swanson, S., Hsu, M.: The effect of recovery locus attributions and service failure severity on word-of-mouth and repurchase behaviors in the hospitality industry. J. Hospitality Tourism Res. **35**(4), 511–529 (2011)
4. Maxham, J.G.: Service recovery's influence on consumer satisfaction, positive word-of-mouth, and purchase intentions. J. Bus. Res. **54**(1), 11–24 (2001)
5. De Matos, C.A., Henrique, J.L., Rossi, C.A.V.: Service recovery paradox: a meta-analysis. J. Serv. Res. **10**(1), 60–77 (2007)
6. Lin, W.: Service recovery model: the integrated view. Serv. Ind. J. **29**(5), 669–691 (2009)
7. Hart, C., Heskett, J., Sasser, W.: The profitable art of service recovery. Harvard Bus. Rev. **68** (4), 148–156 (1990)
8. Van Vaerenbergh, Y., Varga, D., De Keyser, A., Orsingher, C.: The service recovery journey: conceptualization, integration, and directions for future research. J. Serv. Res. **22** (2), 103–119 (2019)
9. Smith, A., Bolton, R., Wagner, J.: A model of customer satisfaction with service encounters involving failure and recovery. J. Market. Res. **36**(3), 356–372 (1999)
10. Priluck, R., Lala, V.: The impact of the recovery paradox on retailer-customer relationships. Managing Serv. Qual. **19**(1), 42–59 (2009)
11. Heskett, J., Jones, T., Loveman, G., Sasser, W., Schlesinger, L.: Putting the service profit chain to work. Harvard Bus. Rev. **72**, 164–170 (1994)
12. Iglesias, V., Varela-Neira, C., Vázquez-Casielles, R.: Why didn't it work out? The effects of attributions on the efficacy of recovery strategies. J. Serv. Theory Practice **25**(6), 700–724 (2015)
13. Pugh, H., Brady, M., Hopkins, L.: A customer scorned: effects of employee reprimands in frontline service encounters. J. Serv. Res. **21**(2), 219–234 (2018)
14. Guchait, P., Simons, T., Pasamehmetoglu, A.: Error recovery performance: the impact of leader behavioral integrity and job satisfaction. Cornell Hospitality Q. **57**(2), 150–161 (2016)
15. Pasamehmetoglu, A., Guchait, P., Tracey, J.B., Cunningham, C.J., Lei, P.: The moderating effect of supervisor and coworker support for error management on service recovery performance and helping behaviors. J. Serv. Theory Practice **27**(1), 2–22 (2017)
16. Heskett, J., Sasser, W., Schlesinger, L.: What Great Service Leaders Know and Do. Berrett-Koehler, Oakland (2015)
17. Ringle, C., Wende, S., Becker, J.: SmartPLS, SmartPLS GmbH, Boenningstedt. Rochester Inst. Technol. **20**(3), 405–420 (2015)
18. Hair Jr., F., Sarstedt, J.M., Hopkins, L., Kuppelwieser, G.V.: Partial least squares structural equation modeling (PLS-SEM) an emerging tool in business research. Eur. Bus. Rev. **26**(2), 106–121 (2014)
19. Webster, N.: Webster's New Collegiate Dictionary, 7th edn. G&C Merriam, Springfield (1967)
20. Gronroos, C.: Service quality: the six criteria of good perceived service. Rev. Bus. **9**(3), 10 (1988)

21. Zeithaml, V., Berry, L., Parasuraman, A.: The behavioral consequences of service quality. J. Market. **60**(2), 31–46 (1996)
22. Simons, T.: Behavioral integrity: the perceived alignment between managers words and deeds as a research focus. Organ. Sci. **13**(1), 18–35 (2002)
23. Simons, T., Leroy, H., Collewaert, V., Masschelein, S.: How leader alignment of words and deeds affects followers: a meta-analysis of behavioral integrity research. J. Bus. Ethics **132** (4), 831–844 (2015)
24. Kannan-Narasimhan, P., Lawrence, B.: Behavioral integrity: how leader referents and trust matter to workplace outcomes. J. Bus. Ethics **111**(2), 165–178 (2012)
25. Palanski, M.E., Vogelgesang, G.R.: Virtuous creativity: the effects of leader behavioural integrity on follower creative thinking and risk taking. Can. J. Adm. Sci./Revue Canadienne des Sciences de l'Administration **28**(3), 259–269 (2011)
26. Leroy, H., Palanski, M., Simons, T.: Authentic leadership and behavioral integrity as drivers of follower commitment and performance. J. Bus. Ethics **107**(3), 255–264 (2012)
27. Bagozzi, R.P.: The self-regulation of attitudes, intentions, and behavior. Soc. Psychol. Q. **55** (2), 178–204 (1992)
28. Babakus, E., Yavas, U., Karatepe, O.M., Avci, T.: The effect of management commitment to service quality on employees' affective and performance outcomes. J. Acad. Market. Sci. **31** (3), 272–286 (2003)
29. Bass, B.M., Avolio, B.J.: Transformational leadership and organizational culture. Public Adm. Q. 112–121 (1993)
30. Schein, E.H.: Organizational Culture and Leadership, vol. 2. Wiley, Hoboken (2010)
31. Van Dyck, C., Frese, M., Baer, M., Sonnentag, S.: Organizational error management culture and its impact on performance: a two-study replication. J. Appl. Psychol. **90**, 1228–1240 (2005)
32. Zhao, B., Olivera, F.: Error reporting in organizations. Acad. Manag. Rev. **31**(4), 1012–1030 (2006)
33. Palanski, M.E., Yammarino, F.J.: Integrity and leadership: a multi-level conceptual framework. Leadersh. Q. **20**(3), 405–420 (2009)
34. Bowling, N.A., Hammond, G.D.: A meta-analytic examination of the construct validity of the Michigan Organizational Assessment Questionnaire Job Satisfaction Subscale. J. Vocat. Behav. **73**, 63–77 (2008)
35. Fornell, C., Larker, D.F.: Evaluating structural equation models with unobservable variable sand measurement error. J. Market. Res. **18**, 39–50 (1981)
36. Dormann, T., Frese, M.: Error training: replication and the function of exploratory behavior. Int. J. Hum.-Comput. Interact. **6**(4), 365–372 (1994)
37. Judge, T., Piccolo, R.: Transformational and transactional leadership: a meta-analytic test of their relative validity. J. Appl. Psychol. **89**(5), 755–768 (2004)
38. Podsakoff, P.M., Organ, D.W.: Self-reports in organizational research: problems and prospects. J. Manag. **12**, 69–82 (1986)
39. Podsakoff, P.M., MacKenzie, S.B., Lee, J.Y., Podsakoff, N.P.: Common method biases in behavioral research: a critical review of the literature and recommended remedies. J. Appl. Psychol. **88**(5), 879 (2003)
40. Marôco, J.: Análise de equações estruturais: Fundamentos teóricos, software e aplicações (2014)

Author Index

Printed in the United States
By Bookmasters